Handbook of Industrial and Business Applications with Digital Twins

A digital twin represents the indistinguishable digital counterpart of the physical object to simulate, monitor and test with real-time synchronization. This book presents a framework and important key aspects of digital twins, including various technologies with coverage of the digital twins in various industry and business applications. It provides a background of modeling and simulation, computer sensor technology and other areas required to create the next wave of digital twins.

Features:

- Presents exclusive material on industrial and business applications of digital twins
- Includes diversified digital twin applications with use cases
- Focuses on tools and methods for digital twins, platforms, application domains and industries
- Emphasizes advances and cutting-edge technologies throughout
- Reviews artificial intelligence (AI), fog/edge computing, industrial automation, blockchains and the Internet of Things (IoT)

This book is aimed at researchers and graduate students in cloud computing, simulation, the IoT and computer engineering.

Handbook of Industrial and Business Applications with Digital Twins

Edited by
Saravanan Krishnan, A. Jose Anand and
S. Sendhilkumar

CRC Press
Taylor & Francis Group
Boca Raton London New York

CRC Press is an imprint of the
Taylor & Francis Group, an **informa** business

Designed cover image: Shutterstock

First edition published 2025
by CRC Press
2385 NW Executive Center Drive, Suite 320, Boca Raton FL 33431

and by CRC Press
4 Park Square, Milton Park, Abingdon, Oxon, OX14 4RN

CRC Press is an imprint of Taylor & Francis Group, LLC

ISBN: 9781032737232 (hbk)
ISBN: 9781032737829 (pbk)
ISBN: 9781003465904 (ebk)

DOI: 10.1201/9781003465904

Typeset in Times
by Newgen Publishing UK

Contents

Preface

Despite people's perception of the agricultural process, the fact is that today's agriculture industry is more data-centric, reliable, and smarter than ever before. The purpose of this book is to cover the applications of digital twins in the Industry 4.0 era. Many technologies, such as cloud/fog computing, Internet of Things (IoT), big data, machine learning (ML), and blockchains are initiated to be adopted in smart agricultural engineering. This book deals with all aspects of smart agriculture with state-of-the-art cloud IoT systems in the complete 360-degree view spectrum. It presents the rapid advancement of the technologies in the existing Agri model by applying IoT techniques. Novel architectural solutions in smart agricultural engineering are the core aspects of this book. Several use cases with IoT and smart agriculture will also be incorporated. This book can be used as a handbook and reference book. Every reader will walk away with a deep understanding of complementary features of the cloud and IoT paradigm in smart agriculture applications.

Automated farming is an engineering problem to be solved with agro technology. The book deals with automating the data collection process using agricultural drones and robots, extending remotely obtained parameters of crops and accessing real-time data from any device at any time. Precision agriculture is the application of information technologies. Strength of this edited book comprises the concept of precision agriculture to make the practice of farming more accurate and controlled inter and intra-field variability of crops. The book also presents Agro IoT tools, open source platforms, protocols and techniques that will help researchers/students who are interested in developing applications to make precision farming.

The book includes the technologies and applications of digital twins for the business world. A digital twin can simulate the best real-world experience for business customers. Also, for the business, predicting customer behavior is quite simple. This book will examine the digital twin industrial and business applications with real-time use cases. A digital twin is ideal for mirroring product development and predicting potential downtimes of high-cost industrial machines. The wide variety of topics presented makes the book ideal for students, academicians and researchers who are studying and implementing the digital twin applications.

Editor Biographies

Saravanan Krishnan is Associate Professor in the Department of Computer Science & Engineering at the College of Engineering, Guindy, Anna University, Chennai, Tamil Nadu. He has done an ME in Software Engineering and a PhD in Computer Science Engineering. His research interests include cloud computing, software engineering, the Internet of Things (IoT), and smart cities. He has published papers in 14 international conferences and 30 international journals. He has also written 16 book chapters and edited eight books with international publishers. He has done numerous consultancy work for municipal corporations and smart city schemes. He is an active researcher and academic. He is also a reviewer for many reputed journals. He is a member of the ISTE, IEI, ISCA, ACM, and others.

A. Jose Anand is Professor in the Department of Electronics and Communication Engineering, KCG College of Technology, Chennai, Tamil Nadu. He has one year of industrial experience and 24 years of teaching experience. He has presented several papers at national and international conferences. He has published several papers in national and international journals, as well as books on polytechnic and engineering subjects. He is a member of the CSI, IEI, IET, IETE, ISTE, INS, QCFI, and EWB. His current research interest is in wireless sensor networks, embedded systems, IoT, machine learning, and image processing.

S. Sendhilkumar is Professor in the Department of Information Science and Technology, CEG, Anna University, with 18 years of teaching experience in data mining, machine learning, data science and social network analytics. His research interests include personalized information retrieval, bibliometrics and social network mining. He has authored the book *Concepts, Techniques and Applications*. He received the CTS Best Faculty Award for 2018 and was awarded with Visvesvaraya Young Faculty Research Fellowship by the Ministry of Electronics and Information Technology (MeitY), Government of India, for 2019–2021. He has authored/co-authored textbooks on the fundamentals of computing and machine learning. He has developed e-content (40 modules with all four quadrants) for the course on data analytics for the E-Pathsala project of MHRD-UGC under its national mission on education through ICT. He has delivered live cast lecture series Edusat Programme on Artificial Intelligence for the engineering colleges affiliated with Anna University.

Contributors

Sayed Abdulhayan
P.A. College of Engineering
Mangalore, India

Senan Ali Abd
University of Anbar Iraq – Al Anbar – Ramadi
Iraq

Elif Alptekin
Istanbul Medeniyet University
Istanbul, Turkey

R. Amuthakkannan
College of Engineering, National University of Science and Technology
Sultanate of Oman, Muscat

Anupama
Geeta University
Panipat, Haryana, India

J. Baskaran
Sri Krishna College of Engineering and Technology,
Coimbatore, Tamil Nadu, India

Rodrigo Bonacin
Renato Archer Information Technology Center (CTI)
Campinas, São Paulo – Brazil

Ahmet Harun Çelik
Ipragaz A. S
Istanbul, Turkey

Elif Cesur
Istanbul Medeniyet University
Istanbul, Turkey

Muhammet Raşit Cesur
Istanbul Medeniyet University,
Istanbul, Turkey

Ananya Chenat
Vellore Institute of Technology
Chennai, Tamil Nadu, India

S.K. Devatharani
Kongu Engineering College
Perundurai, Erode, Tamil Nadu, India

Reinaldo Padilha França
Renato Archer Information Technology Center (CTI)
Campinas, São Paulo – Brazil

Naveenkumar J.
Vellore Institute of Technology
Vellore, Tamil Nadu, India

Shrikant Jadhav
San Jose State University
California, United States of America

Rajkumar Jhapte
SVKM Institute of Technology
Dhule, Maharashtra, India

R. Jeyabharath
K.S.R. College of Engineering
Namakkal, Tamil Nadu, India

A. Jose Anand
KCG College of Technology
Chennai, Tamil Nadu, India

Sumanth Ratna Kandavalli
Tandon School of Engineering
New York University, Brooklyn, New York, USA

Saravanan Krishnan
College of Engineering, Guindy
Anna University, Chennai, Tamil Nadu, India

V.T. Krishnaprasath
Nehru Institute of Engineering and Technology
Coimbatore, Tamil Nadu, India

Nitaigour Premchand Mahalik
California State University
Fresno, USA

Anita Mohanty
Silicon Hills, Patia
Bhubaneswar, Odisha, India

Subrat Kumar Mohanty
Einstein Academy of Technology and Management
Bhubaneswar, Odisha, India

Ambarish G. Mohapatra
Silicon Institute of Technology
Patia, Bhubaneswar, Odisha, India

Ana Carolina Borges Monteiro
Renato Archer Information Technology Center (CTI)
Campinas, São Paulo – Brazil

Saptarshi Mukherjee
Maulana Abul Kalam Azad University of Technology
Makaut, West Bengal, India

S. Muthukaruppasamy
Velammal Institute of Technology
Panchetti, Tamil Nadu, India

Soorya M. Nair
Jyothi Engineering College
Thrissur, Kerala, India

J. Nandha Gopal
Velammal Institute of Technology
Panchetti, Tiruvallur Dt, Tamil Nadu, India

Nguyen Thi Bich Ngoc
Hue University, Thua Thien Hue Province
Vietnam

E. Parimalasundar
Mohan Babu University
Tirupati, India

R. Parvathi
Vellore Institute of Technology
Katpadi, Vellore, Tamil Nadu, India

Narayanan Prasanth
Vellore Institute of Technology
Vellore, Tamil Nadu, India

E.B. Priyanka
Kongu Engineering College
Perundurai, Erode, Tamil Nadu, India

Shrinwantu Raha
Bhairab Ganguly College, Belgharia, Kolkata
West Bengal, India

M. Sundar Raj
Panimalar Engineering College
Chennai, Tamil Nadu, India

Geeta Rani
Jagannath University, Delhi-NCR,
Bahadurgarh, Haryana, India

Ch. Venkata Krishna Reddy
Chaitanya Bharathi Institute of Technology
Hyderabad, India

Divya S.
Vellore Institute of Technology
Chennai, Tamil Nadu, India

Saurabh Saoji
Nutan Maharashtra Institute of Engineering & Technology
Pune, Maharashtra, India

Mayakannan Selvaraju
Vidyaa Vikas College of Engineering and Technology
Namakkal, Tamil Nadu, India

Riyaz Abdullah Sheikh
Department of MIS, College of Business
Jazan University, Jazan, Saudi Arabia

Priya Singh
Institute of Science, Banaras Hindu University
Varanasi, Uttar Pradesh, India

Abhinav Singhal
Christ (Deemed to be University)
Ghaziabad, Uttar Pradesh, India

M. Sucharitha
Vellore Institute of Technology
Amaravati, Andhra Pradesh, India

S. Thangavel
Kongu Engineering College
Perundurai, Erode, Tamil Nadu, India

Duong Thanh Tung
Nguyen Tat Thanh University
Ho Chi Minh, Vietnam

Joshva Devadas T.
Vellore Institute of Technology
Vellore, Tamil Nadu, India

G. Arun Sampaul Thomas
J.B. Institute of Engineering and Technology
Hyderabad, Telangana, India

Sankata Tiwari
Indian Institute of Technology (BHU)
Varanasi, Uttar Pradesh, India

Part 1

Concepts and Architecture

1 A Perspective on Digital Twins as an Emerging Technology

Ana Carolina Borges Monteiro, Rodrigo Bonacin and Reinaldo Padilha França

1.1 INTRODUCTION

The technology concept of a digital twin is a digital reproduction of a real asset—i.e., it is a form or dynamic digital representation of a physical process, service, or product, as well as a virtual replica of a physical process, thing, part, instrument, apparatus, artifact, device, merchandise, product, item, utensil, unit, volume, manufacture or even any object (Figure 1.1). In this case, the physical item can then be qualified, classified, analyzed, and worked on based on the analysis of the technology digital twin in different work environments. In its current condition, with historical data or with a mathematical representation of a physical thing, it can even be a system made of subsystems, all virtualized, simulating, and even representing [1]–[2].

A digital twin basically mimics the behaviors of a real item and interprets it in the online environment, working as if it were a mirror. In this way, it is possible to make tests with much more precision, simulating in several ways to understand what will bring the best result. Otherwise, when simulations are not carried out, the range of possibilities is much smaller, resulting in more rework and a waste of raw materials [1], [3].

Digital twins will help an organization to gain robust, improved customer service, and more accurate information about product performance, as well as accurately predict the current analysis of the digital counterparts and future state of physical assets simply. Virtual assets can be optimized, and even modified, with more ease and practicality, with greater speed and at a lower cost, until the most suitable result is found for the purpose sought [2], [4].

As a technology that goes beyond artificial intelligence (AI), the digital twin uses data from different systems to mimic the behaviors of the physical twin in the face of various factors. Due to this system integration, it is very faithful in the simulation [5]–[6].

Digital twins can be divided into a few groups, such as product twins. Given our current reality, where new products appear every day and previous models become

DOI: 10.1201/9781003465904-2

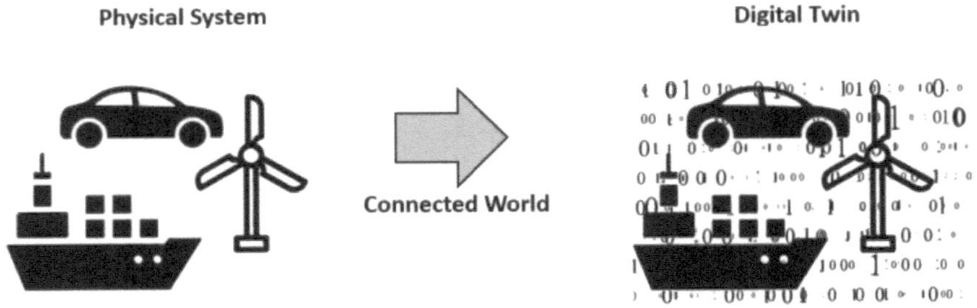

FIGURE 1.1 Digital Twins.

disposable, new challenges arise for industries, such as producing quickly, with fewer errors and saving [7]–[8].

Process twins are a type of digital twin that are used to simulate different manufacturing processes, facilitate decision-making and visualize which process will have more security. In processes that use different equipment, the production flow needs to be well integrated so that it works correctly and each step complements the other. It is possible to create, for example, a digital twin of the entire factory floor and, thus, have a macro view of the environment. This may help to understand the best industrial layout so that employees can move freely and access necessary places easily, avoiding accidents at work and locomotion difficulties [7]–[9].

Equipment or optimization twins are the virtual mirrors of equipment, which simulate existing equipment and, thereby, calculate how the originals will react to certain variables or seasonality that may occur. For this, sensors are used in the physical object, which captures information and feeds the digital twin in real time. Then, it analyzes how several external factors can impact production, aiming for continuous improvement of processes. This anticipates problems such as equipment overload, for example, enabling a power adjustment or preventive maintenance to be carried out. Solving a problem in advance prevents greater expenses [9]–[10].

Digital twins are virtual models that can be combined with new emerging technologies, such as the Internet of Things (IoT) and AI, among diverse others. They can be used in the most diverse applications, from product development to process or logistics improvement, and even more. With this, the fabrication of real prototypes becomes obsolete. Today, instead of investing an enormous amount of time and money in testing to apply changes and move projects forward, a virtual twin is created that can be tried virtually and instantaneously [1], [10].

In practice, within Industry 4.0, a digital twin is the same as a copy of a part or of all the elements that are involved in its production, such as the environment, material, machinery, processes, etc. With it, it is possible to validate all points and replicate behaviors in a virtual environment that mirrors the real world [11]–[13]. Thus, it is possible to simulate situations, changes, and different external factors while monitoring all of this in real time, using the results reports and all performance data on the test done on the digital twin [3]–[4].

Digital twins consist of three primary elements such as real-time presentation of objects like historical data, the performance of process, and the systems. Real-time data is related to receiving outputs from digital platforms or continuous updates from equipment sensors and systems, such as manufacturing, customer service, and even purchasing equipment. Future data, including AI approaches such as deep learning and machine learning (ML), will also receive contributions from engineers [1], [14].

Therefore, this chapter aims to provide an updated overview of digital twins, showing the fundamentals of this disruptive technology. In that regard, it demonstrates a landscape view of the applied aspect, with a concise bibliographic background to point out key concerns and challenges, along with the potential of technologies.

1.2 FUNDAMENTALS OF DIGITAL TWINS AND THEIR RELATIONSHIP WITH EMERGING TECHNOLOGIES

Digital twins are based on creating and maintaining accurate digital replicas of objects, processes, or even biological entities (Figure 1.2). They have emerged as a crucial technological innovation in the Digital Era, transcending the boundary between the virtual and the physical, representing a technological revolution. These digital avatars have been instrumental in numerous sectors, from industry to healthcare, transforming the way we interact with the world around us [15]–[17].

Digital twins are characterized by their ability to accurately reflect the behavior and characteristics of their real-world counterparts and to provide detailed simulations of production processes. This is made possible through the integration of real-time data, advanced sensors, and algorithm ML complexes in industrial sectors; for example, optimizing efficiency and identifying areas for improvement [16], [18].

In healthcare, digital twins of patients enable personalized simulations. This modeling capability creates opportunities for significant advances in personalized medicine, enabling professionals to test different treatment approaches virtually before implementing them in the real world [19]–[20].

The synergistic interaction between digital twins and modern reality drives the accuracy and usefulness of these digital avatars in diverse applications, from smart

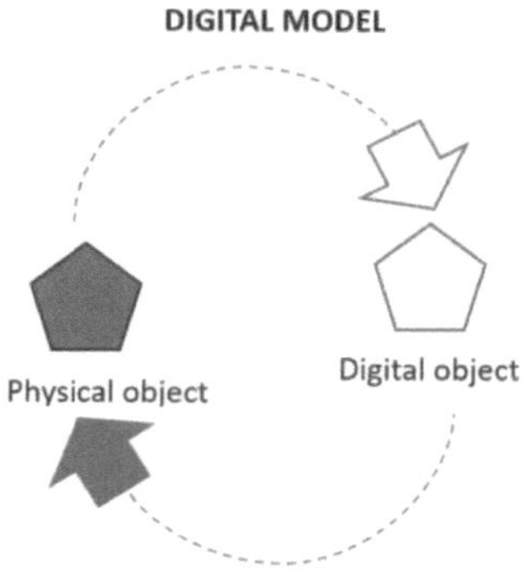

FIGURE 1.2 Digital Twins Concept.

cities to advanced manufacturing. The ability to collect real-time data to power digital twins becomes more robust as the IoT and 5G connectivity become more widespread [21]–[23].

AI can analyze large sets of data generated by these twins, as digital twins play a vital role through AI and provide valuable insights for decision-making and continuous improvements in various sectors. However, the advancement of the digital twin is not without challenges. Ensuring that the information fed into digital twins is accurate and secure is essential to avoid negative consequences. These issues of cybersecurity, privacy, and data reliability are crucial concerns [1], [24].

Ethics and information security are also important considerations; questions about who controls and accesses these digital avatars, and how they are used, require careful attention, especially when digital twins are applied in sensitive contexts such as medicine [24]–[27].

The backbone of the digital twin is accurate three-dimensional modeling, or advanced 3D modeling. Advanced techniques, such as laser scanning and photogrammetry, are often employed to capture geometries and details. This involves creating highly detailed 3D models, which can represent everything from industrial machines to biological organisms [3].

Digital twins rely on real-time data to stay in sync with their physical counterparts; this constant data input is crucial for ensuring that it accurately reflects real-world conditions. This is achievable through the IoT, which is implemented to collect operational, environmental, and performance data [11], [28].

AI and ML algorithms enable digital twins to learn, adapt, and predict behavior based on historical patterns, playing a central role in interpreting and analyzing collected data. These technologies come from ML models that are often trained on large data sets to improve the accuracy of simulations [29]–[30].

Given the data-intensive nature of digital twins, cloud computing enables large data sets to be processed in real time. It is often used for efficient processing and storage, facilitating dynamic simulations and complex interactions [31]–[35].

Simulation capability is one of the fundamental pillars of the digital twin, whereby mathematical, physical, and statistical models are incorporated to simulate specific scenarios. Advanced simulation tools enable you to replicate the dynamic behavior of complex systems, resulting in accurate predictions and performance analysis [36]–[38].

In environments where real-time interaction is crucial, minimal latency is essential to ensure that actions in the digital twin instantly correspond to changes in the real world. As industrial automation, communication networks play a vital role. In this context, efficient communication protocols are frequently implemented. For example, message queuing telemetry transport (MQTT) is a lightweight messaging protocol used in transmission control protocol/Internet protocol (TCP/IP) networks to optimize sensors and small mobile devices. Open platform communications unified architecture (OPC UA) is another protocol that enables information and data to be exchanged in internal devices of machines, between machines and from machines to systems, available free of charge and specifically designed for industrial automation [39].

Digital twins are incorporated into predictive maintenance strategies, considering that real-time sensors, historical data, and ML algorithms converge, using advanced analytics to predict failures and optimize the uptime of physical systems, and even to identify patterns that indicate potential problems before they occur [24].

The choice of operating systems and development platforms, such as integrated development environments (IDEs), and specific platforms for simulation and 3D modeling, play a crucial role in the effectiveness and scalability of digital twins, providing the tools necessary to create and enhance them [3].

Federated learning is an ML technique where generated data can be trained locally at the edge of the network. This enables organizations, or groups within the same organization to train their own data. Only the training results are passed out of the equipment, ensuring that all locally generated data is kept. By sharing these results, organizations can collaboratively and iteratively improve a shared global ML model [40, 41].

In scenarios where data privacy is paramount, such as in healthcare, federated learning techniques can train ML models locally on distributed data, without the need for centralization. By implementing these in digital twins, the confidentiality of the data is preserved [25]–[26].

Digital twins often integrate data from multiple sources to create a holistic representation. The efficient integration of this data is essential to the accuracy and usefulness of the digital twin, and can include geospatial data, weather information, and industry-specific sensor data, among others [1]–[2].

Collaborative development and the open-source approach are gaining prominence in building digital twins, as they promote innovation and rapid dissemination of best practices. Communities of developers contribute to frameworks and libraries that facilitate the implementation and advancement of this technology. Given the critical dependence of many sectors on digital twins, systems are designed to be robust, with the ability to quickly recover in the event of failures, resilience, and redundancy are essential technical aspects, to ensure operational continuity [7].

Advanced pattern recognition algorithms through computer vision play a central role in collecting and interpreting visual information for digital twins, where they are employed to process visual data coming from cameras and other image capture devices. These provide crucial details for the accurate representation of the physical environment. In certain contexts, such as creating digital twins to represent human beings, which may involve the use of facial recognition, fingerprints, or other biometric methods, biometric authentication techniques are applied to ensure unique and secure identification [42].

Digital twins rely heavily on efficient real-time data management to ensure that the information that is fed into digital twins is always up-to-date and accurate. In this context, specialized database systems, optimized data structures, and low-cost data transmission protocols are required, and latency is essential [43].

The creation of natural user interfaces (NUIs) is crucial for efficient interaction with digital twins, especially in environments where human intervention is required. This includes the development of gestures and other intuitive forms of communication, or even voice commands, to facilitate collaboration between humans and digital twins [44].

To deal with the increasing complexity of data and the sophistication of simulations, these approaches enable the creation of more complex and adaptive models, neural networks, and deep learning techniques incorporated into digital twins, enabling continuous learning and improving accuracy throughout time [7].

Interoperability is critical, especially in industries where multiple systems need to collaborate, thereby ensuring that digital twins can function efficiently in complex ecosystems, as well as adopt industry standards such as the Digital Twin Consortium and the Industrial Internet Consortium (IIC) [45].

For a more immersive experience, some digital twins incorporate haptic feedback, which is derived from sensory realism. This is a technology that enables you to feel tactile sensations when touching electronic devices, simulating them. This is particularly relevant where physical sensation is crucial to understanding and learning, such as in medical and training applications [46].

Given the dynamic nature of physical environments, digital twins need to continually evolve and be implemented to ensure that they remain aligned with changes in physical reality, as dynamic update mechanisms, whether through continuous online learning or software updates in real time [1], [3].

Although quantum computing is still in its early stages, as it becomes more viable, the evolution of this technology is an area to watch for substantial improvements in data processing to improve the efficiency of digital twins significantly [47]–[48].

Cybersecurity is a critical component in implementing digital twin. Encryption, multi-factor authentication, and proactive security practices are incorporated to safeguard the integrity of digital twins, and robust security protocols are required to protect against threats, such as unauthorized access, manipulation of data, and interruptions in the flow of information [49]–[51].

Scalability is a key consideration, and effective scalability strategies, including the use of microservice architectures and container technologies, are necessary, especially when digital twins are implemented on a large scale, such as in smart cities or Industry 4.0. This ensures that digital twins can handle growing data volumes and operational complexity [52].

In addition to augmented and virtual reality, mixed reality is integrated into some digital twins, and devices such as smart glasses are used to provide a mixed-reality experience, enabling for more immersive and accurate interactions. This approach combines elements of the physical world with virtual elements [53].

Edge computing is often incorporated into digital twins to reduce latency and improve data transmission efficiency. Derived from distributed processing close to the data source, it helps to minimize network overhead and ensure fast real-time responses, whether on IoT devices or local servers [54]–[55].

Digital twins can be designed to be autonomous in self-actualization, adapting to changes in real-world conditions without direct human intervention. It is this ability that enables digital twins to evolve based on continuous learning and dynamic adjustments. Data provenance is critical to the reliability of digital twins, which is essential for auditing, informed decision-making, and problem-solving. Traceability mechanisms are implemented to ensure the origin and quality of the data that feeds the digital twin. Some digital twin systems explore the use of blockchains, not only for security, which provides transparency and reliability in the evolution of digital twins, but also to maintain an immutable record of changes over time, ensuring the integrity of historical data [1]–[2], [4], [7].

The implementation of recommendation systems and contextual intelligence enriches the interaction between users and digital twins, improving their effectiveness and adaptability. Intelligent algorithms analyze usage patterns from recommendation systems and contextual intelligence, such as user preferences and operational context, to provide personalized suggestions [38], [52].

As AI plays a significant role in digital twins, transparency in decision-making and equity in treatment should be incorporated into development processes. Ethical consideration becomes imperative with issues such as algorithmic bias to ensure that digital twins are fair and socially responsible [56]–[57].

The technical approach of digital twins reflects the integration of a wide range of advanced technologies. This technical interdisciplinarity is essential for the effectiveness and continued success of the technology in different domains, each playing a specific role in the creation, maintenance, and use of these complex digital representations [4].

1.3 APPLICATIONS CATEGORIZED INTO DIFFERENT DOMAINS, ISSUES, AND CHALLENGES

Applications of the digital twin can be seen in manufacturing societies. In terms of Industry 4.0, digital twins can be used to predict equipment failures, facilitate predictive maintenance, optimize processes, and perform production simulations. In medicine, digital twins are used to model organs for personalizing treatments, simulating surgical procedures, and training healthcare professionals [11], [58]–[60].

The digital twin is applied in smart cities for waste management, environmental monitoring, urban simulations, traffic forecasting, and infrastructure improvements. In the energy sector, digital twins are used to monitor and simulate the performance of plants, optimize the operation of electrical grids, and facilitate the transition to renewable sources [61]–[63].

In the aerospace industry, digital twins are applied for flight simulation, aircraft maintenance, component testing, and more efficient design. Logistics and supply chains can use digital twins to optimize transportation routes, manage inventories, simulate logistics operations, and predict future demands [64].

In construction, digital twins are employed to simulate designs, optimize workflow, predict environmental impacts, and improve construction efficiency. In educational environments, they are used for interactive simulations, virtual laboratories, and personalized training [11].

The challenges to digital twins consist of integrating data from the collection and integration of data from different sources, which require standardized protocols and solutions to deal with heterogeneous formats (Figure 1.3). Protecting sensitive data

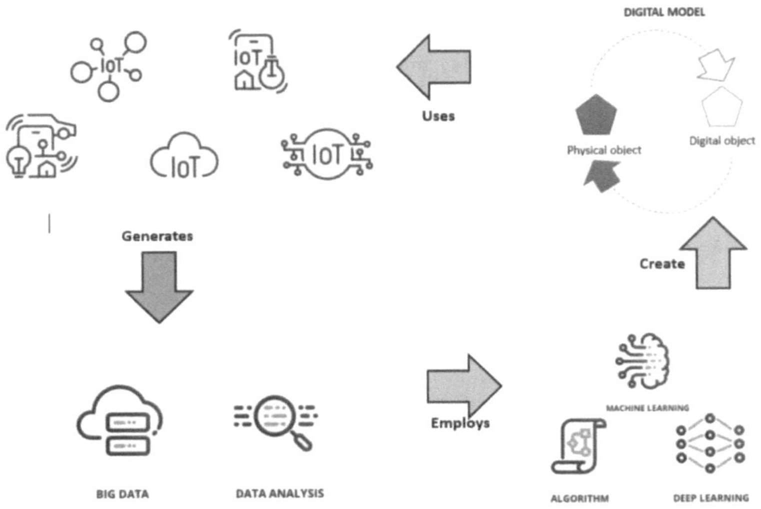

FIGURE 1.3 Technologies Using Digital Twins.

and ensuring privacy, especially in sectors such as healthcare and smart cities, are crucial challenges that require stringent security measures [11].

The use of digital twins in conjunction with digital image processing (DIP) for blood cell segmentation—considering that images of blood samples are acquired through optical microscopy or other imaging methods may vary in terms of resolution, quality, and lighting—is a promising application in areas, such as hematology, medical diagnosis, and biological research [65].

Captured images may undergo a pre-processing process, which may include corrections for brightness, contrast, noise removal, and histogram equalization to improve quality and facilitate segmentation. A digital twin of the blood sample can be created, including information, such as cell arrangement, morphological characteristics, and optical properties, representing a virtual version of the physical sample. DIP algorithms are applied to the digital twin to segment blood cells, which identify and isolate cells in the image through techniques such as thresholding, morphological operators, and edge detection [65]–[68].

The segmentation process can be refined by considering adjustments to the segmentation model parameters, which can be made to improve accuracy, and validated by comparing the results obtained in the digital twin with those in the real image. Once segmented, specific cell characteristics (such as shape, size, and color) can provide valuable information for medical diagnosis or biological research, which can be analyzed in both the digital twin and the real image. The system can be integrated with ML techniques to increase segmentation accuracy over time, since this enables the model to adapt to different types of blood samples, such as automatic cell counting, segmentation, diagnosis of blood diseases, and imaging conditions [65]–[68].

The digital twin can be dynamically updated as new samples are introduced. It is important to highlight that the effective implementation of this process requires an interdisciplinary approach. This enables the system to learn from a variety of cases, especially when it comes to medical diagnoses based on automated systems, making it more robust and adaptable to different conditions [3]–[4], [69]. In this sense, the path to digital twin maturity involves overcoming these obstacles [70]–[71].

From the point of view that considers the future developments and emerging challenges of digital twins, it is worth highlighting the issue of sustainability and environmental impact. As digital twins become more widespread, developing sustainable practices in the implementation and operation of this technology is essential. The environmental impact resulting from increased data processing and energy consumption becomes a critical concern [70]–[71].

The need for communication standards and collaboration protocols between different digital twins becomes crucial as digital twins become more prevalent across different industries; this will ensure a holistic and integrated view across complex environments [72].

In applications involving digital representations of human beings (i.e., ethics) and the manipulation of biometric data, ensuring the privacy and security of this data is fundamental to mitigating ethical risks. Therefore, ethics in the collection and manipulation of biometric data is an area of increasing concern [70]–[71].

Improving the interaction between humans and digital twins is an ongoing challenge, considering that more intuitive interfaces are areas that require constant development to provide a more engaging and efficient user experience. In this sense, advanced sensory feedback and improved personalization are necessary [73]–[74].

As digital twins become more valuable assets, developing advanced security measures to protect against targeted attacks and malicious manipulation is an ongoing challenge as the sophistication of cyber threats increases [1]–[2].

Integrating digital twins with emerging technologies and harmonizing these technologies to optimize efficiency is a constantly evolving field of research and development. Quantum computing, more advanced edge computing, and advanced AI systems present opportunities and challenges. Developing ethical standards and specific regulations for different sectors and applications is a complex task. Even though regulations around digital twins are still evolving, it is necessary to ensure the responsible and safe use of these technologies [75]–[76].

Effectively managing large volumes of data that are distributed across complex environments through strategies to optimize real-time data collection, processing, and storage is a technical and operational challenge, and will be crucial to ensuring the accuracy of digital twins [75]–[76].

The ability to continually train and educate digital twins to adapt to changes in the physical environment comes from developing systems that learn continuously, and staying up-to-date and relevant is an ever-evolving challenge [1], [75].

Promoting social acceptance of digital twins through creating public awareness and understanding is crucial to mitigating concerns and resistance. This involves comprehensive public education about their benefits, limitations, and ethical implications.

As technology evolves, the future of digital twins is inextricably linked to overcoming these challenges and developing innovative solutions, so the global community will face complex issues that will require collaborative and multidisciplinary approaches to ensure an ethical and efficient future for digital twins [70].

From the point of view of an expansion of the concept of digital twins, it is possible to highlight the concept of social digital twins, which is an expansion of the concept of digital twins that can include digital representations, not only of physical objects, but also digital representations of social entities, such as communities, cultures, and even entire societies [71].

Planetary digital twins refer to an idea of extending digital twins to represent planetary systems, and this could be applied to environmental simulations, advanced climate predictions and strategies for global sustainability, or even digital representations of global ecosystems, which could become an area of research [77].

Biological digital twins serve as digital representations of living organisms for purposes such as genetic studies, ecosystem modeling, and advances in personalized medicine. Digital twins can be applied in machines and industrial processes, as well as in biological systems [78].

Space digital twins could contribute to astrophysical research, space exploration, and understanding of complex cosmic phenomena, by exploiting digital twins to simulate and understand space systems, such as planets, stars, and even galaxies [79].

From the perspective of ethical and philosophical challenges, the advancement of digital twins raises questions about digital consciousness and the identity of these representations. As digital twins become more advanced, they pose philosophical challenges about what it means to exist digitally [70]–[71].

Avoiding the creation of socioeconomic disparities and ensuring that everyone has access to technological advances is crucial in ensuring that the benefits derived from digital twins are distributed equitably, which is an ethical challenge. As digital twins transcend borders, international agreements on ethics, security, and standards for the use of digital twins on a global scale become involved, as the need to develop global control and governance structures becomes more pressing [1], [4], [70]–[71].

As digital twins become more advanced, preventing malicious manipulation and ensuring trust in the information that is generated by digital twins are ongoing challenges, and preserving authenticity in digital interactions and representations becomes crucial. As digital twins continue to expand into new domains and challenges emerge, the reflective and collaborative approach will be essential in shaping a future in which society will face not only technical issues, but also deep ethical and philosophical dilemmas, whereby digital twins benefit humanity in an equitable and responsible manner [9]–[10].

1.4 DISCUSSION

As digital twins continue to evolve, continuous adaptation becomes essential, and emerging challenges also stand out. With the need for flexible frameworks that can evolve as demands and technologies change, the increasing complexity of interconnections between systems and dependence on large volumes of data introduce management and governance challenges.

Collective intelligence, which facilitates a more effective approach to the interdisciplinary challenges that we face, where digital twins from different domains collaborate, emerges as a promising area, creating a synergy between the digital twins of smart cities, industries, and healthcare that can provide a holistic understanding.

From a social and cultural impact point of view, in addition to the technological aspects, digital inclusion and education initiatives are crucial. It is also vital to consider the social and cultural impact of digital twins, to ensure that they extend to all layers of society. Transparency in the use and management of this data is essential between digital twin creators and society to build the necessary trust.

Innovation, however, continues to drive the expansion of digital twins into previously unexplored areas, such as manipulation of the human mind or invasive surveillance, through the integration of neuroscience techniques and brain-machine interfaces. However, such advances require a careful approach to avoid harmful ethical implications, promising to take digital human-twin interaction to a new level.

The digital twin revolution has profound implications for workforces and education. The development of adaptive educational programs, supported by digital twins, while optimizing industrial operations, becomes imperative to prepare future generations for an ever-evolving work landscape, but it also raises concerns about the replacement of traditional jobs with advanced automation.

From the perspective of workforce and education implications, a collaboration between humans and digital twins in the workplace with a focus on the ability to harness the human cognitive potential in conjunction with the analytical efficiency of digital twins also becomes a crucial narrative consisting of a delicate balance to achieve optimal productivity.

From the point of view of future perspectives and sustainable development, the ability of digital twins to model scenarios and simulate environmental impacts offers a powerful tool in the search for sustainable solutions. However, it is essential to consider their role in sustainable development, including a constant assessment of the environmental impact of digital twins through responsible implementation.

Digital twins are not just a technological revolution; they are mirrors that reflect our collective ethics. Collecting data to personalize experiences can become intrusive if adequate safeguards are not implemented, and our commitment to directing the course of technological advancement for the benefit of all is necessary.

Ultimately, digital twin technology is a narrative in the making, but ensuring that this narrative is guided by ethical principles is up to us. As architects of this Digital Era, we must value digital privacy and look critically toward a future of sustainable and equitable digital development.

The continued development of digital twins is intrinsically linked to the evolution of adjacent technologies. For example, detailed simulations of industrial processes enable operators to interact directly with digital twins. The fusion of augmented reality (AR) and virtual reality (VR) with digital twins promises an even more enhanced immersive experience, offering unparalleled levels of training and problem-solving.

In the digital twin era, digital twins are more than a mere technological innovation; they are the architects of the digital future. Their cross-sector impact redefines how we understand and interact with the world around us. It is our responsibility to shape this technology for the common good. However, technology is, ultimately, a tool shaped by humanity. As with any technological advancement, a proactive approach to regulatory compliance is essential to mitigate legal risks, but ethical responsibility must also guide its development and implementation.

1.5 CONCLUSION

Based on all the content previously presented, digital twins can be characterized as a fascinating technological frontier, as their application has the capacity to cover several sectors. Therefore, the adoption of digital twin technology is very important to boost innovation, ensure interoperability, and facilitate collaboration between different systems and sectors. Furthermore, there is also the ability to improve efficiency in various fields, providing substantial benefits, through the junction and intersection between the physical worlds and digital environments, even if this seems challenging today. However, investments in technological training are fundamental for maximizing your potential and solving problems.

Other important points to consider are ethical approaches and cybersecurity, as providing transparency in the use of digital twins is fundamental. Ensuring the authenticity and integrity of data associated with digital twins, along with blockchain integration, can play a vital role in data protection and the ethics of AI. Such measures are directly proportional to the trust in the communities in which digital twins are implemented.

As we conclude this chapter, digital twins, which represent a complex synergy of technologies, uniting advanced modeling, reveal themselves as catalysts for profound changes in society and technology. The intersection between ethics, AI, cybersecurity and social impact, innovation, and real-time data analysis can transform entire industries, defining the path forward for creating digital replicas. The continued development of these diverse technical promises to open new frontiers at the intersection of the digital and the physical. As we commit to developing digital twins, sustainable responsibility falls on the shoulders of everyone involved, as well as shaping a narrative of inclusive progress.

These technical aspects form a complex fabric that defines the infrastructure and functioning of digital twins for an era in which technology not only advances but also elevates the human condition. An integrated approach respecting the ethical principles that underpin our modern world with these elements is essential for creating robust, constantly evolving systems that must be responsive and ethical at the forefront of digital transformation.

REFERENCES

[1] VanDerHorn, E., & Mahadevan, S. (2021). Digital twin: Generalization, characterization, and implementation. *Decision Support Systems*, *145*, 113524.

[2] Singh, M., Fuenmayor, E., Hinchy, E. P., Qiao, Y., Murray, N., & Devine, D. (2021). Digital twin: Origin to future. *Applied System Innovation, 4*(2), 36.
[3] Tao, F., Xiao, B., Qi, Q., Cheng, J., & Ji, P. (2022). Digital twin modeling. *Journal of Manufacturing Systems, 64*, 372–389.
[4] Boschert, S., & Rosen, R. (2016). Digital twin—The simulation aspect. In Hehenberger, P., Bradley, D. (eds.) *Mechatronic Futures: Challenges and Solutions for Mechatronic Systems and Their Designers* (pp. 59–74). Springer.
[5] Monteiro, A. C. B., França, R. P., Arthur, R., & Iano, Y. (2021). An overview of artificial intelligence technology directed at smart sensors and devices from a modern perspective. *Smart Sensor Networks: Analytics, Sharing and Control* (vol. 92, pp. 3–26). Springer.
[6] Monteiro, A. C. B., Franca, R. P., Arthur, R., Iano, Y., Carnielli, G. P., & de Godoy, H. A. (2022). Fundamentals of machine learning and blockchain technologies for applications in cybersecurity. *Machine Learning, Blockchain Technologies and Big Data Analytics for IoTs: Methods, Technologies and Applications, 16*, 165.
[7] Wu, J., Yang, Y., Cheng, X. U. N., Zuo, H., & Cheng, Z. (2020, November). The development of digital twin technology review. In *2020 Chinese Automation Congress (CAC)* (pp. 4901–4906). IEEE.
[8] Grieves, M. W. (2019). Virtually intelligent product systems: Digital and physical twins. In Shannon Flumerfelt, Katherine G. Schwartz, Dimitri Mavris and Simon Briceno (eds.) Complex Systems Engineering: Theory and Practice (pp. 175–200). ARC.
[9] Stavropoulos, P., & Mourtzis, D. (2022). Digital twins in Industry 4.0. In *Design and Operation of Production Networks for Mass Personalization in the Era of Cloud Technology* (pp. 277–316). Elsevier.
[10] Reiche, L. T., Gundlach, C. S., Mewes, G. F., & Fay, A. (2021, September). The digital twin of a system: A structure for networks of digital twins. In *2021 26th IEEE International Conference on Emerging Technologies and Factory Automation (ETFA)* (pp. 1–8). IEEE.
[11] Pires, F., Cachada, A., Barbosa, J., Moreira, A. P., & Leitão, P. (2019, July). Digital twin in industry 4.0: Technologies, applications, and challenges. In *2019 IEEE 17th International Conference on Industrial Informatics (INDIN)* (vol. 1, pp. 721–726). IEEE.
[12] Vachálek, J., Bartalský, L., Rovný, O., Šišmišová, D., Morháč, M., & Lokšík, M. (2017, June). The digital twin of an industrial production line within the Industry 4.0 concept. In *2017 21st International Conference on Process Control (PC)* (pp. 258–262). IEEE.
[13] Monteiro, A. C. B., França, R. P., Arthur, R., Iano, Y., Segatti, A. C., Carnielli, G. P., ... & Fernandes, E. C. (2022). A look at IIoT: The perspective of IoT technology applied in the industrial field. In R. Anandan, S. Gopalakrishnan, S. Pal and N. Zaman (eds.) *The Industrial Internet of Things (IIoT)* (pp. 1–29). Wiley.
[14] Monteiro, A. C. B. Proposta de novas metodologias de análise de células sanguíneas por meio dos métodos BSCM (Blood Smear Computacional Method) e BSIM (Blood Smear Intelligence Method): informática médica de baixo custo aplicada a saúde pública. 2023. 1 recurso online (153 p.) Tese (doutorado) – Universidade Estadual de Campinas, Faculdade de Engenharia Elétrica e de Computação, Campinas, SP. Disponível em: www.hdl.handle.net/20.500.12733/14033. Acesso em: 30 nov. 2023.
[15] Zhou, J., Zhang, S., & Gu, M. (2022). Revisiting digital twins: Origins, fundamentals, and practices. *Frontiers of Engineering Management, 9*(4), 668–676.

[16] Khan, L. U., Han, Z., Saad, W., Hossain, E., Guizani, M., & Hong, C. S. (2022). Digital twin of wireless systems: Overview, taxonomy, challenges, and opportunities. *IEEE Communications Surveys & Tutorials*.
[17] Palensky, P., Cvetkovic, M., Gusain, D., & Joseph, A. (2022). Digital twins and their use in future power systems. *Digital Twin, 1*, 4.
[18] Rathore, M. M., Shah, S. A., Shukla, D., Bentafat, E., & Bakiras, S. (2021). The role of AI, machine learning, and big data in digital twinning: A systematic literature review, challenges, and opportunities. *IEEE Access*, 9, 32030–32052.
[19] Estrela, V. V., Monteiro, A. C. B., França, R. P., Iano, Y., Khelassi, A., & Razmjooy, N. (2018). Health 4.0: Applications, management, technologies, and review: Array. *Medical Technologies Journal, 2*(4), 262–276.
[20] França, R. P., Iano, Y., Monteiro, A. C. B., & Arthur, R. (2020). Potential proposal to improve data transmission in healthcare systems. In *Deep Learning Techniques for Biomedical and Health Informatics* (pp. 267–283). Academic Press.
[21] Monteiro, A. C. B., França, R. P., Arthur, R., & Iano, Y. (2021). An overview of medical Internet of Things, artificial intelligence, and cloud computing employed in health care from a modern panorama. *The Fusion of Internet of Things, Artificial Intelligence, and Cloud Computing in Health Care*, 3–23.
[22] França, R. P. Metodologia de codificação de "bits" por entidades por meio de eventos discretos (CBEDE): uma proposta de transmissão aplicada em canais AWGN, com desvanecimento Rayleigh e Rician e OFDM. 2023. 1 recurso online (147 p.) Tese (doutorado) – Universidade Estadual de Campinas, Faculdade de Engenharia Elétrica e de Computação, Campinas, SP. Disponível em: www.hdl.handle.net/20.500.12733/10889. Acesso em: 30 nov. 2023.
[23] Zhang, Z., Wen, F., Sun, Z., Guo, X., He, T., & Lee, C. (2022). Artificial intelligence-enabled sensing technologies in the 5G/internet of things era: From virtual reality/augmented reality to the digital twin. *Advanced Intelligent Systems, 4*(7), 2100228.
[24] Eckhart, M., & Ekelhart, A. (2019). Digital twins for cyber-physical systems security: State of the art and outlook. In Robert M. Lee and Tom Gilb (eds.) *Security and Quality in Cyber-Physical Systems Engineering: With Forewords* (pp. 383–412). Springer International Publishing.
[25] Holmes, D., Papathanasaki, M., Maglaras, L., Ferrag, M. A., Nepal, S., & Janicke, H. (2021, September). Digital twins and Cyber Security–solution or challenge? In *2021 6th South-East Europe Design Automation, Computer Engineering, Computer Networks and Social Media Conference (SEEDA-CECNSM)* (pp. 1–8). IEEE.
[26] Cellina, M., Cè, M., Alì, M., Irmici, G., Ibba, S., Caloro, E., ... & Papa, S. (2023). Digital twins: The new frontier for personalized medicine? *Applied Sciences, 13*(13), 7940.
[27] El Saddik, A. (2018). Digital twins: The convergence of multimedia technologies. *IEEE Multimedia, 25*(2), 87–92.
[28] França, R. P., Monteiro, A. C. B., Arthur, R., & Iano, Y. (2021). An overview of deep learning in big data, image, and signal processing in the modern digital age. In Vincenzo Piuri, Sandeep Raj, Angelo Genovese and Rajshree Srivastava (eds.) *Hybrid Computational Intelligence for Pattern Analysis, Trends in Deep Learning Methodologies* (pp. 63–87). Academic Press.
[29] Monteiro, A. C. B., França, R. P., Arthur, R., & Iano, Y. (2021). An overview of the technological performance of deep learning in modern medicine. *Deep Learning for Personalized Healthcare Services, 7*, 225.
[30] Qiu, C., Zhou, S., Liu, Z., Gao, Q., & Tan, J. (2019). Digital assembly technology based on augmented reality and digital twins: A review. *Virtual Reality & Intelligent Hardware, 1*(6), 597–610.

[31] França, R. P., Monteiro, A. C. B., Arthur, R., & Iano, Y. (2020). An overview of blockchain and its applications in the modern digital age. *Security and Trust Issues in Internet of Things*, 185–207.
[32] Monteiro, A. C. B., França, R. P., Iano, Y., & Arthur, R. (2021). An overview of blockchain technology concepts from a modern perspective. In Sheikh Mohammad Idrees, Parul Agarwal, M. Afshar Alam (eds.), *Blockchain for Healthcare Systems* (pp. 27–41). CRC Press.
[33] Scheibmeir, J., & Malaiya, Y. (2019, July). An API development model for digital twins. In *2019 IEEE 19th International Conference on Software Quality, Reliability and Security Companion (QRS-C)* (pp. 518–519). IEEE.
[34] Issa, R., Badr, M. M., Shalash, O., Othman, A. A., Hamdan, E., Hamad, M. S., ... & Imam, S. M. (2023). A data-driven digital twin of electric vehicle Li-ion battery state-of-charge estimation enabled by driving behavior application programming interfaces. *Batteries*, *9*(10), 521.
[36] Bonney, M. S., De Angelis, M., Dal Borgo, M., Andrade, L., Beregi, S., Jamia, N., & Wagg, D. J. (2022). Development of a digital twin operational platform using Python Flask. *Data-Centric Engineering*, *3*, e1.
[37] Blair, G. S., & Henrys, P. A. (2023). The role of data science in environmental digital twins: In praise of the arrows. *Environmetrics*, *34*(2), e2789.
[38] Boyes, H., & Watson, T. (2022). Digital twins: An analysis framework and open issues. *Computers in Industry*, *143*, 103763.
[39] Agavanakis, K., Cassia, J., Drombry, M., & Elkaim, E. (2022, August). Telemetry transformation towards industry 4.0 convergence – A fuel management solution for the transportation sector based on digital twins. In *AIP Conference Proceedings* (vol. 2437, no. 1). AIP Publishing.
[40] Li, L., Fan, Y., Tse, M., & Lin, K. Y. (2020). A review of applications in federated learning. *Computers & Industrial Engineering*, *149*, 106854.
[41] Mammen, P. M. (2021). Federated learning: Opportunities and challenges. *arXiv preprint arXiv:2101.05428*.
[42] Pan, Y., Braun, A., Brilakis, I., & Borrmann, A. (2022). Enriching geometric digital twins of buildings with small objects by fusing laser scanning and AI-based image recognition. *Automation in Construction*, *140*, 104375.
[43] Kuftinova, N. G., Ostroukh, A. V., Maksimychev, O. I., Vasil'ev, Y. E., & Klimenko, V. A. (2022). Digital twins in smart data management at a manufacturing enterprise. *Russian Engineering Research*, *42*(2), 162–164.
[44] Schluse, M., Priggemeyer, M., Atorf, L., & Rossmann, J. (2018). Experimentable digital twins—Streamlining simulation-based systems engineering for industry 4.0. *IEEE Transactions on Industrial Informatics*, *14*(4), 1722–1731.
[45] Harper, K. E., Malakuti, S., & Ganz, C. (2019). Digital twin architecture and standards.
[46] Mayer, I. S. (2022). *Digital Twins for the Real World.* Tilburg University.
[47] Padilha França, R., Borges Monteiro, A. C., Arthur, R., & Iano, Y. (2019, October). Understanding quantum computation concepts and applications. In Yuzo Iano, Rangel Arthur, Osamu Saotome, Guillermo Kemper, Reinaldo Padilha França (eds.) *Proceedings of the 5th Brazilian Technology Symposium* (pp. 351–359). Cham: Springer International Publishing.
[48] Lv, Z., Cheng, C., & Song, H. (2022). Digital twins based on quantum networking. *IEEE Network*, *36*(5), 88–93.
[49] França, R. P., Monteiro, A. C. B., Arthur, R., & Iano, Y. (2020). An overview of the Internet of Things technologies focusing on disaster response. In Iman Rahimi, Amir H. Gandomi, Simon James Fong, M. Ali Ülkü (eds.) *Big Data Analytics in Supply Chain Management* (pp. 151–168). CRC Press.

[50] Monteiro, A. C. B., Franca, R. P., Arthur, R., Iano, Y., Carnielli, G. P., & de Godoy, H. A. (2022). Fundamentals of machine learning and blockchain technologies for applications in cybersecurity. *Machine Learning, Blockchain Technologies and Big Data Analytics for IoTs: Methods, Technologies and Applications*, *16*, 165.

[51] Hearn, M., & Rix, S. (2019). Cybersecurity considerations for digital twin implementations. *IIC J. Innov*, *10*, 107–113.

[52] Almasan, P., Ferriol-Galmés, M., Paillisse, J., Suárez-Varela, J., Perino, D., López, D., ... & Barlet-Ros, P. (2022). Digital twin network: Opportunities and challenges. *arXiv preprint arXiv:2201.01144*.

[53] Tu, X., Autiosalo, J., Jadid, A., Tammi, K., & Klinker, G. (2021). A mixed reality interface for a digital twin-based crane. *Applied Sciences*, *11*(20), 9480.

[54] França, R. P., Monteiro, A. C. B., Arthur, R., & Iano, Y. (2021). An overview of the edge computing in the modern digital age. *Fog/Edge Computing For Security, Privacy, and Applications* (vol. 83, pp. 33–52). Springer.

[55] Do-Duy, T., Van Huynh, D., Dobre, O. A., Canberk, B., & Duong, T. Q. (2022). Digital twin-aided intelligent offloading with edge selection in mobile edge computing. *IEEE Wireless Communications Letters*, *11*(4), 806–810.

[56] Angelov, P. P., Soares, E. A., Jiang, R., Arnold, N. I., & Atkinson, P. M. (2021). Explainable artificial intelligence: An analytical review. *Wiley Interdisciplinary Reviews: Data Mining and Knowledge Discovery*, *11*(5), e1424.

[57] Confalonieri, R., Coba, L., Wagner, B., & Besold, T. R. (2021). A historical perspective of explainable Artificial Intelligence. *Wiley Interdisciplinary Reviews: Data Mining and Knowledge Discovery*, *11*(1), e1391.

[58] Stavropoulos, P., & Mourtzis, D. (2022). Digital twins in Industry 4.0. In *Design and Operation of Production Networks for Mass Personalization in the Era of Cloud Technology* (pp. 277–316). Elsevier.

[59] Schroeder, G. N., Steinmetz, C., Rodrigues, R. N., Henriques, R. V. B., Rettberg, A., & Pereira, C. E. (2020). A methodology for digital twin modeling and deployment for industry 4.0. *Proceedings of the IEEE*, *109*(4), 556–567.

[60] Schroeder, G. N., Steinmetz, C., Rodrigues, R. N., Henriques, R. V. B., Rettberg, A., & Pereira, C. E. (2020). A methodology for digital twin modeling and deployment for industry 4.0. *Proceedings of the IEEE*, *109*(4), 556–567.

[61] Mylonas, G., Kalogeras, A., Kalogeras, G., Anagnostopoulos, C., Alexakos, C., & Muñoz, L. (2021). Digital twins from smart manufacturing to smart cities: A survey. *IEEE Access*, *9*, 143222–143249.

[62] Farsi, M., Daneshkhah, A., Hosseinian-Far, A., & Jahankhani, H. (Eds.). (2020). *Digital Twin Technologies and Smart Cities*. Berlin/Heidelberg, Germany: Springer.

[63] França, R. P., Monteiro, A. C. B., Arthur, R., & Iano, Y. (2021). An overview of the machine learning applied in smart cities. In Khan, M.A., Algarni, F., Quasim, M.T. (eds.) *Smart Cities: A Data Analytics Perspective* (pp. 91–111). Springer.

[64] Li, L., Aslam, S., Wileman, A., & Perinpanayagam, S. (2021). Digital twin in aerospace industry: A gentle introduction. *IEEE Access*, *10*, 9543–9562.

[65] Evangeline, P. (2020). Digital twin technology for "smart manufacturing". In *Advances in Computers* (vol. 117, no. 1, pp. 35–49). Elsevier.

[66] Monteiro, A. C. B., Iano, Y., França, R. P., & Arthur, R. (2021). A metaheuristic algorithm for classification of White Blood Cells in healthcare informatics. In *Metaheuristics in Machine Learning: Theory and Applications* (pp. 219–238). Springer International Publishing.

[67] Monteiro, A. C. B., Iano, Y., & França, R. P. (2017, December). Detecting and counting of blood cells using watershed transform: An improved methodology. In *Brazilian Technology Symposium* (pp. 301–310). Springer International Publishing.

[68] dos Santos, C. H., Gabriel, G. T., do Amaral, J. V. S., Montevechi, J. A. B., & de Queiroz, J. A. (2021). Decision-making in a fast fashion company in the Industry 4.0 era: A digital twin proposal to support operational planning. *The International Journal of Advanced Manufacturing Technology*, *116*, 1653–1666.

[69] Latif, H., & Starly, B. (2020). A simulation algorithm of a digital twin for manual assembly process. *Procedia Manufacturing*, *48*, 932–939.

[70] Deng, T., Zhang, K., & Shen, Z. J. M. (2021). A systematic review of a digital twin city: A new pattern of urban governance toward smart cities. *Journal of Management Science and Engineering*, *6*(2), 125–134.

[71] Jeong, D. Y., Baek, M. S., Lim, T. B., Kim, Y. W., Kim, S. H., Lee, Y. T., ... & Lee, I. B. (2022). Digital twin: Technology evolution stages and implementation layers with technology elements. *IEEE Access*, *10*, 52609–52620.

[72] Sun, W., Ma, W., Zhou, Y., & Zhang, Y. (2022). An introduction to digital twin standards. *GetMobile: Mobile Computing and Communications*, *26*(3), 16–22.

[73] Ramasubramanian, A. K., Mathew, R., Kelly, M., Hargaden, V., & Papakostas, N. (2022). Digital twin for human–robot collaboration in manufacturing: Review and outlook. *Applied Sciences*, *12*(10), 4811.

[74] Pairet, È., Ardón, P., Liu, X., Lopes, J., Hastie, H., & Lohan, K. S. (2019, March). A digital twin for human-robot interaction. In *2019 14th ACM/IEEE International Conference on Human-Robot Interaction (HRI)* (pp. 372–372). IEEE.

[75] Chaudhary, G., Khari, M., & Elhoseny, M. (Eds.). (2021). *Digital Twin Technology*. CRC Press.

[76] Kuchemüller, K. B., Pörtner, R., & Möller, J. (2021). Digital twins and their role in model-assisted design of experiments. In Herwig, C., Pörtner, R., Möller, J. (eds) *Digital Twins: Applications to the Design and Optimization of Bioprocesses* (vol. 177, pp. 29–61). Springer.

[77] Teixeira, R., Puccinelli, J., de Vargas Guterres, B., Pias, M. R., Oliveira, V. M., Botelho, S. S. D. C., ... & Paris, M. (2022, April). Planetary digital twin: a case study in aquaculture. In Jiman Hong, Miroslav Bures, Juw Won Park, Tomas Cerny (eds.) *Proceedings of the 37th ACM/SIGAPP Symposium on Applied Computing* (pp. 191–197). Association for Computing Machinery, New York, NY, USA.

[78] Bruynseels, K., Santoni de Sio, F., & Van den Hoven, J. (2018). Digital twins in health care: Ethical implications of an emerging engineering paradigm. *Frontiers in Genetics*, *9*, 31.

[79] Kim, M. U. (2022, October). A survey on digital twin in aerospace in the new space era. In Jiman Hong, Miroslav Bures, Juw Won Park, Tomas Cerny (eds.) *2022 13th International Conference on Information and Communication Technology Convergence (ICTC)* (pp. 1735–1737). IEEE.

2 Architecture of Digital Twins and Applications

Ch. Venkata Krishna Reddy,
Sumanth Ratna Kandavalli, M. Sundar Raj,
R. Amuthakkannan, J. Baskaran,
Mayakannan Selvaraju and A. Jose Anand

2.1 INTRODUCTION

A number of nations' advanced manufacturing development plans have emerged in response to the proposed notions of "intelligent manufacturing" and "manufacturing power". Projects, including Germany's Industry 4.0, the United States's Industrial Internet, and Made in China 2025, are all part of these ambitions [1]. The field of intelligent manufacturing, which makes use of cutting-edge information and artificial intelligence (AI) technologies that are already in use in the manufacturing sector, is actively developing, with the ultimate goal of assisting in the enhancement and transformation of the real economy. Integrating and interacting with the physical and digital worlds is a basic challenge that the modern intelligent manufacturing firm is trying to solve. The idea of a digital doppelganger has reawakened old fears among experts [2].

A digital twin is an electronic copy of a physical item that acts similarly to the real thing in real life by simulating its data. Through various methods, such as data fusion analysis, iterative decision-making, virtual and real-world data interaction, and optimization, it is possible to enhance and augment physical entities' capabilities [3].

During a 2003 University of Michigan course on product life cycle management (PLM), Professor Grieves first introduced the idea of digital twins. However, the proposal was not given much thought at the time because of cognitive and technological limitations [4]. The concept of a digital twin for the future aircraft design was initially proposed in 2011 by the US Air Force Research Laboratory (AFRL) and NASA, before it gained traction and was further developed by academics [5]. Using digital twin technologies, several authors [6] investigated health management and the prediction of equipment malfunctions. "Digital twin-driven fault prediction and health management" (PHM) is the abbreviation for our next suggestion. In their study, authors [7] utilized femtolasers to simulate and test the material removal process. A digital twin was created by combining the simulation results with physics, decision-making algorithms, and machine learning (ML) approaches. By comparing the simulation

DOI: 10.1201/9781003465904-3

results with the experimental data that was received from the femtolaser ablation, the viability of the digital twin concept was confirmed. A computer numerical control (CNC) machine tool-based digital twin model was developed and shown by authors [8]. CNC machine tools (CNCMTs) were found to be more stable, function in an optimal mode, and have a lower chance of unexpected failure, according to the results.

Digital twin technology enables the efficient synchronization of physical and virtual goods by integrating intelligent sensing, data analysis, co-simulation, and intelligent control innovations. As time passes, digital twin technology finds more and more uses across the whole product life cycle, ranging from design to production to service [9].

A key component of processing and producing mechanical equipment in the intelligent manufacturing sector is strongly correlated with levels of intelligence, production efficiency, product quality, and overall intelligence. One of the most ubiquitous parts, the cam mechanism, is ideal for a wide variety of automatic tools, instruments, and control systems, due to its ability to handle complex reciprocating linear motion. In a traditional cam mechanism, the cam profile curve determines the follower's motion law; in other words, any desired motion can be attained by simply designing the cam profile curve [10]. Cam mechanisms such as these have their limitations, though, and they can accomplish only a limited range of motion. A new cam design and production run are required whenever the motion law is modified. In response to the limitations of conventional cam mechanisms, a novel electronic cam was developed, which, when controlled by a computer program, can mimic the mechanical cam's operation [11]. An electronic cam's straightforward design, adaptability, control performance, and accuracy make it preferable to a conventional mechanical cam mechanism. An electronic cam eliminates the need to swap out mechanical components when making a wide variety of goods, which boosts manufacturing productivity and paves the way for integrated management system (IMS) advancements.

On the other hand, a direct drive electronic cam does not include any kind of intermediary connection between the two. The follower's electronic cam is immediately pressured by external disturbance and load force, which makes control more difficult. An important step in creating intelligent manufacturing is figuring out how to make electronic cam motion control as resilient, flexible, and smart as possible.

This study employs digital twin technology to examine a five-dimensional (5D) structural model-based electronic cam servo motion controller. By combining digital twins with multi-dimensional simulation software, we can precisely control the electronic cam's movements, monitor its state, and plan its course. This study proves, by testing and simulation, that a digital twin-based electronic cam servo motion controller is both practical and easily accessible. The system not only satisfies the intelligence, speed, and precision performance requirements, but it also increases production efficiency, decreases costs, and meets the market expectations for product diversification by producing diverse motion laws. Electronic cameras can be better planned, built, and serviced with the help of this paper, which details best practices for digital twin technology.

To summarize the article's contributions, consider the following:

1. A description of the digital twin model's building process, its establishment utilizing multi-dimensional simulation software for the electronic cam, and the actualization of the entity's interaction and fusion with the virtual system
2. Using a genetic algorithm (GA) to optimize a multi-objective function, we present a method for electronic camera motion trajectory planning that is based on the non-uniform rational B-splines modeling (NURBS) curve
3. Creating a digital twin-based motion controller and comparing it to the standard feed-forward + feedback configuration for two degrees of freedom (DOF):
4. Testing the scheme's viability via simulation and experimentation

2.2 A DIGITAL CAMERA'S DIGITAL TWIN SYSTEM (DTS)

2.2.1 Structure of a DTS

Models and data hold essential importance within the DTS. The DTS has progressed from its initial three-dimensional (3D) state to its present 5D state, as depicted in Figure 2.1 [12]. It has a real-life component, data from two entities, a digital representation (co-simulation models), a network of service providers, and connections for information exchange.

1. Physical entity: Sensors, actuators, control systems, and similar components are usually part of the hardware component.

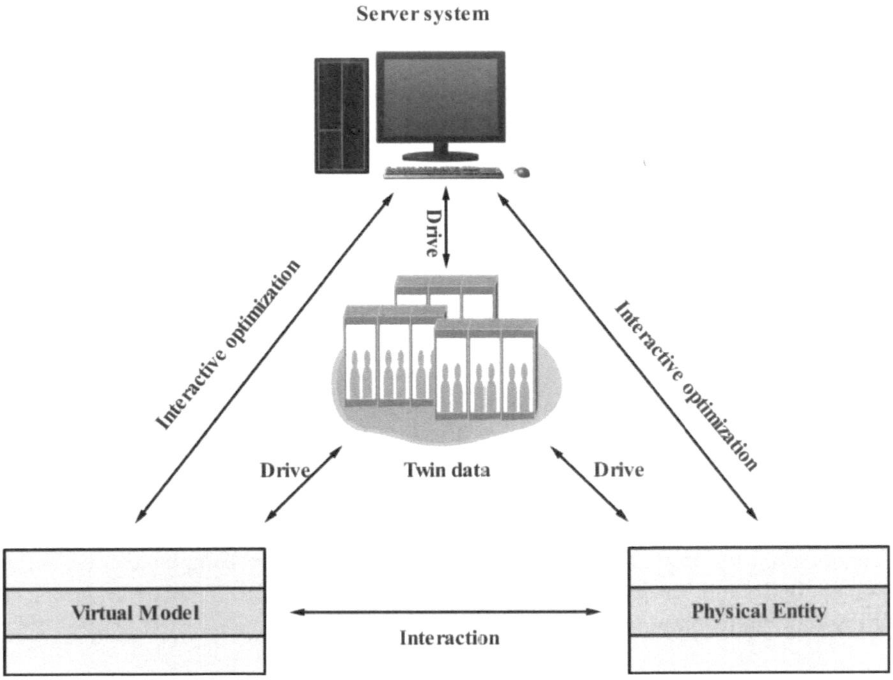

FIGURE 2.1 Digital Twin Model with Five Dimensions.

2. Virtual model: A kinematic model of mechanical goods and a three-dimensional model of the physical entity are both included in this digital representation. It has features for evaluating, optimizing, and predicting, and it can faithfully mimic the physical entity's optimum motion process in real time.
3. Twin data: Data pertaining to real entities, virtual models, service systems, knowledge, etc.; these are the backbone of digital twins.
4. Service system: Information systems for assessment, control, and optimization that are based on virtual models and physical entities are integrated to enable intelligent operation, accurate control, and optimization of activities.
5. Communication connection: The four components listed above are linked in a way that allows for the transmission of data in real time.

An extensive understanding of computer science and intelligent control is necessary to overcome the challenges inherent in the aforementioned 5D structure model of digital twins, which centers on digital twin construction and simulation technologies, as well as virtual-real system coordination and control.

2.2.2 System for Controlling Electronic Cam Servo Motion Powered by DTS

The inherent dynamic nature of the electronic cam makes its movement susceptible to a myriad of external influences. To improve the performance of the electronic camera in real-time monitoring, motion trajectory planning, and control, we can utilize real-time data interaction and virtual-real mapping. As seen in Figure 2.2, the digital twin controls the digital cam servo motion controller.

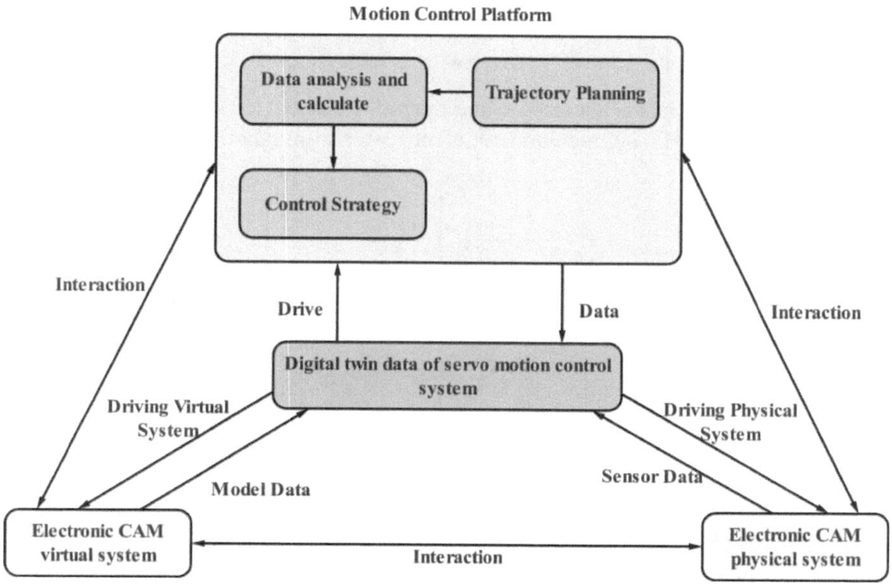

FIGURE 2.2 A DTS of a Cam Servo Motion Controller's Structural Model.

The five main elements of a digital twin-driven electronic cam servo motion controller are a physical system for electronic cams, a virtual system for electronic cams, a platform for motion control, a platform for data from digital twins, and a way of communication.

2.3 PRINCIPLE AND STRUCTURE OF ELECTRONIC CAM

2.3.1 Principle and Structure of Electronic Cam

Woelfel M was the first to suggest the idea of an electronic cam, and he first used one in electronic motion control systems. In Figure 2.3, we can see the electronic cam's schematic. The control system, drive unit, and actuator are its key components. To mimic the behavior of a mechanical cam, the control system's programming can direct the drive unit to operate the actuator. The most common types of electronic cams used today are servo and stepping motors, but these solutions require the translation of rotary into linear motion.

This study details an electronic cam that is a coil-moving, self-developed electromagnetic linear actuator. Figure 2.4 depicts the configuration of a linear motor, also known as an electromagnetic linear actuator, that efficiently converts electrical energy into linear mechanical motion. The primary components of the actuator are an electromagnetic coil, a permanent magnet, a skeleton, and an interior and external yoke, among others. By manipulating the size and direction of the driving current, one may generate the exact pattern of the valve's motion. This is possible because the electromagnetic coil experiences an axial electrostatic field within the magnetic field of the air gap [13]–[14]. An electromagnetic linear actuator's straightforward design, compact size, and lack of a middle conversion link set it apart from competing electronic cams.

2.3.2 An Electronic Cam's Mathematical Model

Many moving parts, electrical circuits, and magnetic circuits make up an electromagnetic linear actuator. Hence, mechanical, electrical, and magnetic circuit subsystems

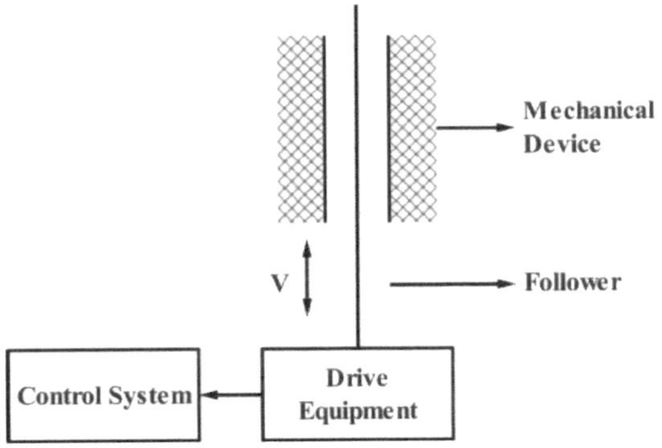

FIGURE 2.3 Electronic Cam Structural Diagram.

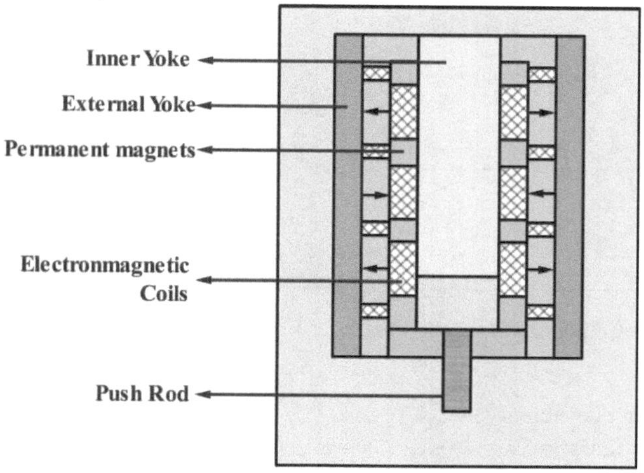

FIGURE 2.4 Physical Entity and Self-Developed Electromagnetic Linear Actuator Structure.

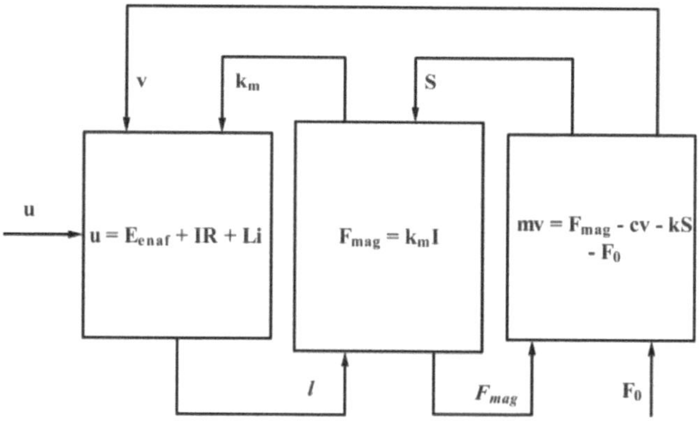

FIGURE 2.5 Relationship of Coupling between Electronic Camera Subsystems.

were developed, and the intersubsystem interaction was thoroughly examined. Figure 2.5 displays the mathematical model.

An electronic camera's differential equations are

$$\begin{cases} \dot{I} = -\frac{R}{L}I - \frac{k_e}{L}v + \frac{u}{L}v = \frac{k_m}{m}I - \frac{c}{m}v - \frac{F_0}{m}\dot{S} = v \end{cases} \tag{2.1}$$

The variable "m" denotes the total weight of the electromagnetic linear actuator, encompassing the electromagnetic coils, framework, and connections [15].

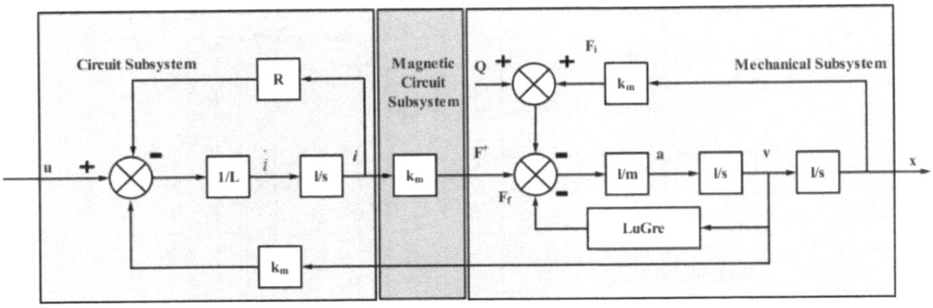

FIGURE 2.6 Computing Dynamic Models of Electronic Cam.

c: Damping coefficient
v: Valve speed, $v = \dot{S}$
u: Voltage
R: Coil resistance
I: The coil's current
L: The coil's inductance
$k_m = k_b B_\delta l N$: The force constant of the actuator
$k_e = BlN$: The back electromotive force (EMF) constant
F_0: The external load

According to formula (2.1), information on the control current, driving voltage, motion displacement, and external load is necessary for monitoring and regulating the electronic cam's motion.

2.3.3 An Electrical Camera Model in Virtual Simulation

The system was modeled using a combination of Unity3D, Pro-E, and MATLAB.

2.3.3.1 Model for Dynamic Simulations

The electronic cam's circuit, mechanical subsystem and magnetic circuit were analyzed in detail, and the results informed the development of the dynamic simulation model in MATLAB/Simulink. Figure 2.6 depicts the structure.

2.3.3.2 Model for Digital Twins

Researchers used Pro-e and Unity3D to build the DTS and evaluate the mapping relationship between an object's physical and virtual versions. The data received from physical entities allows for the real-time tracking of their states, creating a digital twin that can perceive, analyze, and make decisions.

2.4 METHODS FOR ELECTRONIC CAMERA CONTROL AND TRAJECTORY PLANNING

Quite a few things influence how the electronic cam moves. Constant optimization and adjustment of the motion trajectory and state are required to guarantee accurate

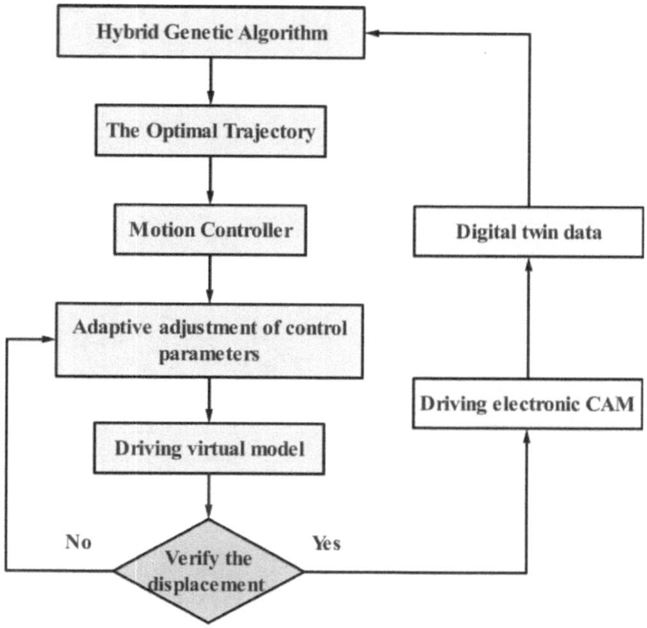

FIGURE 2.7 Flow Diagram for Electronic Cam Motion Control.

cam motion. The motion control stage begins with the necessary steps to establish the platform and gather and send real-time data from the electronic camera. While this happens, the GA in MATLAB is responsible for trajectory planning [16]. Second, the virtual system is used to replicate and examine the electronic cam's motion state. The third phase involves comparing the data generated by simulation with the data obtained in the control system to assess and adjust the control parameters of the virtual model. Lastly, the modified control system propels the actual electronic cam. In Figure 2.7, we can see the model for the motion control system.

2.4.1 Planning an Electronic Cam's Motion Trajectory

Using a GA, this study optimized the trajectory of an electronic camera's motion [17]. In 1975, Professor J. Holland initially proposed the GA. The key benefit of GAs is their ability to search complex landscapes with several peaks simultaneously by reducing the likelihood of iterative convergence to local solutions [18]. Without limiting the derivative or the continuation of the function, GAs can be directly applied to structural objects. Automatic optimization search space retrieval and guidance via the probabilistic optimization method is possible, as is adaptive search direction adjustment free of rule-based constraints.

Figure 2.8 shows the path of an electronic cam that simulates a mechanical cam. It is made up of four stages: the holding ($\delta 4$), the descending ($\delta 3$), the holding ($\delta 2$), and the ascending ($\delta 1$).

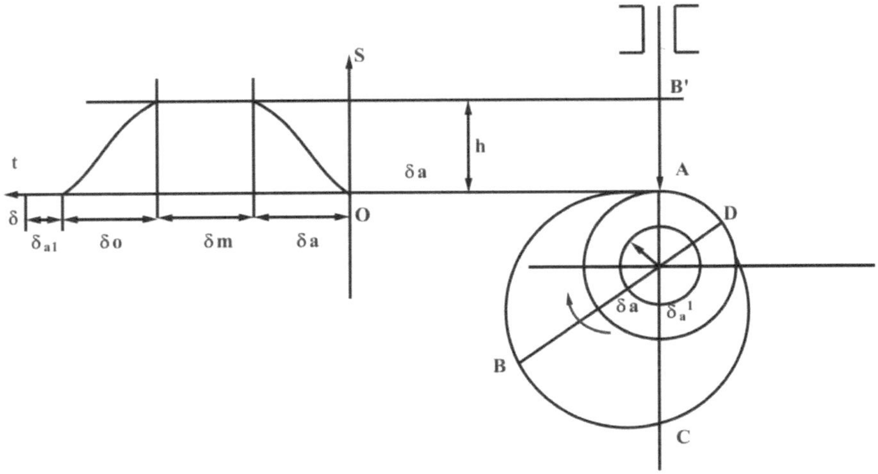

FIGURE 2.8 A Digital Cam's Trajectory that Simulates a Mechanical Cam.

2.4.2 A Multi-Objective GA Optimizes the Flexible Curve of a NURBS-Based Electronic Cam

Designing high-speed cam systems generally involves the use of high-order polynomial curves. The design must account for limitations and complicated linear equations, but assuming a high enough power number, the endpoints and accompanying high-order derivatives are continuous. The cam curve is optimized during design, and the spline function is increasingly utilized due to advancements in computer-aided design (CAD) technology. The spline function represents the generic cam curve, and the optimization approach is used to build the cam curve. The NURBS method's strong local regulatory properties and outstanding expression ability have led to its widespread application. The flexible cam curve is represented in this study using the NURBS curve. The building and solving of a multi-objective optimization model utilizing a hybrid GA was predicated on satisfying the kinematic and dynamic performance criteria while also meeting the basic motion limits [25].

2.4.2.1 The NURBS Curve Mathematical Model

The kth NURBS curve can be mathematically represented as:

$$\begin{cases} S(x) = \sum_{i=1}^{n} d_i R_{i,k(x)} \\ R_{i,k(x)} = \dfrac{w_i N_{i,k(x)}}{\sum_{j=1}^{n} w_j N_{j,k(x)}} \end{cases} \tag{2.2}$$

Where:

n: The number of node vectors
d: A control vertex's vector
wi: The weight factor
$N_{j,k(x)}$: A node vector, which defines the basic functioning of a k-th order B-spline
$R_{i,k(x)}$: The rational basis function that is k-th
$U = [u_0, u_1, \cdots u_{n+k+1}]$: Determined by the De Boor-Cox equation

Three parameters—the weight factor, the control vertex, and the node vector—define the NURBS curve. Modifying the weight factor, relocating the control vertex, and adjusting the node vector all enable one to alter the NURBS curve's shape. The NURBS curve can have its local shape changed when a parameter is changed without impacting the overall performance. The NURBS curve outperforms the polynomial curve by a wide margin because of this.

2.4.2.2 Applying the NURBS Curve

It is usual practice to use a dimensionless form when describing the cam motion curve. The following is the expression for the dimensionless displacement with respect to a dimensionless time T:

$$S = S(T), (0 \leq T \leq 1, 0 \leq S \leq 1) \tag{2.3}$$

The following is an expression for the NURBS cam curve's jerk, acceleration, velocity, and dimensionless displacement:

$$\begin{cases} S(T) = \sum_{i=1}^{n} d_1 R_{i,k(T)} \\ V(T) = \sum_{i=1}^{n} d_1 R'_{i,k(T)} \\ A(T) = \sum_{i=1}^{n} d_1 R''_{i,k(T)} \\ J(T) = \sum_{i=1}^{n} d_1 R'''_{i,k(T)} \end{cases} \tag{2.4}$$

The g-i,(x) derivative of the K-th rational basis function is known at a specific time T. satisfies

$$\sum_{i=1}^{n} R_{i,k}^{(m)}(T) = 0, (m = 1, 2, 3, \cdots) \tag{2.5}$$

We minimized calculation by optimizing the NURBS cam curve using the weight factor and control vertex location, and we characterized the node vector interval as:

$$U = \left[0,\cdots,0,\frac{1}{n-k+1},\frac{2}{n-k+1},\cdots,\frac{n-k}{n-k+1},1,\cdots,1\right] \tag{2.6}$$

The nodes at either end have a repeatability of k+1. Because the k-th NURBS curve is continuous up to the k−1 order, the jump of the cam curve is made continuous by setting (Ck−1), k to 5.

2.4.2.3 Depiction of Multi-Objective Optimization in Mathematics

Cam curve optimization aims to improve dynamics and kinematics performance while keeping geometric and natural limits intact. Cam curve design is thus an example of a common engineering use of multi-objective optimization. The optimization objects in this paper are five NURBS curves' control vertices and weight factors. Pareto optimum solution, merge and exchange target are the three most used approaches to problem optimization with multiple objectives. Due to its strong global search capabilities and rapid convergence speed, the Pareto optimal solution-based method is strongly recommended for multi-objective optimization problems [19].

The node vectors define the basic function of a B-spline of kth order.

$$\begin{cases} \min y = f(x) = \left[f_1(x), f_2(x),\cdots,f_m(x)\right]^T \\ \text{s.t. } g_i(x) \le 0 (i = 1,2,\cdots,p) \\ h_j(x) = 0 (j = 1,2,\cdots,q) \end{cases} \tag{2.7}$$

Where:

x decision vector, $x \in R^n$

For optimization problems with several goals, the Pareto optimum solution is the one that maximizes the value of each sub-objective. Sometimes, the Pareto optimal solution is the best group of solutions to a problem of multi-objective optimization.

2.4.2.4 NSGA-II GA

In this study, the cam curve was optimized using the non-dominated sorting genetic algorithm (NSGA-II GA). Figure 2.9 shows the NSGA-II GA flow. To guarantee the method's rapid convergence, the hybrid GA's general framework included an elitist NSGA-II.

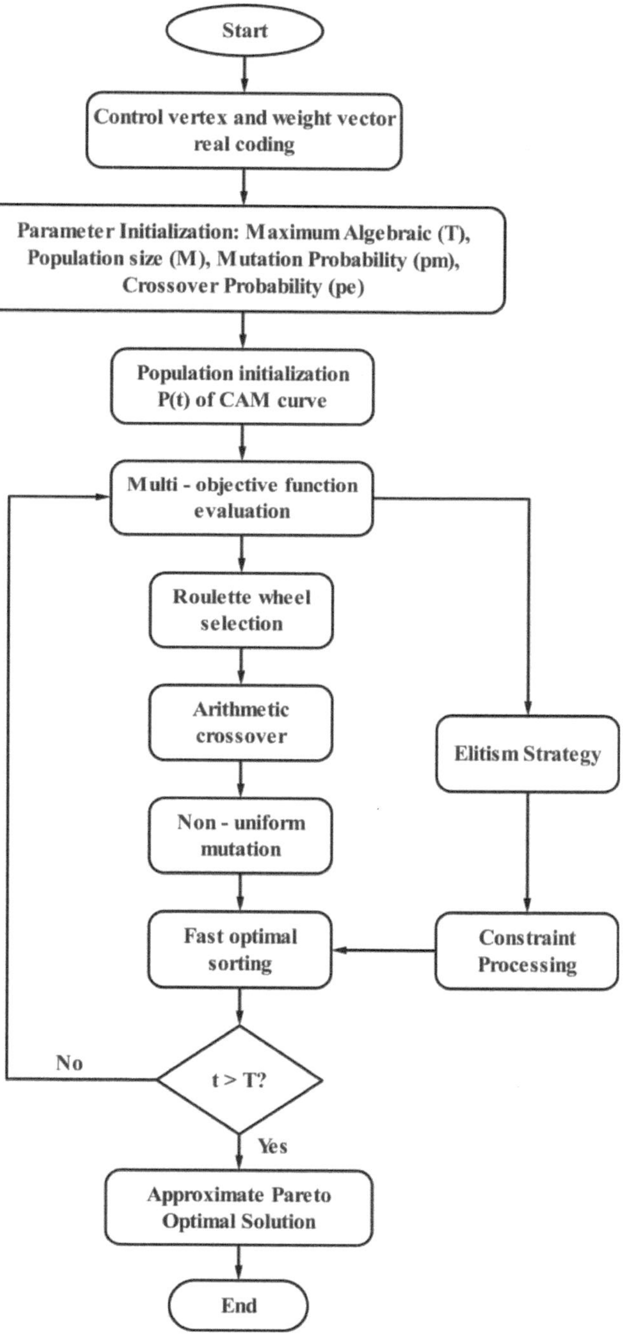

FIGURE 2.9 An NSGA-II GA Flow Diagram.

2.4.2.5 Parameter Initialization Setting

An equation displaying constraints and characteristic values, such as maximum acceleration Am and maximum jerk Jm(T), is used as the goal function for optimizing the cam curve (T).

$$\begin{cases} \min A_m(T) \\ \min J_m(T) \\ V_m(T) \leq 10.0 \\ h(T) = [S(0), V(0), A(0), S(1) - 1, V(1), A(1)]^T = 0 \end{cases} \tag{2.8}$$

Utilizing the 5th NURBS curves with k=5, n = 15, d = 20, and wi = 20, we established f* = [4,50] as the optimal acceleration and jump characteristic values. Additionally, the weight factor and number of control vertices were both set to 20. Here are the parameters that were chosen for the hybrid GA:

- 50 for the population size,
- 1000 for the maximum evolution algebra,
- 0.5 for the crossover probability, and
- 0.015 for the mutation probability.

It was a hybrid GA that optimized the two objective functions.

2.4.3 Techniques for Controlling an Electronic Camera's Motion

By doing away with the middle link in the power transmission chain, the electronic cam greatly simplifies the system structure. Its benefits include quiet operation, quick response time, minimal energy loss, and efficient transmission. The electronic cam's direct connection to the load makes its motion control susceptible to a number of uncertainties and disturbance sources, which makes system management harder [20]–[21]. Common non-linear systems that exhibit time-varying behavior are electronic cam servo motion control systems. The two key aspects that can be used to summarize the uncertainty and disruption factors are:

2.4.3.1 System Variable Uncertainty

During the course of actual operation, the system settings are frequently modified. Air gap magnetic density and coil resistance are two examples of fundamental factors that can be impacted by variations in saturation and temperature. Among other things, changes in loads and voltage can impact operating characteristics, such as driving voltage and motion quality. Finally, the friction coefficient can change after a long period of operation. Depending on their surroundings and internal conditions, these factors can shift within a specific range. A robust and adaptive control system design is essential for ensuring the system's control performance.

2.4.3.2 Disturbance in the External Load Resistance

During operation, the load force acts directly on the actuator, reducing control performance. Nevertheless, changes in the load resistance led to fluctuations in motion. For that reason, suppression capabilities are required of the control system.

By eliminating the need for an intermediate link in the power transfer, the electronic cam drive simplifies the overall design of the system. System control becomes more challenging when electronic components are directly coupled to loads, making actuator motion control vulnerable to a host of uncertainties and disturbance sources. There are essentially two sides to the uncertainty and disturbance elements [22], [23]: (1) the unpredictability of system parameters, which can occur for a variety of reasons (e.g., as a result of saturation or rising temperatures), as well as changes to fundamental parameters (e.g., coil resistance and air gap magnetic density), operational factors (e.g., motion quality, driving voltage, and load), and parameters (e.g., friction coefficient) after extended motion. The system's control performance can only be ensured by an electronic control system that is both resilient and flexible. (2) Changes to the outer resistance to load: The actuator must overcome external load resistance at each stage of the motion process in order to perform the operation. Because of the potential for control performance degradation owing to fluctuating motion caused by changes in the load resistance, it is imperative that the actuator control system has the ability to suppress the load while in motion.

This study presents a comparison of two motion control strategies that are built upon the aforementioned two points.

2.4.3.3 Digital Twin-Based Method for Electronic Cam Motion Control

The physical object uses an open-loop PID control to monitor its own displacement in real time, while the control strategy determines the distance the electronic camera should travel in order to create a virtual model. The physical object's trajectory can be precisely controlled by seamlessly integrating the virtual model with it in real time. Because of this, we can monitor the electronic cam's mobility status. Figure 2.10 shows the schematic of the control system.

2.4.3.4 Applying the Conventional Two-DOF Method of Motion Control

We validated the benefits of digital twin-based electronic camera motion control by comparing it to a conventional two-DOF motion control system.

Designed inside the motion control platform was a control system with two DOFs [24, 25]. An error proportional-derivative (PD) feedback control and a feed-forward unit inverse system are parts of the system. The inverse system's feed-forward control received its input signal from the projected trajectory. Figure 2.11 indicates that this is the control system's schematic.

This is the inverted system that the feed-forward control makes use of:

$$u_{ff} = \frac{mL}{K_m} a_r + \frac{cL}{K_m} a_r + K_e v_r + \frac{RF_0}{K_m} \quad (2.9)$$

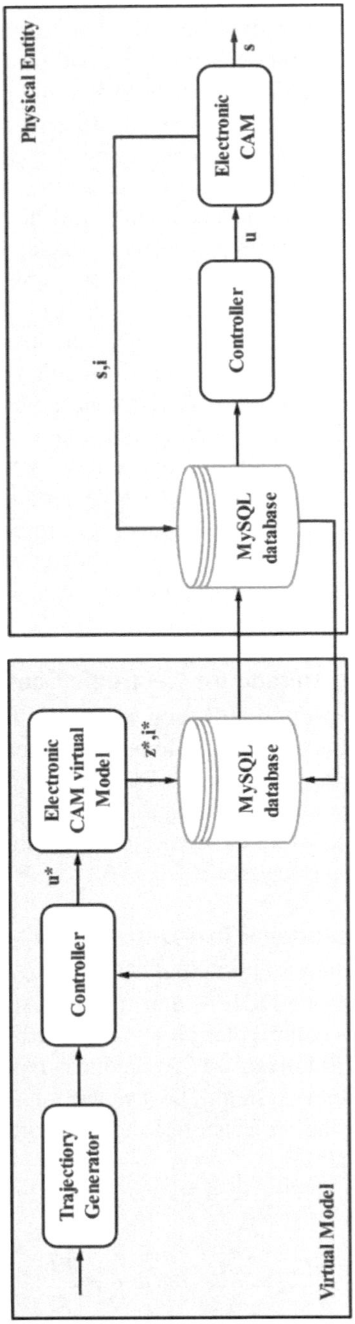

FIGURE 2.10 Flowchart for Motion Control System Using DTS.

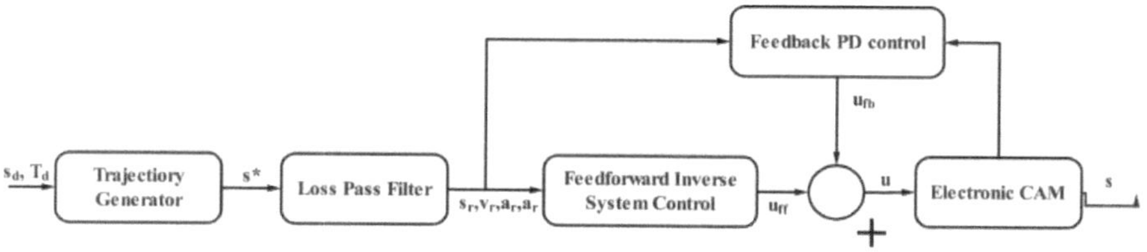

FIGURE 2.11 Motion Control System Block Diagram.

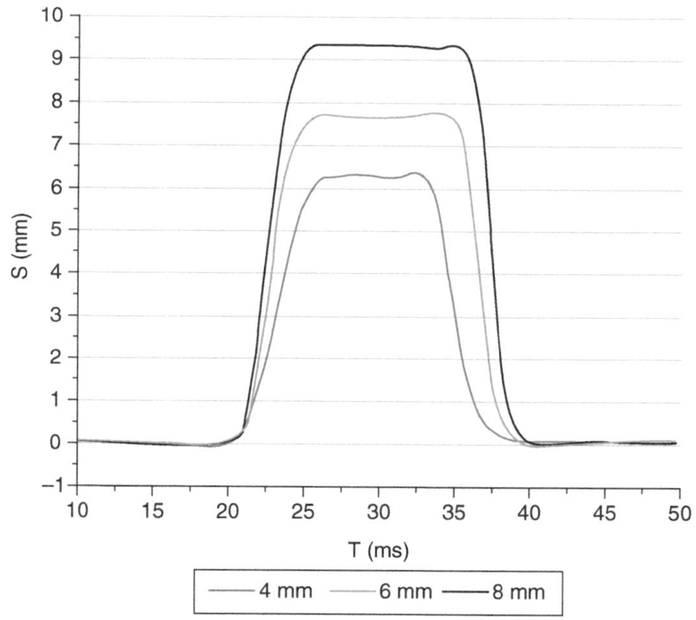

FIGURE 2.12 Various Target Displacements were Analyzed in the Simulation.

This is the following expression of PD control, which is used in feedback control:

$$u_{fb} = k_s(s_r - s) + k_v(v_r - v) \tag{2.10}$$

2.5 EXPERIMENTATION AND VIRTUAL SIMULATION

2.5.1 Electronic Cam Motion Control Simulation Using Digital Twins

Both the response time and the displacement were fine-tuned by adjusting the electronic cam's motion control parameters in real time.

2.5.1.1 Variable Displacement

Figure 2.12 displays the outcomes of the virtual testing of the electronic cam's motion in response to varying displacements of the target. Four, six, and eight millimeters

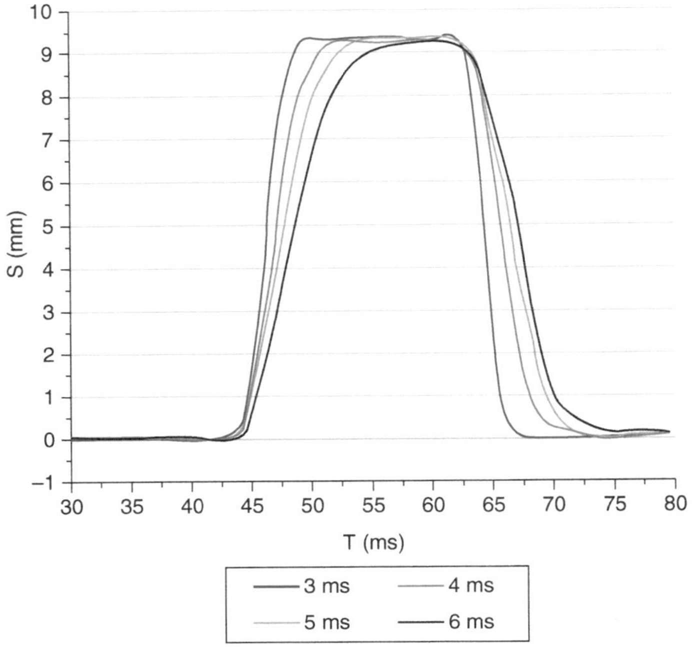

FIGURE 2.13 Various Response Times were Considered in the Simulation Analysis.

of displacement were selected as targets, with a response time of three and a half milliseconds. A continuous range of 0 to 8 mm was available for the motion displacement.

2.5.1.2 Variable Response Time

Modifying the state feedback controller's parameters allowed for a highly adjustable reaction time. The results of the simulation can be seen in Figure 2.13. The electromagnetic linear actuator was specifically designed to meet the time demands of high-speed motion devices, with a minimum reaction time of 3 ms.

2.5.1.3 Controlling and Planning Trajectory

Figure 2.14 shows the results of optimizing the electronic cam's displacement for various displacements in order to achieve motion trajectory planning. The electronic cam's trajectory was fine-tuned within the same motion period (20 ms) to enable its movement under varied situations, according to the system's needs.

2.5.2 Experiment on DTS-based Movement Control for an Electronic Cam

Figure 2.15 shows the experimental setup that was used to validate the palpable effects of the digital twin-based electronic cam servo controller. The major components are a

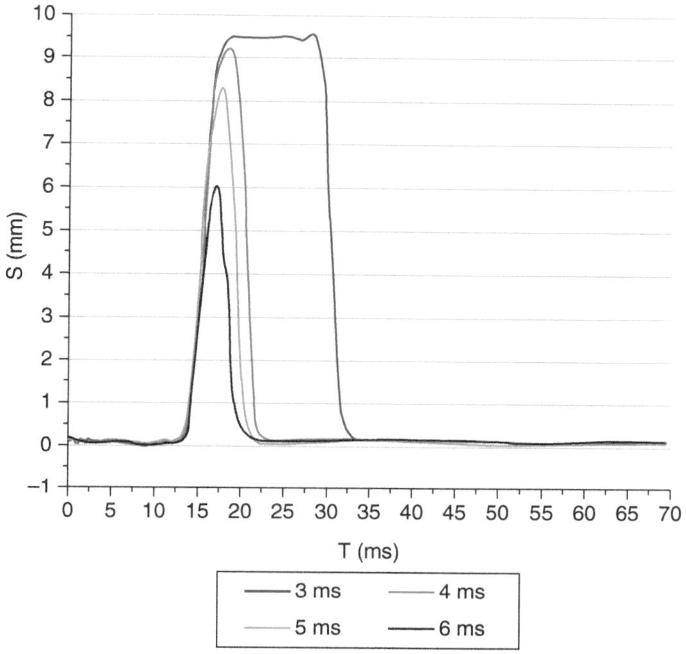

FIGURE 2.14 Various Displacements' Effects on the Electronic Cam's Motion Trajectory.

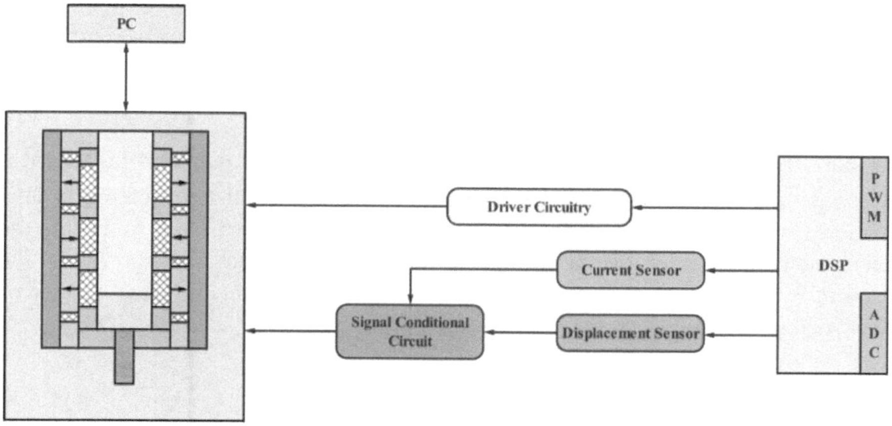

FIGURE 2.15 An Electronic Cam Motion System Block Diagram.

personal computer (PC), a digital signal processor controller, sensors for current and displacement, an electronic camera, and a drive circuit.

In this case, the digital controller was a TMS320F2812 digital signal processor (DSP) with a fixed-point operation running at 150 MHz. The input voltage

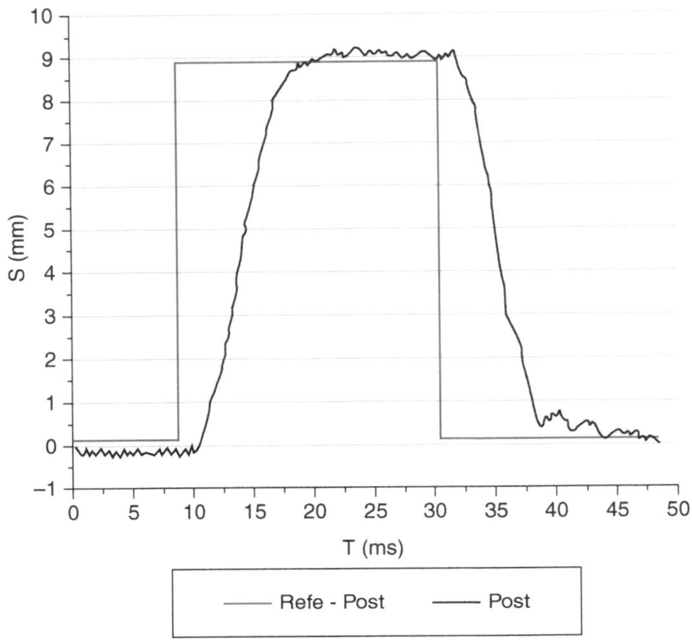

FIGURE 2.16 The Experimental Outcome with an 8 mm Displacement.

and current of an electronic cam were detected using a voltage sensor (SMIV ±50DCE) and a closed-circuit Hall-effect current detector (TBC10SY). In order to display the motion and controlling process, data was received from the DSP and sent to a PC.

2.5.2.1 Experimental Control of the Electronic Cam's Linear Displacement

Pol was the experimental outcome, and Refpos was the ideal-state reference data, as shown in Figure 2.16 At its fastest, the system's response time was 5 ms, and the electronic cam's motion displacement was set at 8 mm. Figure 2.17 shows that the electronic cam's motion displacement error was minor, matching the results of the virtual simulation.

2.5.3 Analysis of Control Performance Comparison

The black curve in Figure 2.18 displays a solution based on digital twin technology, whereas the pink curve depicts the standard two-DOF approach of motion control. After the target displacement was reached, the digital twin-based motion control system showed better stability and faster reaction times than the conventional two-DOF system, according to the results.

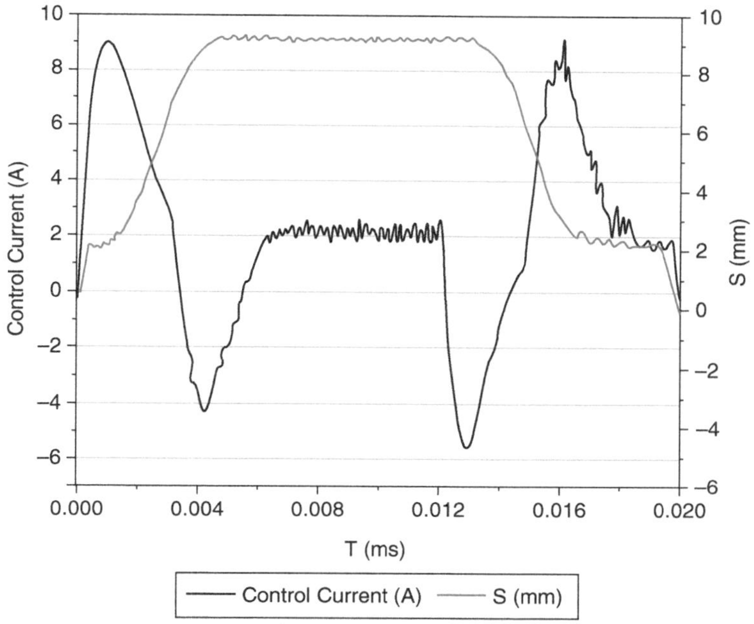

FIGURE 2.17 Current Displacement Diagram and Trajectory Planning for Targets within 8 mm.

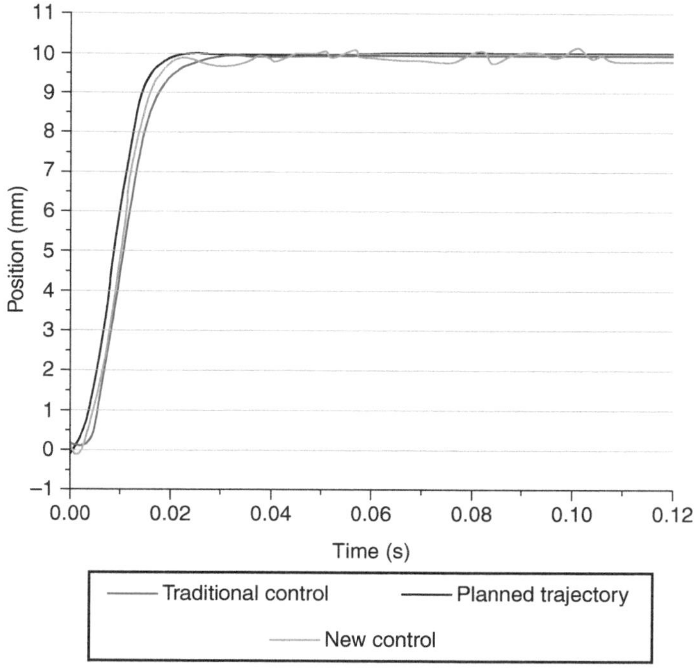

FIGURE 2.18 Performance Comparison of Two Motion Control Systems.

2.6 CONCLUSIONS

This work proposes a digital twin drive-based electronic cam servo control system after reviewing the technical applications of digital twin technology and conducting preliminary research.

1. The digital twin model's construction and simulation methods are described in detail. With the help of sensor technology and data connections, the physical entity and virtual simulation system can be fused interactively. Data from the physical system can be monitored and analyzed online in real time.
2. The GA is employed to execute the electronic cam's trajectory planning in accordance with the plans created by NURBS for various motion displacements as part of the system's motion trajectory planning.
3. The technique for managing the system's motion incorporates an examination of the open-loop PID controller utilizing digital twins, as well as the conventional feedback + feed-forward two-DOF control approach. With its digital twin as its foundation, the motion control system outperforms its conventional counterpart in terms of reaction time and real-time performance. The control method enhances the motion control system's accuracy, robustness, and adaptability.
4. Experiments and simulations confirm that the suggested plan is feasible. We are still in the early stages of developing and implementing this system, which can serve as a model for the wider application of digital twin technology in intelligent manufacturing.

REFERENCES

[1] J. Anand and K. Sivachandar, "Performance analysis of ACO-based IP Traceback," *International Journal of Computing Applications*, vol. 59, no. 1, pp. 1–5, 2012.

[2] D. Srinath, J. Janet, and J. Anand, "A survey of routing instability with IP spoofing on the internet," *Asian Journal of Information Technology*, vol. 9, no. 3, pp. 154–158, 2010.

[3] G. Ramesh, S. S. Vadivelu, and J. Anand, "A survey on wavelength division multiplexing (WDM) networks," *ICTACT Journal Of Communication Technology*, no. 01, 2010.

[4] J. Anand and D. Srinath, "Performance analysis of shared tree based multicast routing protocols for ad-hoc wireless networks," *International Journal of Engineering and Technology*, vol. 2, no. 1, pp. 78–91, 2009.

[5] R. Vaishnavi, J. Anand, and R. Janarthanan, "Efficient security for Desktop Data Grid using cryptographic protocol," in *2009 International Conference on Control, Automation, Communication and Energy Conservation*, IEEE, 2009, pp. 1–6.

[6] J. Anand, D. Srinath, R. Janarthanan, and C. Uthayakumar, "Efficient security for desktop data grid using fault resilient content distribution," *International Journal of Engineering Research and Industrial Applications*, vol. 2, no. VII, pp. 301–313, 2009.

[7] W. Ren, X. Yang, Y. Yan, Y. Hu, and L. Zhang, "A digital twin-based frame work for task planning and robot programming in HRC," in *Procedia CIRP*, Elsevier, pp. 370–375, 2021.

[8] X. Wei, W. Li, Z. Guo, B. Liu, J. Wang, and T. Wang, "Digital twin robot and its motion control for surface cleaning of carbon block," *Jisuanji Jicheng Zhizao Xitong/ Computer Integrated Manufacturing Systems, CIMS*, vol. 29, no. 6, pp. 1950–1964, 2023.

[9] J. Xu, X. Song, X. Liu, B. Chen, and H. Yan, "Modeling and system realization of assembly robot based on digital twin," *Xitong Fangzhen Xuebao / Journal of System Simulation*, vol. 35, no. 7, pp. 1497–1507, 2023.

[10] T. Guo, S. Chang, Y. Wu, and J. Xu, "Digital twin-driven motion trajectory tracking control of electromagnetic linear actuator," in *Proceedings of SPIE – The International Society for Optical Engineering*, 2023.

[11] H. Shi, J. Li, L. Guo, and X. Mei, "Control Performance evaluation of serial urology manipulator by virtual prototyping," *Chinese Journal of Mechanical Engineering (English Edition)*, vol. 34, no. 1, pp. 1–15, 2021.

[12] Y.-H. Chen, W.-T. Yang, B.-H. Chen, and P.-C. Lin, "Manipulator trajectory optimization using reinforcement learning on a reduced-order dynamic model with deep neural network compensation," *Machines*, vol. 11, no. 3, pp. 1–24, 2023.

[13] A. Chaloulos, N. Nikolakis, and K. Alexopoulos, "A model-based digital twin for adaptive trajectory planning of a robot for mixed packaging process and active collision avoidance," in *Lecture Notes in Mechanical Engineering*, Springer, 2024, pp. 583–591.

[14] Z. Zhu, Z. Lin, J. Huang, L. Zheng, and B. He, "A digital twin-based machining motion simulation and visualization monitoring system for milling robot," *International Journal of Advanced Manufacturing Technology*, vol. 127, no. 9–10, pp. 4387–4399, 2023.

[15] X. Zhang *et al.*, "Challenges and developing of the intelligent remote controlon roadheaders in coal mine," *Meitan Xuebao/Journal of the China Coal Society*, vol. 47, no. 1, pp. 579–597, 2022.

[16] X. Li, G. Liu, S. Sun, W. Yi, and B. Li, "Digital twin model-based smart assembly strategy design and precision evaluation for PCB kit-box build," *Journal of Manufacture Systems*, vol. 71, pp. 206–223, 2023.

[17] Z. Wang *et al.*, "Digital Twin System for Propulsion Design of UAVs," in *Proceedings – VRCAI 2022: 18th ACM SIGGRAPH International Conference on Virtual-Reality Continuum and Its Applications in Industry*, 2022.

[18] N. Boros, G. Kallós, and Á. Ballagi, "Implementation of trajectory planning algorithms for track serving mobile robot in ROS 2 ecosystem," *Tehnicki Vjesnik*, vol. 30, no. 4, pp. 1020–1028, 2023

[19] J. Morales, I. Castelo, R. Serra, P. U. Lima, and M. Basiri, "Vision-based autonomous following of a moving platform and landing for an unmanned aerial vehicle," *Sensors*, vol. 23, no. 2, pp. 829, 2023.

[20] W. Xia, X. Liu, C. Yue, H. Li, R. Li, and X. Wei, "Tool wear image on-machine detection based on trajectory planning of 6-DOF serial robot driven by digital twin," *International Journal of Advanced Manufacturing Technology*, vol. 125, no. 7–8, pp. 3761–3775, 2023.

[21] T. Ni, X. Sun, D. Li, Y. Zhao, P. Zhang, and Y. Deng, "Compliance control strategy of parallel robot based on external force estimation," *Nongye Jixie Xuebao/Transactions of the Chinese Society for Agricultural Machinery*, vol. 53, no. 8, pp. 443–451, 2022.

[22] R.-J. Hao, J.-Z. Wang, J.-B. Zhao, and S.-K. Wang, "Adaptive robust control for electrical cylinder with friction compensation using modified LuGre model," *Journal of Beijing Institute of Technology (English Edition)*, vol. 23, no. 3, pp. 358–367, 2014.

[23] A. Kanak *et al.*, "Verification and validation of an automated robot inspection cell for automotive body-in-white: A use case for the VALU3S ECSEL project," *Open Research Europe*, vol. 1, pp. 1–23, 2021.

[24] H. Li *et al.*, "Phenomenological modelling of camera performance for road marking detection," *Energies (Basel)*, vol. 15, no. 1, pp. 1–10, 2022.

[25] C. Wang, Y. Wan, and X. Liang, "Model establishment with Stribeck friction and fuzzy PID control for Cartesian picking robot system," in *Proceedings – 2017 Chinese Automation Congress, CAC 2017*, IEEExplore, 2017, pp. 3070–3074.

3 Unlocking Business Potential with Human Digital Twins

A Journey Toward Human-Centricity, Sustainability and Resilience

Riyaz Abdullah Sheikh

3.1 INTRODUCTION

Currently, there has been a noticeable shift in the industrial sector toward Industry 5.0, which is characterized by the integration of digital technology into operational procedures in order to boost efficiency, environmental friendliness, and robustness [1]–[3]. Figure 3.1 [3] illustrates the progression from Industry 1.0 to Industry 5.0. Improved efficiency and process automation are hallmarks of Industry 4.0 [4]–[6]. Equally neglected, however, are environmental and social metrics, as well as human health and safety. Industry 5.0 endeavors to address these constraints through its emphasis on human-centricity. This method acknowledges and capitalizes on the cognitive abilities, defect tolerance, and inventiveness of individuals. The integration of human and system capabilities enhances adaptability, resilience, and agility. Ensuring the development of labor skills as a top priority is imperative for sustaining competitiveness and relevance within an industry. An industry that adopts a human-centric approach prioritizes the interests and requirements of individuals, including their mental and physical health [7]–[8]. The paramount concerns are enduring contributions to society and the promotion of sustainable economic development. It is crucial to emphasize that advancements in technology should not undermine the protection of human rights [9].

The human-centric approach is critical in achieving sustainability and resilience in society [10]. It emphasizes placing humans at the center of production processes, meeting various human needs from safety to personal growth [4], [11]. Humans offer unique and irreplaceable values in smart manufacturing systems, such as creativity, their decision-making ability, cognition, and dexterity [12]. By working with machines and technology, human workers can achieve higher productivity in manufacturing

DOI: 10.1201/9781003465904-4

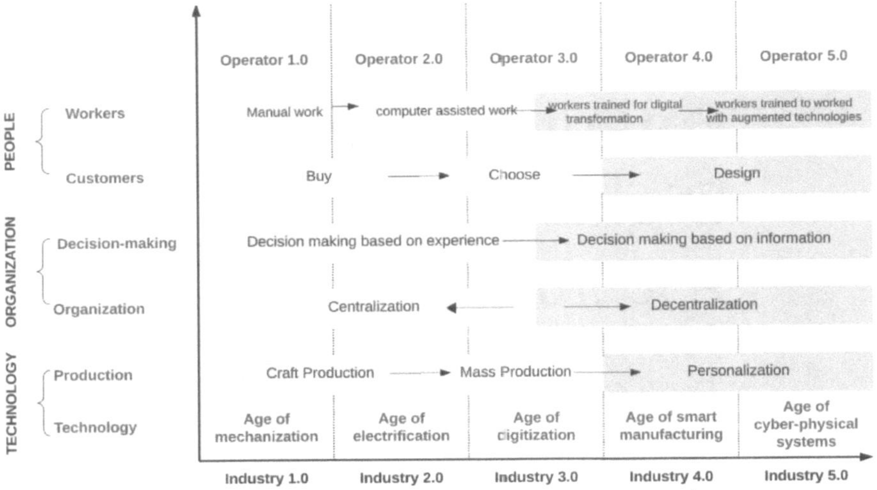

FIGURE 3.1 Transition of Industrial Revolutions from 1.0 to 5.0.

tasks, such as inspections, assembly, logistics, and maintenance, compared to manual work or automation alone [13].

Despite the significant attention devoted to the concept of Industry 5.0, further clarification regarding its form and implementation is still required. Because the human-centeredness vision put forward by researchers at EC and other institutions is so general, it needs to be looked at more closely in terms of how it can be put into practice, and more specific case studies should be included [14]. It is crucial to improve collaboration and capitalize on the respective strengths and deficiencies of humans and machines in order to progress toward a scenario where humans are prioritized [10]. To accomplish this, it is necessary to devise novel approaches that facilitate seamless and effective cooperation between human beings and machines, thereby permitting reciprocal situational support and cooperation [9]. In this new context, machines must be designed to recognize users, remember their capabilities, skills, and preferences, and adapt accordingly, eliminating the need for workers to adapt to machine requirements as is typical in traditional shop floors [15]. To achieve such a paradigm shift, extensive research is required that combines various enabling technologies, including human digital twins (HDTs). These digital twins are the building blocks of future manufacturing used to virtually represent assets, products, and other resources, facilitating the simulation of performance, prediction of failures, or investigation of problems within the factory [12]–[13]. The Industry 5.0 concept requires further clarification and more concrete case studies for its implementation. Collaboration between humans and machines is required to achieve a human-centric scenario, with HDT technology playing a pivotal role [14]–[16]. HDTs are digital representations of humans that aim to complement their strengths and prioritize their well-being in smart manufacturing systems [6]–[7], [17]. HDTs are central to realizing the goals of Industry 5.0 and are instrumental in actualizing the principles of human-centricity, sustainability, and resilience [18]–[20]. In real time, organizations can leverage a virtual representation of physical processes and assets to gain deeper insights, optimize performance, and make informed decisions [21]. Furthermore, HDTs facilitate the creation of safer and more ergonomic work environments by leveraging sensors and data analytics. They also help organizations to build robust contingency plans by modeling scenarios and assessing potential vulnerabilities. By monitoring resource utilization and environmental impact, HDTs drive sustainability [22] while providing a platform for predictive maintenance, risk mitigation, and adaptive strategies, which are essential for resilience [12]–[13]. The potential of HDTs lies in their ability to bring humans and smart devices together in smart manufacturing systems of the future, making them a crucial driver of prosperity for all involved human groups. These systems can be made human-centric, sustainable, and resilient, in line with the principles of Industry 5.0 [23]–[24].

The utilization of HDT technology is essential to the realization of Industry 5.0's goals [24]. HDTs are digital depictions of individuals that augment their capabilities, boost their efficiency, enable them to reach their full potential and place emphasis on their psychological and physical welfare within smart manufacturing systems [6]–[17]. HDTs are one of the most critical technologies that will enable Industry 5.0. They make clear and direct connections between people and technologies so that each can benefit from the other and work seamlessly with smart devices in advanced manufacturing systems [25]. HDTs revolutionize the practice of human-system integration, improve the system's design and performance, and sustainably produce personalized human skills [18]–[20] through real-time sensing, analysis, and automated feedback on humans. Smart manufacturing systems (HDTs) facilitate the development of innovative and complex industrial systems by integrating human cognitive abilities and natural senses. This contributes to the manufacturing sector's advancement and enhances prospective employees' worth [22]. Furthermore, HDTs are pivotal in driving sustainability in Industry 5.0 by monitoring and analyzing resource utilization and environmental impact [26]–[27]. Resilience is another defining characteristic of Industry 5.0, acknowledging the inevitability of disruptions and uncertainties in the global landscape. HDTs emerge as critical enablers of resilience by providing a platform for predictive maintenance, risk mitigation, and adaptive strategies. They help organizations to build robust contingency plans by modeling scenarios and assessing potential vulnerabilities.

In the context of Industry 5.0, HDTs are critical in promoting prosperity for all stakeholders involved. Unlike the previous industrial revolutions, which focused solely on cost-efficiency and profit maximization, HDTs are attracting much attention from various domains [23]–[24]. These technologies are designed to embody human-centricity, sustainability, and resilience principles. They enable organizations to visually represent physical processes and assets in real time, providing deeper insights, optimizing performance, and facilitating informativity [21]. By leveraging sensors and data analytics, HDTs also help to create safer and more ergonomic work environments. In recent years, HDTs have been the subject of extensive research and comparison on account of their immense impact [20]. Table 3.1 shows the total number

TABLE 3.1
Number of Publications Based on the Keyword Number of Publications (Year Wise and Country Wise)

Year	Documents	Country/Territory	Number of Documents
2010	12	Russian Federation	75
2011	13	Netherlands	77
2012	10	Australia	78
2013	10	India	116
2014	16	France	122
2015	9	Italy	191
2016	16	United Kingdom	219
2017	28	Germany	277
2018	51	United States	398
2019	145	China	409
2020	238		
2021	420		
2022	690		
2023	815		
2024	18		
Total	**2491**		

Source: Scopus Database

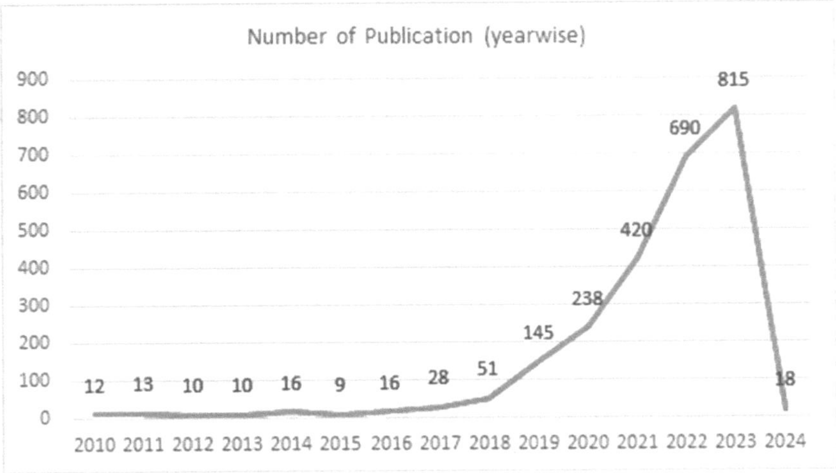

FIGURE 3.2(a) Publication of Research Year Wise.

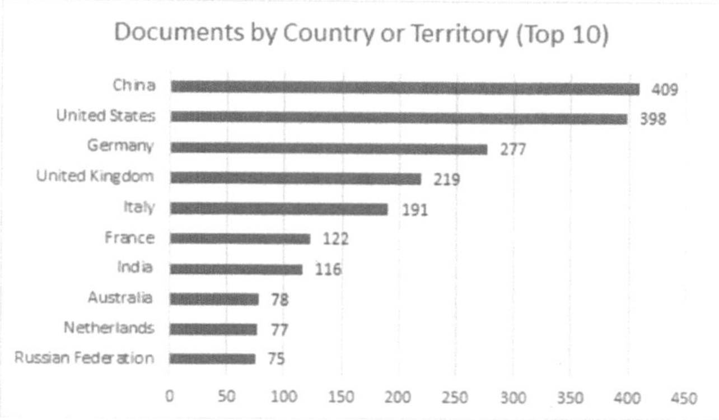

FIGURE 3.2(b) Publication of Research Country Wise (Top 10). (Source: Scopus Database.)

of documents published since 2010 on HDTs and the top ten countries in publication. They are further presented graphically in Figure 3.2.

Publications on HDTs have been on an upward trend since 2010, with a notable spike from 2018 to 2024, reaching their zenith in the latter year, as shown in Figure 3.2(a). China and the United States lead in the field, with Germany, the United Kingdom, and Italy following (Figure 3.2(b)). The varying publication numbers may indicate differing levels of emphasis or resources allocated to this field in each country. The growing interest in HDTs indicates significant potential for practical application in various industries and sectors.

The research on the HDT is expanding and has become extensive. However, some gaps in the current knowledge still require sufficient attention. This chapter's primary objective is to present a comprehensive study of an HDT, which includes its definition, framework, architecture, enabling technologies, and industrial applications. The chapter is structured as follows: Section 2 introduces the evolution of the HDT. Next, Section 3 covers the framework of the HDT, including its components and structure. Sections 4 and 5 discuss the representative applications and enabling technologies, respectively. Our observations and challenges are presented in Section 6. This study aims to provide an in-depth understanding of an HDT, its impact, and specific research areas. By accurately summarizing HDTs' meaning, we aim to bridge the gap in existing knowledge. Our analysis of the framework, architecture, enabling technologies, and industrial applications of HDT will offer valuable insights into the subject matter. In conclusion, this chapter presents a comprehensive study of HDTs, essential for a deeper understanding of their impact and significance.

3.2 HUMAN DIGITAL TWIN: FOUNDATIONS, CONCEPTS, DEFINITION, AND FEATURES

3.2.1 HDT Concepts and Definitions

Michael Grieves, a University of Michigan researcher specializing in product life cycle management (PLM), conceived the initial framework for the digital twin [28].

A digital twin is a simulation that replicates the operational characteristics of a constructed vehicle or system by incorporating various probabilistic, physics, and scale factors [29]. The HDT functions as an evolving human-centric system, facilitating bidirectional communication between the physical and electronic realms through constant updates and adjustments. HDT is a multidimensional database of humans that takes into account human interactions with other resources when digitally describing a specific entity. The development of HDTs is the result of an exhaustive examination of human representations, from the exterior to the interior. Eliminating the barrier that exists between the actual and virtual worlds is the ultimate objective.

3.2.2 Features of HDTs

The objective of this study is to establish a clear distinction between HDTs and digital twins, two terms that have garnered considerable interest in the context of Industry 5.0. HDTs focus on the development of virtual human replicas that emulate both intrinsic and extrinsic attributes, including but not limited to personality, cognition, sensibilities, and capabilities. Simultaneously, digital twins place emphasis on intangible physical entities. Intelligent interpretation and analysis, a visualization of industrial work floors, and a virtual representation of humans are integrated into HDTs to provide actionable insights [18]. Intelligent prediction and decision-making, real-time reflection, and high fidelity are what set HDTs and digital twins apart. The objective of the study is to ascertain the supplementary characteristics that are necessary for HDTs and to emphasize the anticipations for HDT technologies in the context of Industry 5.0.

According to the extant literature, social representation refers to integrated models that show characteristics associated with one or more individuals. Generally, the industrial system functions as a single-agent system. HDTs, on the other hand, are a sociomorphic multi-agent system that makes it easier for agents to connect and interact with each other. This makes it easier for agents to complete group tasks and achieve overall optimization. Within distinct contexts and circumstances, every agent comprising the HDT system endeavors to achieve common goals while preserving individual attributes [30].

The HDT system demonstrates its prowess through its capacity to amalgamate human intelligence with numerous intelligence sources that are oriented toward both human and technological perspectives. Knowledge management is iterated and revised in accordance with what is understood about single-agent and multi-agent systems [31]. The interconnection among them facilitates the exchange of resources and their mutual support, thereby rendering the aggregation of knowledge pertinent to affiliated departments and members. The HDT system must incorporate human-centric personalization as it takes into account the diverse and complex attributes of humans, which are in contrast to the physical system [32]–[33]. Personality attributes consist of traits such as trust propensity, inclination toward suspicion, and brain electrophysiologic signals. Physiological characteristics include blood oxygen levels, heart rate, and anthropometric attributes; perceptual abilities include kinematics, biomechanics, and human kinematics; and emotional states include happiness, depression, and anxiety.

It is essential, prior to implementing digital twin and HDT systems, to thoroughly assess how well they fulfill the requirements or goals of individuals, groups, and production [34]. Digital twin systems place a premium on performance-driven objectives in order to attain large-scale production and enhance economic advantages. Conversely, the human-centered paradigm of HDTs seeks to enhance the welfare of individuals through the recognition and valuing of their roles, needs, abilities, and rights. HDTs expand the range of metrics and measurements that can be employed to assess systems beyond the conventional performance metrics utilized in digital twins, including but not limited to efficiency, productivity, efficacy, and profitability. HDTs take into account metrics such as user experience, usability, and the prevention of damage during usage.

3.3 HDT CONCEPTUAL FRAMEWORK

3.3.1 Theoretical Framework of HDT

Wang *et al.* [35] conceptualized HDTs, which create a sentient digital twin in cyberspace on behalf of humans. Previous research has recognized that people are an important part of the HDT framework. However, more attention should be paid to how the different parts work together, the human-centered value that is emphasized, and other important factors [36]. The HDT theoretical framework is depicted in Figure 3.3.

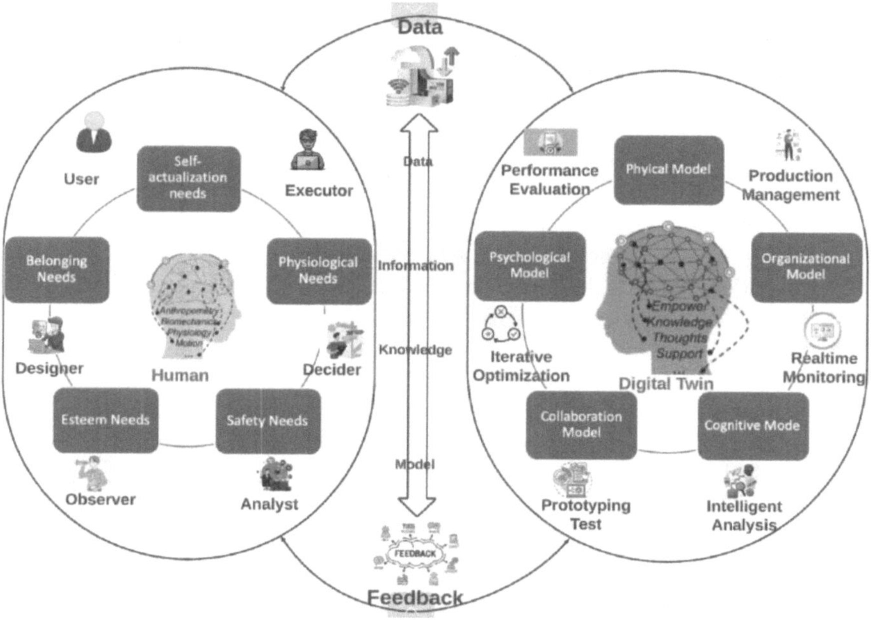

FIGURE 3.3 Theoretical Framework of HDT. (Source: Derived from [35].)

The formulation of the three-dimensional (3D) HDT model is HDT = HE, VE, and IS. IS denotes the interactive system, VE signifies the virtual entity, and HE signifies the human entity. The following sections describe each component:

1. HE: Human participation is crucial to the achievement of HDT's mission and vision. In accordance with Maslow's hierarchy of needs, there are five levels to the industrial human needs paradigm. The aforementioned levels symbolize the incremental advancement from fundamental manufacturing requirements to fully-fledged progress and self-actualization [37]. The hierarchy places physiological needs at the bottom, followed by needs for safety, belonging, esteem, and self-actualization [38]. Humans assume a multitude of roles within HDTs, including those of a user, observer, analyst, decision-maker, and executor. The designer conceptualizes and fabricates the structure, form, and associated components. The fundamental responsibilities of the observer include perception, representation, and interaction. In order to analyze, interpret, monitor, and diagnose, analysts employ their cognitive and personality attributes. The deciding individual employs human capabilities in cognition and creativity in a collaborative manner, taking into account ethical positions and values. The executor is responsible for executing the predetermined actions within the system. Human actions are categorized as either physical or cognitive. Users ultimately derive pleasure or benefits from the physical asset while contributing feedback that aids in the optimization or refinement of the entire system. The objective of the human entity's bottom-up approach is to ascertain human roles in the context of HDTs, as they pertain to industrial development requirements. Industry 5.0 offers an abundance of resources and envision HDTs performing a multitude of functions in contrast to analogous conceptual blueprints. Facilitating Industry 5.0 through the construction of diverse artificial societies, organizations, and systems requires the collaboration and participation of various positions within the system [39].
2. VE: The subsequent passage explains the notion of a VE, which prioritizes human involvement in its development with the aim of creating a perfect representation of the human entity. There are three discrete representations within the VE: anthropomorphic, materialization, and rule. Anthropomorphic depictions of humans incorporate models that embody both their physical as well as mental characteristics. On the contrary, the depiction of materialization is indicative of the characteristics, attributes, and operations of the tangible system or surroundings that are associated with the human being. Human-driven technology gains cognitive and logical abilities through rule representation, such as reasoning, decision-making, evaluation, strategy updates, and optimization [40]. The development and modification of the interactive attributes of virtual environments (VEs) occur within the framework of human-machine interactions. The VE delineates a multitude of engagements between human beings and diverse physical systems, in which the human entity actively participates in the advancement of the cyber entity platform as a whole. Coexistence, interaction, collaboration, and cooperation

are defining features of the cyber system. When a human entity adopts one of the following roles—participant, interactor, cooperator, collaborator, or user—it engages in a variety of interactions. Human-machine coexistence entails the following responsibilities from the participants: sensing, controlling, and interacting with other physical systems. In contrast to the interactor, which is introduced during the rudimentary phase of human-machine interaction, collaborators and co-operators are classified at higher levels due to their execution of more intricate tasks. Users comprise the final consumers of a given product or service. Intelligent computation, analysis, updated data and metadata, and dynamic models all contribute to the VE's ability to reflect changes in the physical world. The rules and mechanisms governing the construction and maintenance of the VE enable interactive attributes that aid in the evolution of the cyber entity platform as a whole.

3. Interactive System (IS): An IS comprises data, services, and objects that are interconnected and enable interaction between virtual entities and humans [18]. The physical objects, services, and information system data that comprise the virtual entity are contingent upon the hardware platform [41]. Integrating data models with physical, processed, virtual, and consumption data pertaining to human entities, services, and physical objects is the objective of the human-centered cyber-physical system (HCPS). A multitude of algorithms, processing techniques, and data models implemented on the hardware platform facilitate this integration. These instruments facilitate the system in performing crucial operations associated with the complete data processing cycle, including data collection, transmission, and storage [39]. The service component of the system functions to deliver applications and services to human users. Provisioning encapsulation, matching and browsing for services, modeling and evaluating the quality of services, optimizing and integrating services, and fault-tolerance management are among its pertinent theories. Within this stratum are a multitude of service models and service management models.

Moreover, tangible entities comprise a fundamental component of the HDT. These entities comprise tangible systems that engage in direct and indirect interactions with human beings, including miniature biosensors and manufacturing equipment. These objects fulfill diverse functions, such as providing leadership and assistance [41]. Assembled products or industry machinery that are involved in the product life cycle constitute the subject component, which holds the dominant position. Assembled intelligent sensors that are supplementary to manufacturing processes and products are examples of accessory components, which are additional tangible objects or natural surroundings. Adjacent components are responsible for gathering data, while an interactive system (IS) is composed of interconnected objects, services, and data that facilitate interaction between virtual entities and human beings [18]. Virtual entities have the capability to encompass physical objects, information system data, and services, contingent upon the hardware platform in use [42]. HCPSs use data models that include physical, processed, virtual, and consumption data for services, physical objects, and real and virtual people and things. In order to facilitate this integration, a multitude of algorithms, processing methods, and data models are implemented on the hardware platform. Throughout the data processing cycle, these tools empower the system to execute critical operations, including data collection, transmission, and storage [43].

3.3.2 System Architecture

As shown in Figure 3.4, a four-layer model delineates the obligations and cooperation of HDT components. The architectural structure consists of the subsequent layers: (a) the agent layer; (b) the data layer; (c) the inference layer; and (d) the implementation layer [35].

The HDT architecture has four layers with different functions. Foundational in the design, the agent layer connects human entities, physical assets, and dimensions via wearable devices and the Internet of Things (IoT) technology to prioritize human safety [44]. This layer connects the upper layers by capturing physical system parameters and human psychophysical data. The layer handles data and metadata, supporting knowledge management confidentiality and integrity through extraction, storage, and access governance. Math, simulation, and reasoning let the inference layer process facts, predict outcomes, and make judgments, emphasizing human values, autonomy, creativity, and responsibility. It uses advanced AI models to ensure secrecy and provide a human interface. End users consume inference data modules from the implementation layer. This layer keeps physical systems and people working together by taking into account each person's unique traits and helping people to make smart choices about how to use, what the benefits are, and what the risks are across several interfaces, such as product analysis, performance prediction, testing, production planning, equipment status prediction, and plant optimization [45].

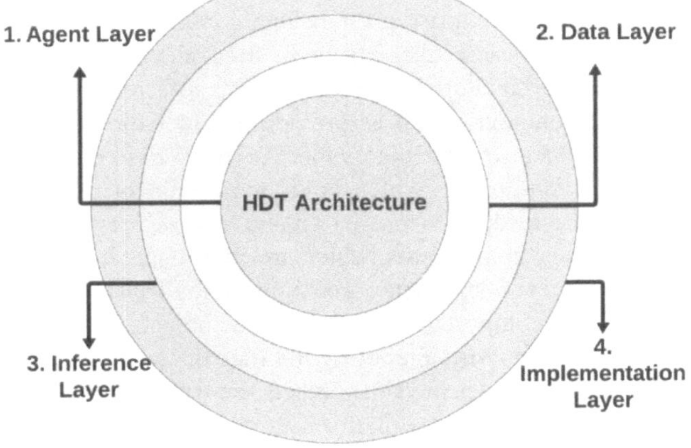

FIGURE 3.4 High-Level HDT Architecture.

3.4 ENABLING HDT TECHNOLOGIES

Within the ever-evolving context of Industry 5.0, the incorporation of HDT technologies presents itself as a revolutionary influence, holding the potential to effortlessly merge the physical and digital domains. The narrative progresses by means of a methodical structure that incorporates pivotal technologies, each making a distinct and valuable contribution toward the achievement of a symbiotic relationship between people and the digital environment [17]. This is shown in Figure 3.5.

3.4.1 Data Collection and Sensing

Utilizing data is crucial for achieving HDT, as it involves using information about humans to create accurate models. Modern sensor technologies provide technical support that enhances both human and physical systems' perception abilities. In general, the nature of the information permits the classification of sensing data into the following categories: physical, physiological, psychological, cognitive, and organizational.

1. *Physical Data:* Physical data includes body measurements, motion conditions, and other aspects collected through scanning systems and motion capture

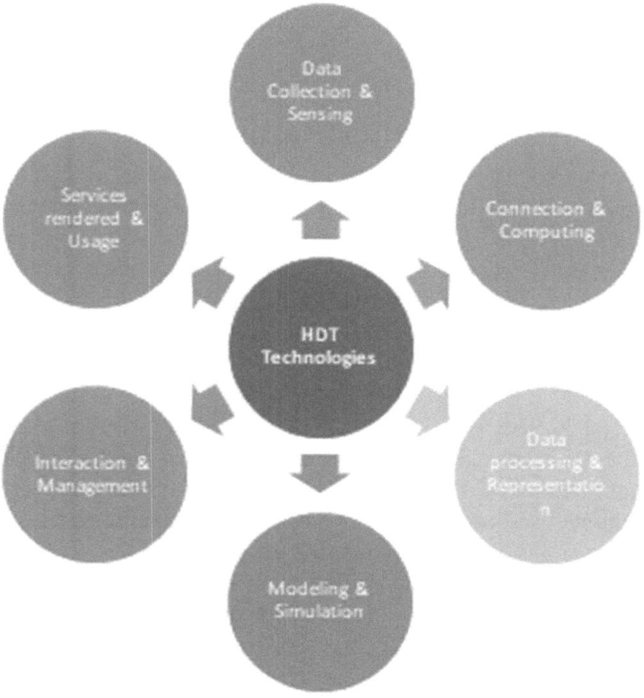

FIGURE 3.5 Categories of HDT Enabling Technologies.

[45]. Biomechanics data, such as haptic and vibration signals, provide information on human intention, comfort, performance, and health [46]. Vision systems estimate 3D human poses and calculate joint angles, while sensors detect position information [47]. Humans can reflect their physical sensations, such as physical exertion fatigue, through scaling methods like Borg RPE and Borg CR10 scales [48].

2. *Physiological Data:* Physiological data pertains to measuring an individual's various physical and mental parameters. These parameters include electroencephalography (EEG), electromyography (EMG), electrocardiogram (ECG), body temperature, breathing rate, skin conductance, and pupil diameter [49]. Non-invasive biosensors are typically employed to keep an eye on these parameters. For instance, a wireless EEG collection cap can gather human brain signals to detect fatigue [50]. To calculate mental and physical workload, physiological data can be analyzed in conjunction with other information, such as joint angles [51].
3. *Psychological and Cognitive Data:* The cognitive and psychological data that are commonly gathered via interviews or questionnaires [85]– [86] can be used to characterize an individual's subjective perception of the external environment. Flow time, error rate, production time, worker utilization, efficiency, and reaction time are a few of the performance metrics that can be assessed using performance measurement data. In contrast, the workers' information data set comprises identification numbers, roles, skill levels, credentials for equipment operation, staffing levels, flexibility levels, and individual learning rates [52].

3.4.2 Connection and Computing

A system driven by HDTs can generate a significant amount of data. This data can include information about the position and velocity of objects obtained through sensors and data generated by the virtual model. This data is sent in real time to cyberspace via the communication network to conduct further analysis. The synchronization of states is essential in establishing HDTs, as illustrated in Table 3.2, and is achieved through bidirectional, real-time communication between the physical space and cyberspace.

1. Internet-of-X: The IoT [53] refers to a network of interconnected objects that detect, collect, and exchange information regarding operational and environmental dynamics. In manufacturing, sensors on tools or workers' bodies gather human resource data for faster decision-making and service delivery. The autonomous interactions of smart devices make real-time scheduling and worker assignment possible [65]. Smart devices like smartphones, Google Glass, iPods, and smartwatches enable users to better integrate into the system [57].
2. Edge Computing and Cloud Computing: Edge computing, which entails the localization of computations in proximity to the data source, can substantially

TABLE 3.2
Major HDT Technologies for Connection and Computing

Technology	Description	Functions	Reference
 Internet of Things	This is a collection of interconnected devices that include actuators, controllers, sensors, and human operators.	Efficiently accomplish comprehensive digital integration across the complete manufacturing enterprise.	[53]
 Industrial Internet of Things	Industrial IoT (IIoT) refers to the implementation of traditional IoT principles and technologies in the business sector.	Enhance industrial operations through real-time data, predictive maintenance, and improved efficiency via interconnected devices and systems.	[53]
 Social Internet of Things	The modeling of Internet-connected devices according to the connections between humans and their daily activities is referred to as SIoT.	Enable seamless communication and collaboration between smart objects, facilitating efficient information exchange and decision-making in the IoT ecosystem.	[54]
 Internet of Services (IoS)	IoS is a paradigm that emphasizes the dynamic and scalable composition of services over the Internet to fulfill diverse user needs.	Dynamically and flexibly compose and deliver digital services over the Internet to meet diverse user requirements.	[55]

(continued)

TABLE 3.2 (Continued)
Major HDT Technologies for Connection and Computing

Technology	Description	Functions	Reference
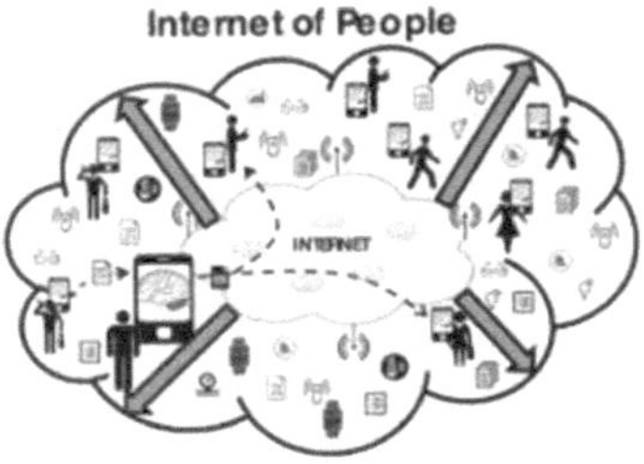 Internet of People	Humans are a part of the network structure and play an active role in its management services.	Enable direct interactions and connections between individuals through digital platforms and devices.	[56]
Internet of All (IoA)	The statement encompasses not solely conventional IoT but also incorporates human beings as an essential constituent.	Integrate and interconnect diverse entities, including devices, data, and people, fostering seamless communication and collaboration across the digital ecosystem.	[57]
 Internet of Everything (IoE)	By establishing connections between entities, data, processes, and humans, this technology improves the well-being and communication of numerous communities.	Enable the comprehensive connectivity and interaction of people, processes, data, and devices, creating a unified and intelligent network.	[58]
 Edge Computing	The text pertains to the computations that take place on the device's terminal and are not sent to the cloud.	Process and analyze data near the source of generation, reducing latency and optimizing network bandwidth.	[59]

TABLE 3.2 (Continued)
Major HDT Technologies for Connection and Computing

Technology	Description	Functions	Reference
 Cloud Computing	Users can access computing services and resources through the cloud.	This platform provides computing and storage resources to meet your needs.	[60]
Wireless Sensor Networks	Data is collected through the extensive use of numerous sensors. This data is then processed, analyzed, and transmitted as information.	Preserving the confidentiality of transmitted and collected data within the HDT system.	[61]
 5G	It is suitable for connecting many nodes in an industrial context due to its high bandwidth, speed, low latency, and reliability.	Wireless communication among HDT system modules facilitates control, real-time interaction, and high-quality simulation, among other vital capabilities.	[62]
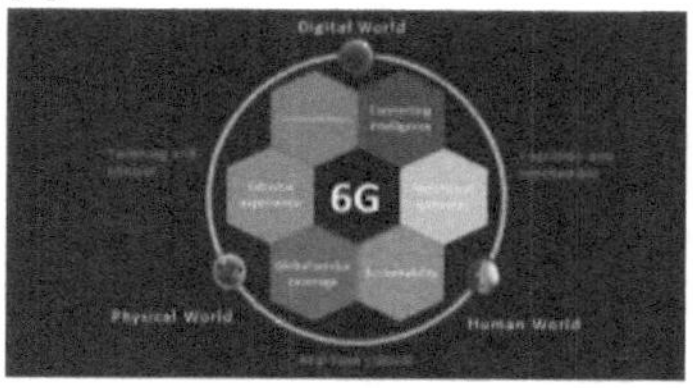 6G	This technology offers numerous advantages, such as low latency and high bandwidth for integration. Sustainability-related energy management is also a top priority.	Enable ultra-fast and reliable wireless communication, offering unprecedented data speeds, low latency, and supporting advanced technologies like holographic communication and seamless integration of physical and virtual worlds.	[27]

(continued)

TABLE 3.2 (Continued)
Major HDT Technologies for Connection and Computing

Technology	Description	Functions	Reference
Wi-Fi	This technology facilitates Internet access and local area networking of devices. It is a technology for wireless networking.	Facilitate wireless connectivity by employing radio waves to enable devices to establish connections with the internet and exchange information over brief distances.	[63]
Bluetooth	A technology standard that enables the exchange of data between mobile and fixed devices over short distances is known as wireless technology.	Enable short-range wireless communication between devices, facilitating data transfer, audio streaming, and device connectivity.	[64]

diminish transmission times that are influenced by network traffic and distance when an AI-embedded edge server is utilized [96]. In the cloud, consumers have access to computing resources and services [60]. An HDT communication architecture that is based on the edge or the cloud could include scalable nodes, such as transducers that can do a number of different computational tasks, IIoT devices, Cobots (collaborative robots), mixed reality (MR) / industrial augmented reality (AR) / virtual reality (VR) devices, embedded systems, and human-machine interfaces (HMIs) [66].

3. Wireless Communication Technologies: The handling of real-time data processing and connected devices is increasingly being done through wireless networks such as 5G, 6G, Wi-Fi, and Bluetooth [27], [62]–[64]. These cutting-edge communication technologies make human-robot collaboration possible and can also be utilized for wearable sensors equipped with camera modules that allow for wireless transmission and video reading [67]. Using multiple computation nodes in wireless networks enables faster system responses and shorter computation times.

3.4.3 Data Processing and Representation

Data obtained from the agent layer constitutes the HDT system's foundation. The aforementioned information is vital for the system to execute evaluation, testing, analysis, monitoring, and optimization operations. Either AI-based or non-AI-based methods may be applied to the data processing in order to extract relevant information. Ordinary differential equations (ODE), state-space models, and distance computation algorithms comprise the latter techniques. However, the emphasis of this section will be solely on algorithms and data pre-processing methods based on AI.

1. *Data pre-processing technologies*: Data pre-processing is the initial step in converting raw data into usable information that can be further processed. Eliminating duplicate, insignificant, deceptive, redundant, and contradictory data is the objective of the procedure. A range of techniques are used for this purpose, including filtering, segmentation, feature dimensionality reduction, augmentation, normalization, and standardization, which are discussed in several studies [68]. These methods ensure that the data is appropriate for analysis, and that meaningful patterns and insights can be extracted. Effective data pre-processing is essential for data analysis in various fields [69], and Figure 3.6 illustrates a few of the most common data pre-processing techniques and their applications in HDT systems.
2. *Algorithms for AI*: AI technology has transformed HDT capabilities by revolutionizing product design, manufacturing, and maintenance. Combining AI and other technologies has provided valuable insights into an increasingly complex and interconnected world.

A subset of AI is machine learning (ML), which employs gathered data to discern patterns and provide assistance in various tasks, including detection, recognition, prediction, optimization, and decision-making. HDTs commonly employ a variety

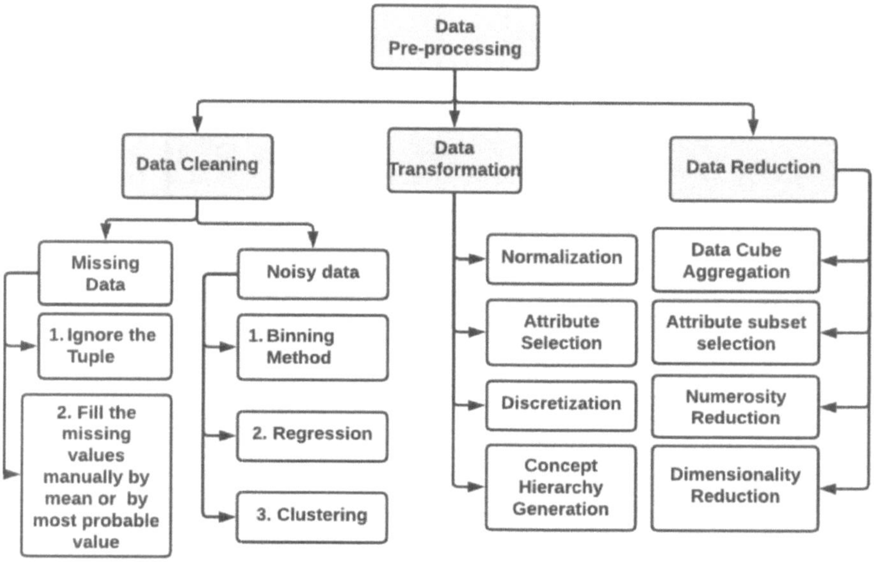

FIGURE 3.6 Common Data Pre-Processing Techniques.

of ML techniques, including K-means, K-nearest neighbors (K-NN), decision trees, Markov model, Bayesian network, and support vector machine (SVM) [69]–[70]. Unlike the manual programming of knowledge into computers, ML enables systems to acquire meaningful relationships and patterns automatically, thereby endowing them with cognitive abilities akin to those of humans.

Deep learning (DL), a subtype of ML, manages large-volume and high-dimensional data through the implementation of representation learning and artificial neural networks. For processing human-related data from multiple sources, DL is frequently more efficient than naive ML algorithms. Many HDT applications have utilized DL, including the inference of human operators' intentions during car engine assembly tasks and the recognition of human movement intentions [71].

A second type of AI, reinforcement learning (RL), discovers the optimal solution for sequential decision-making problems. "State space, action space, transition probability, policy, and reward" functions are all essential components of RL. As an example of RL algorithms utilized to resolve HDT problems, such as assembly tasks and HRC, "Deep Q-Network (DQN), Double DQN, Deep Deterministic Policy Gradient, and Deep Reinforcement Learning (DRL) (DDPG)" [72]–[73] are within this category. DRL facilitates proactive behavior adaptation by robotics, which is advantageous for HDT applications involving human uncertainty, such as assembly assistance and HRC. In addition to assembly sequence and task distribution optimization, DRL has enhanced manufacturing processes. DRL eliminates irrational sequencing and decreases the manager's workload by substituting human decision-making [72]. The algorithms and technologies depicted in Figure 3.7 are typical.

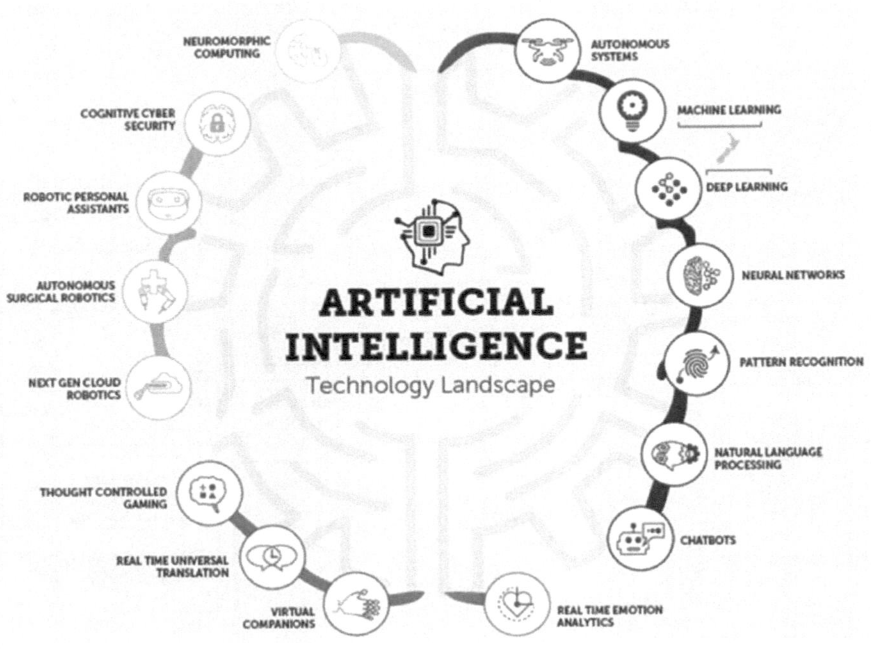

FIGURE 3.7 AI Algorithms.

The integration of deep neural networks and DRL algorithms enables the management of human uncertainty in HDT applications such as assembly assistance and HRC, thereby increasing the efficiency of manufacturing processes. In an ever more interconnected and complex world, these developments in AI technology offer a hopeful transformation of the HDT discipline and impart invaluable insights.

3.4.4 HDT Modeling and Simulation

Incorporating modeling and simulation is crucial in developing HDTs, which utilize digital representations of physical entities, employees, and operations. The data layer makes it easier for different models, such as the physical, psychological, cognitive, and interaction layers, to work together. It does this by enabling complex tasks like task assignment and ergonomic evaluation. The physical framework comprises motion, physical ergonomics, and human physical models; it uses kinematic models, algorithms for motion detection, and methods for evaluating ergonomics [74]–[75]. The psychological framework monitors and tries to guess what operators think using emotion and mental workload models based on physiological signals, voice, body language, and facial expressions [76]–[81]. The cognitive framework incorporates human intention and decision-making models, which facilitate human motion prediction and promote human-robot collaboration [82]–[83]. The interaction framework within the organization layer considers various factors, such as dynamic scheduling, ergonomics, and computer-aided design (CAD) models, when allocating tasks and designing factory layouts to ensure consistent and effective production [76], [84]. Factors relating to people, the environment, and automation affect trust between humans and robots. Dynamic Bayesian network models determine trust based on subjective human feedback [85].

3.4.5 Interaction and Management

Many facilitating technologies are critical components within HDT management and interaction. Human-robot interaction (HRI) is a burgeoning domain that facilitates improved communication between human operators and autonomous systems, specifically focusing on intelligent manufacturing [33]. Collaborative robots, also known as cobots, simplify tasks that require both entities to work together by utilizing sensing technologies that detect and interpret bioelectric, visual, auditory, and haptic information [86]. Multimodal interfaces integrating numerous sensory modalities are becoming increasingly popular to improve communication and perception [87]–[89]. Critical to guaranteeing the safety of HRI, compliance management employs techniques such as hybrid force / impedance control and impedance control to guarantee both adaptability and security [90]–[91]. By incorporating feedback modalities and VR to provide realistic remote control experiences and utilizing human motion as an alternative to conventional input devices, remote management provides a solution for hazardous or unstructured environments. Implementing hierarchical structures that allocate specific duties to humans and robotics, supervision, and shared control improve task execution and decision-making via feedback, intent detection, and arbitration [69], [92]. Subsequent investigations ought to prioritize inventive approaches to enhance collaboration and communication between human beings and robotics, to optimize task execution across diverse domains.

3.4.6 Services Rendered and Usage

Within HDT systems, technologies that facilitate service and consumption are of the utmost importance in assisting with various human responsibilities. AR is a cutting-edge technology that facilitates the performance of manufacturing layout, assembly, maintenance, and product design by superimposing computer-generated data onto the physical environment [93]. VR can replicate real and simulated environments, presenting potential use cases in data-driven decision-making, vehicle maintenance, and health monitoring [94]–[95]. Blending digital and physical components, MR has demonstrated its versatility across multiple production phases, encompassing training and inspection [96]. XR, which consists of AR, VR, and MR, augments engagement and perception with machinery [93]. Knowledge graphs (KGs) provide a structured collection of concepts, entities, and semantic connections that assist manufacturing administrators and decision-makers [97]–[98]. By harnessing the power of cloud computing and modern information technologies, cloud-based services facilitate the commercialization of user-conceived concepts by providing agile platforms that enable resource sharing, collaborative design, and constant information revision [99]. The integration of these technologies results in enhanced efficiency, comprehension, and interaction within HDT systems [100].

3.5 REPRESENTATIVE APPLICATIONS OF HDT

Industry 5.0 is a manufacturing paradigm that aims to deliver tailored, highly satisfactory products while providing a production environment that liberates humans from arduous physical and cognitive tasks. This enables them to undertake more pivotal roles, including those concerning transformation, innovation, and high-level decision-making. Human involvement is a vital consideration in this context, owing to human beings' complex, dynamic, and evolutionary nature. So, using HDT makes it possible to create and improve high fidelity, almost real-time versions of people, which makes it easier for different human roles to get direct optimization and management services. This section examines representative HDT applications throughout the product design, production, optimization, and maintenance phases of the manufacturing life cycle.

3.5.1 HDTs in Design

To tackle this concern, an innovative system utilizing cloud-based design and additive manufacturing (CDAM) has been devised to fabricate prototypes of ankle foot orthosis (AFO). This system is user-centric, placing an emphasis on their requirements. An optical scanning process is employed to generate a three-dimensional geometry of the user's foot and limb during their clinic visit. This data is subsequently converted into a designated file format and transmitted to the cyber design center. Clinicians then have the ability to modify the personalized point cloud data in accordance with the needs of the user by accessing it via a web portal and interface on a cloud server. By making the tool path for validating the HDT design scheme, the additive manufacturing software makes it easier for clinicians to solve difficult engineering analysis and design problems. Continuously executed outside of the clinic utilizing embedded IMU, the performance evaluation determines whether additional modifications are required. An expeditious, effortless, and precise customization has resulted from implementing the CDAM system, thereby enhancing the one-day visitor experience. An HDT [101] study that demonstrated the CDAM system's viability indicates a positive outlook for the system's implementation in clinical environments (Figure 3.8).

3.5.2 HDT in Production: Dynamic Collaboration between Humans and Robots

In the contemporary world of mass-customized production, collaborative robotics with humans engaged in complex manufacturing tasks must be capable of adapting to their surroundings and rendering prompt decisions. A novel approach for forecasting instances in which humans and machines desire collaboration is the fusion-based spiking neural network (FSNN), a multi-channel information processing method [102]. As an instance of real-time monitoring and decision-making, this FSNN-based module initiates robotic responses in response to alterations in the surrounding environment. The inference layer of the system integrates and analyses the HDT data acquired via video capturing, fusion, and processing, including information on assembly part status, robot posture sequence, and human behavior. Decoded output spiking signals pass as instructions to the automaton. In contrast to alternative prediction techniques like long short-term memory (LSTM) and Hidden Markhov Model (HMM), the proposed method has achieved an almost 30% improvement in decision accuracy. Robotic systems enhance productivity and operational efficiency by reducing human intervention and facilitating real-time adaptive robot assistance (Figure 3.9).

In conclusion, the proposed multi-channel information processing method using an FSNN can potentially revolutionize the collaborative task completion process for complex manufacturing tasks. It allows for real-time monitoring and decision-making, increasing accuracy and efficiency.

3.5.3 HDT in Other Life Cycle Stages: Adaptation of the Layout to Ergonomics

At the present moment, there are a multitude of approaches available to evaluate the potential for biomechanical overload; however, these methods are predominantly methodical, time-intensive, and objective in nature. To improve the precision and consistency of measurements and overcome the shortcomings of these approaches, a methodological framework has been suggested and verified in an assembly scenario. As depicted in Figure 3.10 [103], this framework is intended to facilitate monitoring and decision-making regarding the ergonomic performance of manual production lines. The suggested approach entails gathering data about working postures throughout consecutive working cycles using a wearable inertial motion monitoring

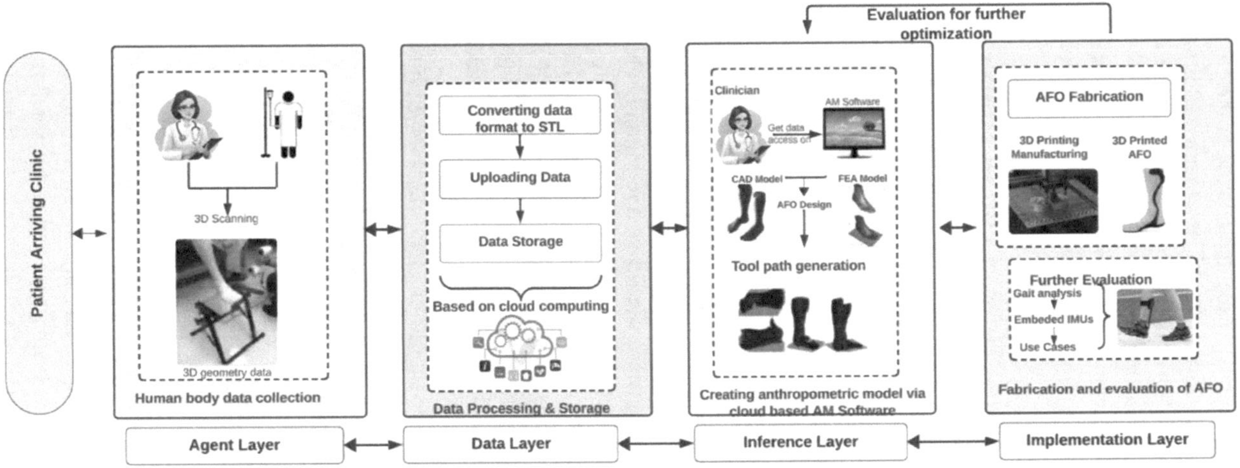

FIGURE 3.8 A Cloud-Based Additive Manufacturing and Design System for Customizing Orthosis Prototypes.

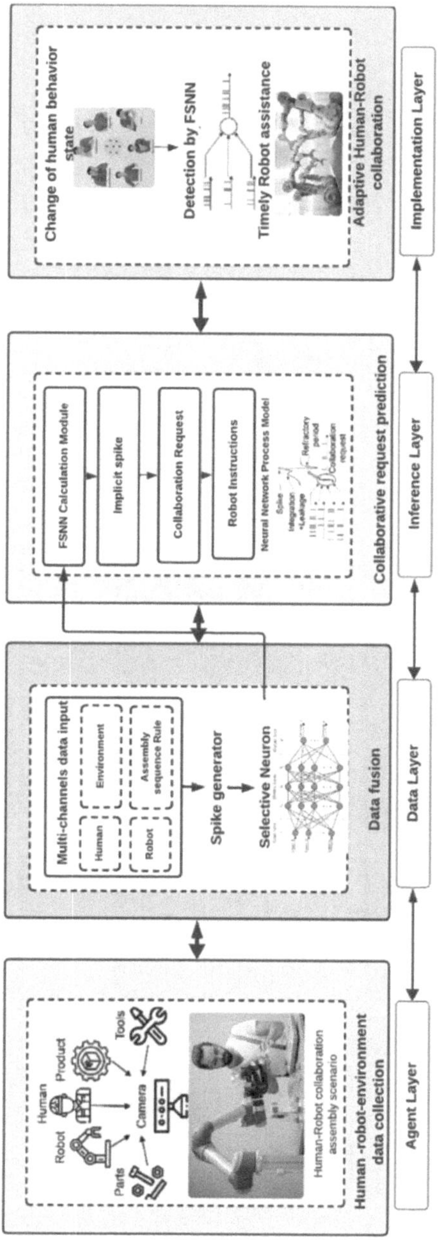

FIGURE 3.9 An FSNN-based Multi-Channel Information Processing Method for Forecasting Collaboration Requests.

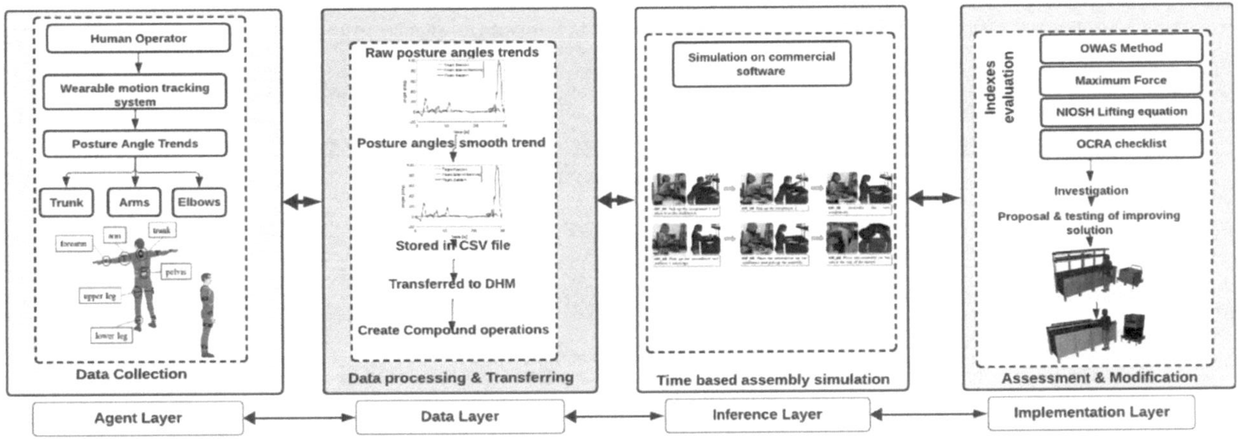

FIGURE 3.10 A Methodological HDT Framework to Facilitate Decision-Making And Monitoring With Respect to Ergonomics Performance.

TABLE 3.3
Summary of Cases Based on HDT

Cases	Data Input	Computing	Modeling	Control & Optimization	Services	References
Human-centric product prototype customization	3D geometry data of the human body	Tailored tool paths and process parameters in additive manufacturing software.	Anthropometric model for orthosis customization	Prototype model optimization through gait analysis in the usage stage.	Accurate, easy, and quick customization for designers and a one-day visit experience for users.	[101]
Dynamic collaboration between humans and robots	Image data of environmental factors affecting human robots	Utilizing the FSNNs methodology to map data to neurons and decode the spike signal into robot instructions.	Cognitive model for predicting collaborative requests	The update of robot decisions occurs concurrently with the alteration of human behavior.	Elimination of repetitious robot monitoring and control tasks and provision of the operator with real-time adaptive robot assistance.	[102]
Operator safety and worker management	Real-time location coordinates and musculoskeletal data	Fatigue analysis and safety evaluation via localization and posture detection.	A physical model including position, ergonomic, and motion detection	Precautions include adjusting work posture, and ensuring productivity in accordance with anticipated outcomes.	Enabling administrators to oversee and render decisions at any moment and in any location.	[104]

(continued)

TABLE 3.3 (Continued)
Summary of Cases Based on HDT

Cases	Data Input	Computing	Modeling	Control & Optimization	Services	References
Adaptation of the layout to ergonomics	Data on posture angles	An ergonomic evaluation using numbers for four distinct risk indices.	Physical ergonomic model and dynamic posture model	Analyzing, implementing, simulating, and assessing modification strategies for identified critical issues.	Physicians and ergonomists are then able to concentrate on problem identification and solution proposal.	[103]
Human knowledge-based decision support for maintenance	Textual maintenance requirements The process of calculating	Analyzing similarities between current cases and those previously recorded in the knowledge base.	A cognitive framework for making maintenance decisions based on knowledge	In order to update the knowledge base, one may implement a model-recommended strategy or generate, retain, and store a potentially viable new solution.	System value enhancement as an intangible knowledge asset and reduction of the cognitive burden associated with maintenance.	[105]

system. Following the transmission of the gathered data to commercial software, it is merged with human avatars and numerically simulated. Four risk indices—working postures, exerted forces, material manual handling, and repetitive actions—are assessed by analyzing the numerical data. Occupational clinicians and seasoned ergonomists should deliberate on the feasibility of layout modification if the analysis reveals critical issues. Case study results demonstrating the framework's efficacy included a cycle time of four seconds and a risk index reduction of 38.5%. This methodology enables ergonomists to concentrate predominantly on problem identification and solution proposition.

In its entirety, the suggested methodology offers a holistic strategy to tackle the obstacles traditional approaches face in evaluating the risk of biomechanical overload. Moreover, it improves measurement precision and repeatability, as well as monitoring and decision-making pertaining to the ergonomic performance of manual production lines.

A summary of the application cases of HDT in different sectors is explained in Table 3.3.

3.6 HDT ADOPTION CHALLENGES

HDT is a fundamental enabling technology that promotes human-centricity through the three pillars of Industry 5.0: digitization, digitization, and intelligentization [18]. To ensure optimal development of HDTs, it is essential to consider philosophical, social, and ethical issues and theories.

1. *Digitization*: The digital aspect of HDTs involves gathering information from multiple sources in the cyber world using multimodal sensory technologies. Nevertheless, in order to optimize the effectiveness of the HDT, it is critical to fortify connections with realism in terms of feedback. Incorporating more data sources and using extended reality (XR) technologies can significantly boost the strength of HDTs, thus enabling more informed decision-making and innovative service delivery.
2. *Networking*: HDTs rely heavily on networking to facilitate the distribution and administration of digital resources. Web 3.0, the forthcoming iteration of the World Wide Web, is characterized by its extreme decentralization and user-friendliness. A fundamental element of HDTs is digital rights management, which encompasses the storage, verification, transaction, and operation of digital assets. This enhances the integrity of HDT data and promotes the openness of the ecosystem. The widespread integration of IoT sensors and Web 3.0 has the potential to significantly improve the digital twin infrastructure in various sectors and regions.
3. *Intelligentization*: By leveraging generative AI technology, such as ChatGPT, which executes a range of functions predicated on the twin model of InstructGPT, including code debugging and responding to inquiries, HDTs can further their intelligentization. By simulating various scenarios and constructing a virtual physical system model, generative AI can predict events while maintaining data on-premises, secure, and private. This technology

has the potential to significantly impact the design and construction of HDT models.

Developing digitization and intelligentization in HDTs is challenging, including technological, application, and organizational societal issues. Addressing these challenges is critical to the optimal development of HDTs, ensuring that it continues to promote human-centricity and meets the demands of the modern world.

3.6.1 HDT Technological Challenges

1. *Standardization of technologies*: HDTs are structured as an all-encompassing development and design platform capable of managing standardization and standard formats across heterogeneous systems. This includes data formats, interfaces, communication protocols, and more. Nevertheless, the adherence of current systems to unique standards and protocols, which vary depending on the specific demands of the project (e.g., document formats, log files, and Excel spreadsheets), hinders the ability to communicate effectively with multiple systems. Furthermore, such heterogeneity can hinder such systems' subsequent development and expansion. Standardization, by contrast, enables the creation of a global collaborative environment among participants for HDTs, facilitating its integration into diverse systems [41].
2. *Seeking transparency and explanation*: Transparency and explainability are fundamental principles in personalized HDTs, as they require algorithmic decisions to be explicit and easily understood by humans. In order to fathom the algorithms that affect our lives, there has been an increasing need for transparency and explanation. It is critical to comprehend the manner in which the system obtains data, assembles conclusions, executes actions, and engages with the cyber system. Unfortunately, the current system is deficient in this respect. Transparency and explainability ought to cultivate a reciprocal relationship of trust between individuals and the physical system. Moreover, it is imperative that consumers or participants possess the necessary authority to fully understand the origins of the gathered data and its ramifications. The credibility of data and the comprehensibility of implemented algorithms are both contingent upon the attributes of transparency and explainability.

3.6.2 HDT Application Challenges

1. *Human-centric Approach*: Although existing systems assert their commitment to human-centricity, significant advancements remain in order to attain genuinely human-centric methodologies. HDTs, as opposed to conventional digital twins, prioritize human rights throughout the production process. In the pursuit of Industry 5.0, HDTs place a human-centric emphasis on sustainability and resilience, employing a direct technological methodology. To ensure sustainability, we must respect our planet's limits and our employees' welfare. By establishing adequately resilient value chains, adaptable production capacities, and flexible business processes, resilience increases the degree of

industry robustness, especially in sectors where value chains cater to fundamental human necessities [106]. Instead of focusing on the capabilities of technology for the system as a whole, HDTs emphasize their potential benefits for human sustainability and resilience [107]. At the moment, existing systems incorporate human factors to a limited extent, rendering them unsuitable for classification as HDT systems. Therefore, it is imperative to establish a precise definition of an intentional substance in order to facilitate future classification and comprehension.

2. *Exhaustive modeling*: Digital human modeling (DHM) must progress in congruence with the HDT's specifications within the context of the manufacturing process. This calls for developing interactive environments with superior fidelity that can effectively support a wide range of physical characteristics and effective communication functionalities. Furthermore, it is crucial to consider HRC in relation to efficiency, security, planning, and operation sequences, as it is an essential development in Industry 5.0 [36]. Evaluating HRC practically and efficiently through the use of modeling is therefore essential. The development of static frameworks with restricted interaction has been the extent of previous DHM research. Further research should incorporate intelligent and dynamic interaction in order to improve the development of electronic systems in HDTs.

3.6.3 Challenges of an Organizational and Societal Nature

1. *Security and privacy considerations*: HDTs are in the process of developing an HCPS-based environment for smart manufacturing component representation. Using HCPS, the goal is to develop a system capable of representing the numerous components of smart manufacturing. Innovative manufacturing systems may encompass various political authorities, cultures, and managerial perspectives due to their operation in various contexts. In such circumstances, technical collaborations are insufficient; collaborations encompassing legal, political, and ethical dimensions are also imperative. Furthermore, HDTs encompass both tangible assets, such as manufacturing systems, and intangible assets, including critical data that is not observable. It is of the utmost importance to safeguard the personal information of those who interact with the systems and the privacy and security of these assets. This is essential for the development and popularization of HDTs. Associated technologies, including sophisticated cyber-security protocols and novel system architecture design patterns, are required to address the obstacles in this context. These technologies play a critical role in protecting intangible assets against cyber-attacks.
2. *Market Awareness within the Trade*: It is our responsibility to guarantee the protection of laborers and to establish them as significant contributors to both society and industry. By integrating socio-centric and human-centric perspectives on HDTs into industrial policies, it is possible to protect the rights of groups and individuals while advancing the implementation of HDTs via value-based policies. In the interim, it is imperative for the sector to develop business models that ensure the reciprocal advantage of HDTs companies and employees.

Restructuring social policies, including health protection and welfare systems, might be necessary to foster substantial collaboration between businesses and institutions of higher learning and training.

3.4 SUMMARY

Exploring Industry 5.0 and the HDT's role represents a transformative journey. This chapter highlights the evolution from Industry 4.0 toward a more human-centric approach in Industry 5.0, emphasizing sustainability, resilience, and prosperity. The study underscores the critical significance of HDTs in fostering collaboration, optimizing performance, and ensuring well-being within smart manufacturing environments. The chapter comprehensively examines the technical aspects of HDTs, including data processing, communication, and computing requirements. The section emphasizes the real-time, low latency, and secure nature of HDT communication systems, incorporating technologies such as 5G, 6G, Wi-Fi, and Bluetooth. The transformative impact of AI, ML, and DL on product design, manufacturing, and maintenance in HDTs and the importance of modeling, simulation, and HRI frameworks are underscored. The chapter further delves into practical applications, illustrating how HDTs revolutionize industries by producing customized products, optimizing labor management, and enhancing decision-making processes. The section emphasizes the challenges associated with digitalization, networking, and intelligence in HDTs, stressing the importance of standardization and transparency and addressing application difficulties for successful integration into diverse systems. Integrating human-centric elements, including emotional and mental workload models, cognitive frameworks, and trust-building measures, is crucial for successfully implementing HDTs. The chapter also explores the emerging field of HRI, emphasizing the need for improved communication and interaction to enhance joint tasks' efficiency and success.

Finally, the chapter explores the convergence of AR, cloud services, and KGs in HDT systems, showcasing their role in enhancing human roles, safety, efficiency, and decision-making. The summary covers all aspects of HDTs, from how they were thought of as a way to change industrial paradigms to how they are used in the real world to transform industries, including all the technical details and human-centered concerns that go into their implementation. Together, these components paint a comprehensive picture of HDTs as critical enablers in the evolution of Industry 5.0 toward sustainable, resilient, and human-centered industrial systems for the future.

REFERENCES

[1] J. Cotta, M. Breque, L. De Nul, and A. Petridis, "Towards a sustainable, human-centric and resilient European industry," *Luxemb. Publ. Off. Eur. Union*, 2021, doi: 10.2777/308407.
[2] S. Nahavandi, "Industry 5.0—A human-centric solution," *Sustainability*, 11(16), 2019.
[3] M. C. Zizic, M. Mladineo, N. Gjeldum, and L. Celent, "Of paradigm shift for the people, organization and technology," *Energies*, 15, 5221, 2022.

[4] J. Alves, T. M. Lima, and P. D. Gaspar, "Is industry 5.0 a human-centred approach? A systematic review," *Processes*, vol. 11, no. 1, pp. 0–15, 2023, doi: 10.3390/pr11010193.
[5] Q. Qi, F. Tao, Y. Zuo, and D. Zhao, "Digital twin service towards smart manufacturing," *Procedia CIRP*, vol. 72, no. March, pp. 237–242, 2018, doi: 10.1016/j.procir.2018.03.103.
[6] B. Wang, P. Zheng, Y. Yin, A. Shih, and L. Wang, "Toward human-centric smart manufacturing: A human-cyber-physical systems (HCPS) perspective," *J. Manuf. Syst.*, vol. 63, 2022, doi: 10.1016/j.jmsy.2022.05.005.
[7] J. Zhou, Y. Zhou, B. Wang, and J. Zang, "Human–cyber–physical systems (HCPSs) in the context of new-generation intelligent manufacturing," *Engineering*, vol. 5, no. 4, 2019, doi: 10.1016/j.eng.2019.07.015.
[8] S. Alam *et al.*, "An overview of blockchain and IoT integration for secure and reliable health records monitoring," *Sustainability*, vol. 15, no. 7, 2023, doi: 10.3390/su15075660.
[9] S. Huang, B. Wang, X. Li, P. Zheng, D. Mourtzis, and L. Wang, "Industry 5.0 and society 5.0—Comparison, complementation and co-evolution," *J. Manuf. Syst.*, vol. 64, 2022, doi: 10.1016/j.jmsy.2022.07.010.
[10] O. Kolade and A. Owoseni, "Employment 5.0: The work of the future and the future of work," *Technol. Soc.*, vol. 71, 2022, doi: 10.1016/j.techsoc.2022.102086.
[11] F. Longo, A. Padovano, and S. Umbrello, "Value-oriented and ethical technology engineering in industry 5.0: A human-centric perspective for the design of the factory of the future," *Appl. Sci.*, vol. 10, no. 12, pp. 1–25, 2020, doi: 10.3390/APP10124182.
[12] M. Lorenzini, M. Lagomarsino, L. Fortini, S. Gholami, and A. Ajoudani, "Ergonomic human-robot collaboration in industry: A review," *Front Robotics AI*, vol. 9, 2023, doi: 10.3389/frobt.2022.813907.
[13] X. Li, Y. Koren, and B. I. Epureanu, "Complementary learning-team machines to enlighten and exploit human expertise," *CIRP Ann.*, vol. 71, no. 1, 2022, doi: 10.1016/j.cirp.2022.04.019.
[14] J. Leng *et al.*, "Industry 5.0: Prospect and retrospect," *J. Manuf. Syst.*, vol. 65, 2022, doi: 10.1016/j.jmsy.2022.09.017.
[15] Y. Lu *et al.*, "Outlook on human-centric manufacturing towards Industry 5.0," *J. Manuf. Syst.*, vol. 62, 2022, doi: 10.1016/j.jmsy.2022.02.001.
[16] G. E. Modoni and M. Sacco, "A human digital-twin-based framework driving human centricity towards Industry 5.0," *Sensors*, vol. 23, no. 13, 2023, doi: 10.3390/s23136054.
[17] B. Wang, H. Zhou, G. Yang, X. Li, and H. Yang, "Human Digital Twin (HDT) Driven human-cyber-physical systems: Key technologies and applications," *Chinese J. Mech. Eng. (English Ed.)*, vol. 35, no. 1, 2022, doi: 10.1186/s10033-022-00680-w.
[18] H. Bomström *et al.*, "Digital twins about humans—Design objectives from three projects," *J. Comput. Inf. Sci. Eng.*, vol. 22, no. 5, 2022, doi: 10.1115/1.4054270.
[19] Kazakov, V. V., Kazakov, V. G., Katsel, E. M., Nechiporenko, A. V., Panina, N. L., Pischik, B. N., & Shershnev, A. A. (2020). Personal digital twins and their sociomorphic networks: Current research trends and possibilities of the approach. In *CEUR Workshop Proceedings* (Vol. 2569, pp. 29–34). CEUR-WS.
[20] S. D. Okegbile, J. Cai, D. Niyato, and C. Yi, "Human digital twin for personalized healthcare: Vision, architecture and future directions," *IEEE Netw.*, vol. 37, no. 2, 2023, doi: 10.1109/MNET.118.2200071.
[21] M. Ghobakhloo, M. Iranmanesh, M. F. Mubarak, M. Mubarik, A. Rejeb, and M. Nilashi, "Identifying industry 5.0 contributions to sustainable development: A

strategy roadmap for delivering sustainability values," *Sustain. Prod. Consum.*, vol. 33, pp. 716–737, 2022, doi: 10.1016/j.spc.2022.08.003.

[22] X. Li, A. Nassehi, B. Wang, S. J. Hu, and B. I. Epureanu, "Human-centric manufacturing for human-system coevolution in Industry 5.0," *CIRP Ann.*, vol. 72, no. 1, 2023, doi: 10.1016/j.cirp.2023.04.039.

[23] A. Agrawal, R. Thiel, P. Jain, V. Singh, and M. Fischer, "Digital Twin: Where do humans fit in?," *Autom. Constr.*, vol. 148, 2023, doi: 10.1016/j.autcon.2023.104749.

[24] M. Ciccarelli, A. Papetti, M. Germani, A. Leone, and G. Rescio, "Human work sustainability tool," *J. Manuf. Syst.*, vol. 62, 2022, doi: 10.1016/j.jmsy.2021.11.011.

[25] E. Montini *et al.*, "An IIoT platform for human-aware factory digital twins," in *Procedia CIRP*, 2022, vol. 107, doi: 10.1016/j.procir.2022.05.042.

[26] S. Shetty, "How to use digital twins in your IoT strategy -Smarter," *Gartner*, 2017. www.gartner.com/smarterwithgartner/how-to-use-digital-twins-in-your-iot-strategy/.

[27] A. Adel, "Future of industry 5.0 in society: Human-centric solutions, challenges and prospective research areas," *Journal of Cloud Computing*, vol. 11, no. 1. 2022, doi: 10.1186/s13677-022-00314-5.

[28] M. Grieves, "Digital Twin: Manufacturing excellence through virtual factory replication," *White Paper*, 1–7, March, 2014.

[29] E. H. Glaessgen and D. S. Stargel, "The digital twin paradigm for future NASA and U.S. Air force vehicles," 2012, doi: 10.2514/6.2012-1818.

[30] J. Sun, Z. Tian, Y. Fu, J. Geng, and C. Liu, "Digital twins in human understanding: a deep learning-based method to recognize personality traits," *Int. J. Comput. Integr. Manuf.*, vol. 34, no. 7–8, 2021, doi: 10.1080/0951192X.2020.1757155.

[31] S. Saniuk, D. Caganova, and A. Saniuk, "Knowledge and skills of industrial employees and managerial staff for the Industry 4.0 implementation," *Mob. Networks Appl.*, vol. 28, no. 1, 2023, doi: 10.1007/s11036-021-01788-4.

[32] B. M. G. Rosa and G. Z. Yang, "A flexible wearable device for measurement of cardiac, electrodermal, and motion parameters in mental healthcare applications," *IEEE J. Biomed. Heal. Informatics*, vol. 23, no. 6, 2019, doi: 10.1109/JBHI.2019.2938311.

[33] A. Bilberg and A. A. Malik, "Digital twin driven human–robot collaborative assembly," *CIRP Ann.*, vol. 68, no. 1, 2019, doi: 10.1016/j.cirp.2019.04.011.

[34] E. Coronado, T. Kiyokawa, G. A. G. Ricardez, I. G. Ramirez-Alpizar, G. Venture, and N. Yamanobe, "Evaluating quality in human-robot interaction: A systematic search and classification of performance and human-centered factors, measures and metrics towards an industry 5.0," *J. Manuf. Syst.*, vol. 63, 2022, doi: 10.1016/j.jmsy.2022.04.007.

[35] B. Wang *et al.*, "Human Digital Twin in the context of Industry 5.0," *Robot. Comput. Integr. Manuf.*, vol. 85, no. April 2023, p. 102626, 2024, doi: 10.1016/j.rcim.2023.102626.

[36] L. Gualtieri, E. Rauch, and R. Vidoni, "Emerging research fields in safety and ergonomics in industrial collaborative robotics: A systematic literature review," *Robot. Comput. Integrated Manufact.*, vol. 67. 2021, doi: 10.1016/j.rcim.2020.101998.

[37] A. H. Maslow, "A theory of human motivation," *Psychol. Rev.*, vol. 50, no. 4, 1943, doi: 10.1037/h0054346.

[38] B. Wang, F. Tao, X. Fang, C. Liu, Y. Liu, and T. Freiheit, "Smart manufacturing and intelligent manufacturing: A comparative review," *Engineering*, vol. 7, no. 6, 2021, doi: 10.1016/j.eng.2020.07.017.

[39] D. Shangguan, L. Chen, C. Su, J. Ding, and C. Liu, "A triple human-digital twin architecture for cyber-physical systems," *C. – Comput. Model. Eng. Sci.*, vol. 131, no. 3, 2022, doi: 10.32604/cmes.2022.018979.

[40] G. Paul, N. D. Abele, and K. Kluth, "A review and qualitative meta-analysis of digital human modeling and cyber-physical-systems in Ergonomics 4.0," *IISE Transactions on Occupational Ergonomics and Human Factors*, vol. 9, no. 3–4. 2021, doi: 10.1080/24725838.2021.1966130.

[41] Asad, U.; Khan, M.; Khalid, A.; Lughmani, W.A. "Human- Centric Digital Twins in Industry: A Comprehensive Review of Enabling Technologies and Implementation Strategies". Sensors, 23, 3938, 2023.

[42] Q. Qi *et al.*, "Enabling technologies and tools for digital twin," *J. Manuf. Syst.*, vol. 58, 2021, doi: 10.1016/j.jmsy.2019.10.001.

[43] W. Shengli, "Is human digital twin possible?," *Comput. Methods Programs Biomed. Updat.*, vol. 1, 2021, doi: 10.1016/j.cmpbup.2021.100014.

[44] S. A. Suhluli and S. M. F. Ali Khan, "Determinants of user acceptance of wearable IoT devices," *Cogent Eng.*, vol. 9, no. 1, 2022, doi: 10.1080/23311916.2022.2087456.

[45] V. M. Manghisi, A. E. Uva, M. Fiorentino, V. Bevilacqua, G. F. Trotta, and G. Monno, "Real time RULA assessment using Kinect v2 sensor," *Appl. Ergon.*, vol. 65, 2017, doi: 10.1016/j.apergo.2017.02.015.

[46] H. Ahmadian, S. R. Hassan-Beygi, B. Ghobadian, and G. Najafi, "ANFIS modeling of vibration transmissibility of a power tiller to operator," *Appl. Acoust.*, vol. 138, 2018, doi: 10.1016/j.apacoust.2018.03.018.

[47] M. Menolotto, D. S. Komaris, S. Tedesco, B. O'flynn, and M. Walsh, "Motion capture technology in industrial applications: A systematic review," *Sensors (Switzerland)*, vol. 20, no. 19. 2020, doi: 10.3390/s20195687.

[48] T. Tashtoush *et al.*, "Human-robot interaction and collaboration (HRI-C) utilizing top-view RGB-D camera system," *Int. J. Adv. Comput. Sci. Appl.*, vol. 12, no. 1, 2021, doi: 10.14569/IJACSA.2021.0120102.

[49] A. Buerkle, H. Matharu, A. Al-Yacoub, N. Lohse, T. Bamber, and P. Ferreira, "An adaptive human sensor framework for human–robot collaboration," *Int. J. Adv. Manuf. Technol.*, vol. 119, no. 1–2, 2022, doi: 10.1007/s00170-021-08299-2.

[50] X. Zhang *et al.*, "Design of a fatigue detection system for high-speed trains based on driver vigilance using a wireless wearable EEG," *Sensors (Switzerland)*, vol. 17, no. 3, 2017, doi: 10.3390/s17030486.

[51] R. K. Khamaisi, A. Brunzini, F. Grandi, M. Peruzzini, and M. Pellicciari, "UX assessment strategy to identify potential stressful conditions for workers," *Robot. Comput. Integr. Manuf.*, vol. 78, 2022, doi: 10.1016/j.rcim.2022.102403.

[52] H. Yang, S. Kumara, S. T. S. Bukkapatnam, and F. Tsung, "The internet of things for smart manufacturing: A review," *IISE Trans.*, vol. 51, no. 11, 2019, doi: 10.1080/24725854.2018.1555383.

[53] S. Latif *et al.*, "Deep learning for the industrial internet of things (Iiot): A comprehensive survey of techniques, implementation frameworks, potential applications, and future directions," *Sensors*, vol. 21, no. 22. 2021, doi: 10.3390/s21227518.

[54] L. Atzori, A. Iera, G. Morabito, and M. Nitti, "The social internet of things (SIoT) – When social networks meet the internet of things: Concept, architecture and network characterization," *Comput. Networks*, vol. 56, no. 16, 2012, doi: 10.1016/j.comnet.2012.07.010.

[55] J. Cardoso, K. Voigt, and M. Winkler, "Service engineering for the internet of services," *Lect. Notes Bus. Inf. Process.*, vol. 19, 2009, doi: 10.1007/978-3-642-00670-8_2.

[56] M. Younis, "Internet of everything and everybody: Architecture and service virtualization," *Comput. Commun.*, vol. 131. 2018, doi: 10.1016/j.comcom.2018.07.008.

[57] D. S. Sousa Nunes, P. Zhang, and J. Sa Silva, "A Survey on human-in-The-loop applications towards an internet of all," *IEEE Commun. Surv. Tutorials*, vol. 17, no. 2, 2015, doi: 10.1109/COMST.2015.2398816.

[58] B. Galitsky and A. Parnis, "Accessing validity of argumentation of agents of the internet of everything," *In Artificial Intelligence for the Internet of Everything*, Elsevier, pp. 187–216, 2019.

[59] S. Svorobej *et al.*, "Simulating fog and edge computing scenarios: An overview and research challenges," *Future Internet*, vol. 11, no. 3. 2019, doi: 10.3390/fi11030055.

[60] G. Adamson, L. Wang, M. Holm, and P. Moore, "Cloud manufacturing–A critical review of recent development and future trends," *Int. J. Comput. Integr. Manuf.*, vol. 30, no. 4–5, 2017, doi: 10.1080/0951192X.2015.1031704.

[61] J. Yang, T. Liu, Y. Liu, and P. Morgan, "Review of human-machine interaction towards Industry 5.0: Human-centric smart manufacturing," in *Proceedings of the ASME Design Engineering Technical Conference*, 2022, vol. 2, doi: 10.1115/DETC2022-89711.

[62] J. Cheng, Y. Yang, X. Zou, and Y. Zuo, "5G in manufacturing: A literature review and future research," *Int. J. Adv. Manuf. Technol.*, 2022, doi: 10.1007/s00170-022-08990-y.

[63] P. S. Henry and H. Luo, "Wi-Fi: What ' s Next?," *IEEE Communication Magazine*, vol. 40, no. 12, pp. 66–72, 2002.

[64] P. McDermott-Wells, "What is Bluetooth?," *IEEE Potentials*, vol. 23, no. 5. 2004, doi: 10.1109/MP.2005.1368913.

[65] S. Strohmeier, "Smart HRM–A Delphi study on the application and consequences of the Internet of Things in Human Resource Management," *Int. J. Hum. Resour. Manag.*, vol. 31, no. 18, 2020, doi: 10.1080/09585192.2018.1443963.

[66] P. Fraga-Lamas, D. Barros, S. I. Lopes, and T. M. Fernández-Caramés, "Mist and edge computing cyber-physical human-centered systems for Industry 5.0: A cost-effective IoT thermal imaging safety system," *Sensors*, vol. 22, no. 21, 2022, doi: 10.3390/s22218500.

[67] F. Chen *et al.*, "WristCam: A wearable sensor for hand trajectory gesture recognition and intelligent human-robot interaction," *IEEE Sens. J.*, vol. 19, no. 19, 2019, doi: 10.1109/JSEN.2018.2877978.

[68] E. Debie *et al.*, "Multimodal fusion for objective assessment of cognitive workload: A review," *IEEE Trans. Cybernetics*, vol. 51, no. 3. 2021, doi: 10.1109/TCYB.2019.2939399.

[69] J. Asensio-Cubero, J. Q. Gan, and R. Palaniappan, "Extracting optimal tempo-spatial features using local discriminant bases and common spatial patterns for brain computer interfacing," *Biomed. Signal Process. Control*, vol. 8, no. 6, 2013, doi: 10.1016/j.bspc.2013.07.004.

[70] R. Sheikh and N. Goje, "Role of big data analytics in business transformation," in *Internet of Things in Business Transformation*, Wiley Online Library, 2021.

[71] A. Buerkle, W. Eaton, N. Lohse, T. Bamber, and P. Ferreira, "EEG based arm movement intention recognition towards enhanced safety in symbiotic Human-Robot Collaboration," *Robot. Comput. Integr. Manuf.*, vol. 70, 2021, doi: 10.1016/j.rcim.2021.102137.

[72] R. Zhang, Q. Lv, J. Li, J. Bao, T. Liu, and S. Liu, "A reinforcement learning method for human-robot collaboration in assembly tasks," *Robot. Comput. Integr. Manuf.*, vol. 73, 2022, doi: 10.1016/j.rcim.2021.102227.

[73] Z. Liu, Q. Liu, L. Wang, W. Xu, and Z. Zhou, "Task-level decision-making for dynamic and stochastic human-robot collaboration based on dual agents deep reinforcement

learning," *Int. J. Adv. Manuf. Technol.*, vol. 115, no. 11–12, 2021, doi: 10.1007/s00170-021-07265-2.
[74] T. Maruyama *et al.*, "Digital twin-driven human robot collaboration using a digital human," *Sensors*, vol. 21, no. 24, 2021, doi: 10.3390/s21248266.
[75] T. Falcari, O. Saotome, R. Pires, and A. B. Campo, "Evaluation of multi-class support-vector machines strategies and kernel adjustment levels in hand posture recognition by analyzing sEMG signals acquired from a wearable device," *Biomed. Eng. Lett.*, vol. 10, no. 2, 2020, doi: 10.1007/s13534-019-00141-9.
[76] M. Faccio *et al.*, "Human factors in cobot era: A review of modern production systems features," *J. Intell. Manuf.*, vol. 34, no. 1, 2023, doi: 10.1007/s10845-022-01953-w.
[77] M. Hita-Gutiérrez, M. Gómez-Galán, M. Díaz-Pérez, and Á. J. Callejón-Ferre, "An overview of reba method applications in the world," *Int. J. Environ. Res. Public Health*, vol. 17, no. 8. 2020, doi: 10.3390/ijerph17082635.
[78] K. Koushik Balaji and M. S. Alphin, "Computer-aided human factors analysis of the industrial vehicle driver cabin to improve occupational health," *Int. J. Inj. Contr. Saf. Promot.*, vol. 23, no. 3, 2016, doi: 10.1080/17457300.2014.992351.
[79] R. Gervasi, F. Barravecchia, L. Mastrogiacomo, and F. Franceschini, "Applications of affective computing in human-robot interaction: State-of-art and challenges for manufacturing," *Proc. Instit. Mech. Eng., Part B: J. Eng. Manufact.*, vol. 237, no. 6–7. 2023, doi: 10.1177/09544054221121888.
[80] D. Kulíc and E. Croft, "Affective state estimation for human-robot interaction," *IEEE Trans. Robotics*, 2007, vol. 23, no. 5, doi: 10.1109/TRO.2007.904899.
[81] J. Oliveira and I. Praca, "On the usage of pre-trained speech recognition deep layers to detect emotions," *IEEE Access*, vol. 9, 2021, doi: 10.1109/ACCESS.2021.3051083.
[82] L. Bi, A. Feleke, and C. Guan, "A review on EMG-based motor intention prediction of continuous human upper limb motion for human-robot collaboration," *Biomed. Signal Process. Control*, vol. 51. 2019, doi: 10.1016/j.bspc.2019.02.011.
[83] K. Sycara, C. Lebiere, Y. Pei, D. Morrison, Y. Tang, and M. Lewis, "Abstraction of analytical models from cognitive models of human control of robotic swarms," In *Proceedings of ICCM 2015 - 13th International Conference on Cognitive Modeling* (pp. 13–18). University of Groningen, 2015.
[84] Q. Lv, R. Zhang, X. Sun, Y. Lu, and J. Bao, "A digital twin-driven human-robot collaborative assembly approach in the wake of COVID-19," *J. Manuf. Syst.*, vol. 60, 2021, doi: 10.1016/j.jmsy.2021.02.011.
[85] A. Xu and G. Dudek, "OPTIMo: Online Probabilistic Trust Inference Model for asymmetric human-robot collaborations," in *ACM/IEEE International Conference on Human-Robot Interaction*, 2015, vol. 2015-March, doi: 10.1145/2696454.2696492.
[86] A. Mohammed and L. Wang, "Brainwaves driven human-robot collaborative assembly," *CIRP Ann.*, vol. 67, no. 1, 2018, doi: 10.1016/j.cirp.2018.04.048.
[87] T. Xue, W. Wang, J. Ma, W. Liu, Z. Pan, and M. Han, "Progress and prospects of multimodal fusion methods in physical human-robot Interaction: A review," *IEEE Sensors J.*, vol. 20, no. 18. 2020, doi: 10.1109/JSEN.2020.2995271.
[88] M. P. Mayer, B. Odenthal, M. Faber, C. Winkelholz, and C. M. Schlick, "Cognitive engineering of automated assembly processes," *Hum. Factors Ergon. Manuf.*, vol. 24, no. 3, 2014, doi: 10.1002/hfm.20390.
[89] C. Cui, G. Bin Bian, Z. G. Hou, J. Zhao, and H. Zhou, "A multimodal framework based on integration of cortical and muscular activities for decoding human intentions about lower limb motions," *IEEE Trans. Biomed. Circuits Syst.*, vol. 11, no. 4, 2017, doi: 10.1109/TBCAS.2017.2699189.

[90] S. G. Khan, G. Herrmann, M. Al Grafi, T. Pipe, and C. Melhuish, "Compliance control and human-Robot interaction: Part 1-survey," *Int. J. Humanoid Robot.*, vol. 11, no. 3. 2014, doi: 10.1142/S0219843614300013.
[91] A. L. Shoushtari, P. Dario, and S. Mazzoleni, "A review on the evolvement trend of robotic interaction control," *Indus. Robot*, vol. 43, no. 5. 2016, doi: 10.1108/IR-02-2016-0073.
[92] S. Music and S. Hirche, "Control sharing in human-robot team interaction," *Ann. Rev. Control*, vol. 44. 2017, doi: 10.1016/j.arcontrol.2017.09.017.
[93] S. Doolani *et al.*, "A review of Extended Reality (XR) technologies for manufacturing training," *Technologies*, vol. 8, no. 4, 2020, doi: 10.3390/technologies8040077.
[94] Z. Guo *et al.*, "Applications of virtual reality in maintenance during the industrial product lifecycle: A systematic review," *J. Manufact. Syst.*, vol. 56. 2020, doi: 10.1016/j.jmsy.2020.07.007.
[95] S. Shah, S. Petluru, R. Singh, and S. Srivastava, "GAR-age: A feedback-enabled blended ecosystem for vehicle health monitoring," 2018, doi: 10.1145/3239060.3239075.
[96] S. Rokhsaritalemi, A. Sadeghi-Niaraki, and S. M. Choi, "A review on mixed reality: Current trends, challenges and prospects," *Appl. Sci. (Switzerland)*, vol. 10, no. 2, 2020, doi: 10.3390/app10020636.
[97] B. Zhou, J. Bao, Z. Chen, and Y. Liu, "KGAssembly: Knowledge graph-driven assembly process generation and evaluation for complex components," *Int. J. Comput. Integr. Manuf.*, vol. 35, no. 10–11, 2022, doi: 10.1080/0951192X.2021.1891572.
[98] L. He and P. Jiang, "Manufacturing knowledge graph: A connectivism to answer production problems query with knowledge reuse," *IEEE Access*, vol. 7, 2019, doi: 10.1109/ACCESS.2019.2931361.
[99] P. Busch and D. Richards, "Modelling tacit knowledge via questionnaire data," in *Lecture Notes in Artificial Intelligence (Subseries of Lecture Notes in Computer Science)*, 2004, vol. 2961, doi: 10.1007/978-3-540-24651-0_27.
[100] S. Mihai *et al.*, "Digital Twins: A survey on enabling technologies, challenges, trends and future prospects," *IEEE Commun. Surv. Tutorials*, vol. 24, no. 4, pp. 2255–2291, 2022, doi: 10.1109/COMST.2022.3208773.
[101] A. Shih, D. W. Park, Y. Y. Yang, R. Chisena, and D. Wu, "Cloud-based design and additive manufacturing of custom orthoses," *Procedia CIRP*, 2017, vol. 63, doi: 10.1016/j.procir.2017.03.355.
[102] R. Zhang, J. Li, P. Zheng, Y. Lu, J. Bao, and X. Sun, "A fusion-based spiking neural network approach for predicting collaboration request in human-robot collaboration," *Robot. Comput. Integr. Manuf.*, vol. 78, 2022, doi: 10.1016/j.rcim.2022.102383.
[103] A. Greco, M. Caterino, M. Fera, and S. Gerbino, "Digital twin for monitoring ergonomics during manufacturing production," *Appl. Sci.*, vol. 10, no. 21, 2020, doi: 10.3390/app10217758.
[104] I. Ahmed, A. M. Medabesh, K. H. Mehrez, N. Shili, C. Socci, and S. Deriu, "Socioeconomic spillovers of COVID-19: case of Saudi Arabia," *Kybernetes*, vol. 52, no. 3, 2023, doi: 10.1108/K-04-2021-0275.
[105] D. Romero and J. Stahre, "Towards the resilient operator 5.0: The future of work in smart resilient manufacturing systems," *Procedia CIRP*, vol. 104, no. March, pp. 1089–1094, 2021, doi: 10.1016/j.procir.2021.11.183.

4 Digital Platforms for Business Applications

R. Jeyabharath, Sumanth Ratna Kandavalli, Shrinwantu Raha, Saptarshi Mukherjee, Abhinav Singhal, Mayakannan Selvaraju and A. Jose Anand

4.1 INTRODUCTION

As digital platforms enable organizations to become more digitized, they are predicted to revolutionize future value generation and are already disrupting nearly every industry. Due to digital platforms' cutting-edge innovation, businesses can now take a more analytical and predictive approach to product and process improvement, which has vast implications for the industry. To be more precise, businesses will not be able to function efficiently enough without technology that bridges the gap between the real and virtual worlds. Those businesses that do not enable and take advantage of these prospects will be left behind by their competitors. Also, digital platforms provide for new ways of interacting and working together, which should fundamentally speed up innovation and alter manufacturing organizations' operations [1]. According to authors, digital platforms mediate between various user groups using a combination of technical components (hardware and software) and related organizational activities. Digital platforms can be categorized according to organizational context, with openness level being the key feature. The main difference between an industry platform—which is accessible to all parties—and an internal platform—which is accessible exclusively to employees from a single company—is that the latter is open to all parties. The degree to which supply-chain platforms are accessible to third parties is a matter of personal decision for the platform owner [2].

Platforms for creating digital twins, exact virtual replicas of processes, products, and performance, have garnered much interest in recent years as a technology. One of the most anticipated technologies of the next several years is digital twins, which has been riding high on the hype cycle for new technology [3]. In the same way that a supply-chain platform allows for the collaboration of numerous external parties, this platform's owner controls which of these parties can access the platform. The development of digital twins that faithfully mimic all aspects of reality enables businesses

DOI: 10.1201/9781003465904-5

to virtually test and commission assembly lines prior to their actual construction. The great degree of detail that an exact reproduction would entail, however, is of little use in present practice [4]. Rather, digital twins differ in specifics according to each client's demands. Digital twins have many potential uses and can shorten the time that it takes to bring new products to a market, lower commissioning costs and times, and increase uptime through predictive analytics. However, there is currently little information available on how to use digital platforms such as digital twins effectively [5].

The prospects presented by these platforms are exciting, but getting people to use them is difficult. Cooperation is critical to the success of the simulation platform because the virtual models rely on specific data that a third party often owns. There is a lot more room for error when using a digital twin than when using an internal platform because data is required from all parties involved in the supply chain [6]–[7]. One major problem that could arise, for instance, is when third parties are eager to divulge sensitive information. For instance, in order to build a digital twin of their operations, a manufacturing organization requires comprehensive data on all machinery located at their site. On the other hand, a machine creator may be reluctant to divulge their machine's inner workings for fear of being exposed. The digital doppelganger loses some of its value because it does not represent reality. More studies are required in this area, but for now, it is clear that all parties involved must work together to reap the benefits of simulation platforms [8]–[9].

Companies still struggle to build relationships with external parties, despite the fact that they know this is going to be crucial. Constantinides and Henfridsson, all point to data-sharing as a big problem [10]–[11]. Another difficulty is that it is not always apparent who owns what data, and some businesses see becoming dependent on others as unsafe. Additionally, digital platforms frequently require substantial investments that must be recouped through the value that is produced, thus complicating the cooperation [12]–[13]. Nevertheless, digital twins' worth is tied to the involvement of all relevant parties; if they do not work together, then their value will be diminished. Thus, it is challenging for a small group of people to launch a digital platform partnership without the full supply buy-in chain's. There is a lot of pressure on the manufacturing sector to digitalize in order to remain competitive, and digital platforms are essential to this effort. Consequently, it is critical to learn how businesses may inspire and oversee teamwork on digital platforms [14].

While there is some literature on digital platforms, most of it is concerned with the technology itself and the difficulties of developing them, rather than the management elements [15]–[16]. A thorough examination of the difficulties associated with collaboration is lacking, even though several scholars have conducted ground-breaking work in the field. Consequently, the topic of potential obstacles and how to handle them has received even less focus. A one-size-fits-all approach is also unlikely to work because every partnership has its own specific requirements and difficulties [17]. Thus, there is a dearth of qualitative studies regarding the management of difficulties associated with digital platform collaboration. Studying and understanding the mechanics behind digital platform collaboration is necessary due to the novelty of the subject and the increasing vitality of industrial digitalization.

With this backdrop in mind, the study's overarching goal is to build a contingency framework that will shed light on the best practices for digital platform collaboration in the manufacturing sector. This will add to the existing body of literature on digital platforms by highlighting the difficulties of managing collaboration on these platforms from a managerial viewpoint. This area has received little attention in prior research. Ultimately, the results will help manufacturing companies to maintain their competitive edge by facilitating the smooth implementation and operation of digital platforms. In order to achieve its goals, this study will offer a comprehensive case study of how manufacturing companies work around platform-type digital twins. Analyzing the data obtained will provide a foundation for finding the right ways to handle the difficulties of certain partnerships.

4.2 BACKGROUND THEORY

4.2.1 The Idea of a Digital Platform

The literature offers a plethora of platform kinds, each with its own unique definition and set of attributes. Authors separate the debate around digital platforms into two camps, economics and engineering design, to aid readers in navigating the field of literature. From an economic point of view, online marketplaces such as Facebook and Twitter bring together buyers and sellers who may not have been able to meet otherwise. Uber is an example of this type of platform because it facilitates transactions between drivers and riders. According to authors, the engineering design perspective is more applicable to this study because it sees digital platforms as a set of technical components (hardware and software) and related organizational processes that mediate between various user groups. An essential principle is that the platform is modular, allowing for the decomposition of complicated systems into more manageable parts that can be linked through interfaces.

The following is a summary of some of the features that differentiate platforms, as shown in Table 4.1. There are three types of platforms: internal, supply chain, and industry. The latter two types are external, while the former two are internal

TABLE 4.1
An Overview of the Many Types of Platforms and Their Attributes

	Internal platform	Supply-chain platform	Industry platform
Interpretation level	Company	Supply chain	Industry network
Level of transparency	Closed-loop systems	Interfaces that are selectively open and closed	accessible interfaces
Innovation talents that are easily accessible	Company's capacities	Supply-chain competencies	An infinite quantity of outside sources
Control mechanisms	Hierarchy of managers	Agreements among supply-chain firms	Governance of ecosystems

to the business. A distinct level of analysis is required for each type of platform. For example, while the industry platform permits the study of a complete industry ecosystem, the internal platform restricts the study to a single business. The accessibility of information on the platform affects actors, but it also affects their ability to innovate. This is because more participants mean more viewpoints, which leads to stronger innovation skills. Also noteworthy is the method of platform control, which may be hierarchical or based on contracts.

Articles that examine internal platforms in the current literature tend to focus on the underlying technology. In contrast, research examining business platforms focuses more on network configurations and pays less heed to technologies. It is not unexpected that debates have focused on industry platforms, which encourage external collaboration, as opposed to internal platforms, which are entirely restricted to outside users. However, both points of view must be considered while thinking about supply-chain platforms. The value created on a supply-chain platform is heavily dependent on the collaboration between its actors, definition, since these platforms are partially available to other actors. Even while actor relationships are obviously very important, the current work on supply-chain platforms primarily addresses underlying technologies. The concept of digital twins, which has gained significant traction in the manufacturing sector in recent years, sheds light on this [18]–[19]. However, as the world becomes more digital, the interdependencies among the players will grow in significance, necessitating further investigation.

4.3 DIGITAL TWINS

4.3.1 Functions and Definitions

One definition of a digital twin is an identical copy of an item, piece of machinery, or procedure. Since an accurate model would necessitate parameters that are not feasible nor desired, practitioners view this description as a utopia. Depending on the region of utilization, it is not guaranteed that the resulting value grows with each added detail, and the cost increases with each detail as well. Thus, digital twins can alternatively be seen as a digital copy that has the required amount of information about the virtual object.

One kind of digital platform where multiple players share data to recreate a physical object is known as a digital twin. Most of the time, there are rigorous regulations and limits on the number of users in the production sector. Considering the flow of data is crucial since the virtual entity relies on data generated by the physical entity. Authors [20] states that there are three distinct types of digital twins distinguished by the degree to which their data is integrated. In the first type, alterations to the physical part have no effect on the digital representation, and vice versa. Consequently, handling the data flow by hand is required. In the second group, there is an automated one-way data flow; any changes made to the physical entity will have an effect on the virtual entity, but the process cannot be turned backward. The last type is a full-scale digital twin, which has a fully automated data flow and can respond to changes

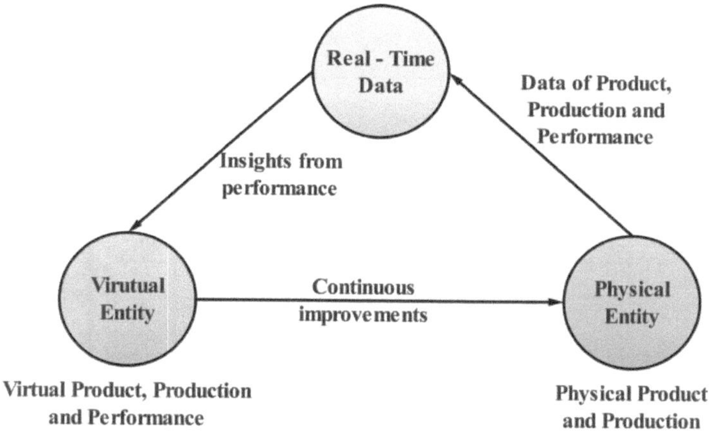

FIGURE 4.1 Digital Twin Structure.

in both the physical and virtual entities. Nevertheless, the idea of a full-scale digital twin is still in its infancy due to its immense complexity. Many businesses are already battling to develop the first kind of digital twin and ensure it accurately reflects reality. Figure 4.1 shows the structure of a digital twin.

Digital twins offer much promise, even though there are still significant technical hurdles to overcome. Companies can optimize production processes and products online before physically testing them with the help of digital twins, which represent a digital factory environment. According to authors [21]–[22], this method shortens the time that it takes to introduce new goods and also decreases the time and money needed for commissioning. Because of its superior analytical capabilities, the digital twin opens the door to more predictive labor techniques. Another crucial feature is sharing a visual representation of the same digital twin with stakeholders who are not in the same physical place. However, both digital platforms and the cooperation surrounding digital twins are understudied.

4.3.2 Digital Twin Data

A crucial component of the digital twin is twin data. Data types that digital twins process include heterogeneous, multi-dimensional, multi-source, and multi-temporal data. Both static attribute data and dynamic condition data are sourced from physical entities to a certain extent. Virtual models generate some data that represents the outcome of the simulation. Information about the service's invocation and execution is gleaned from the data collected from services. Knowledge is a type of data that may be either directly supplied by domain experts or derived from pre-existing data sets. Data that is created by merging all of the data types discussed above is called fusion data [23].

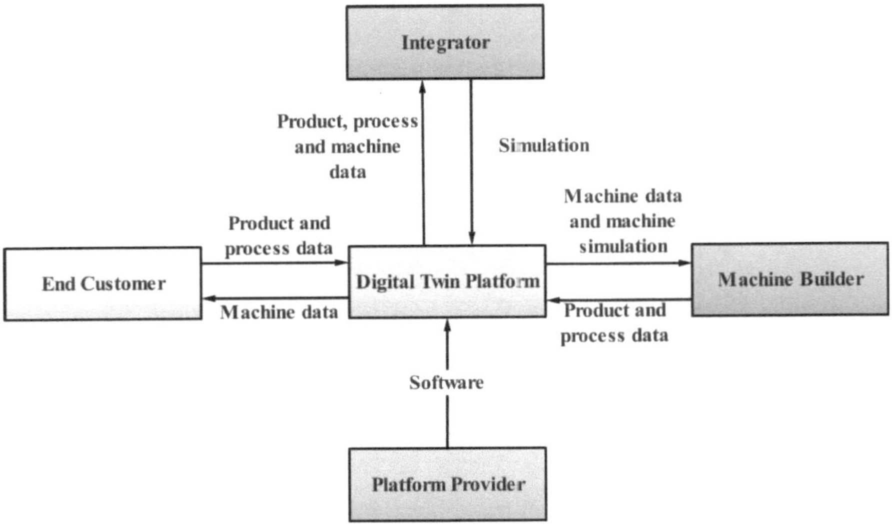

FIGURE 4.2 An Overview of Digital Twin Collaboration.

4.3.3 Actors Collaborating around Digital Twins

Despite the abundance of writing about self-organized map (SOM) and cloud manufacturing, the players in supply-chain platforms have received surprisingly little attention in the literature. Since digital twins are complex structures, all parties involved must work closely together to create them. Forming agreement-based strategic alliances with reliable parties is a common way to keep everyone's interests aligned. According to authors, there are four basic kinds of actors involved with digital twins. Figure 4.2 shows a high-level view of the collaboration round to digital twins.

Customers, or the business housing the production line and the actors using it, make up the first category of actors. The final consumer must buy machines from a machine builder, whose job it is to design, manufacture, and ship out machines, in order for the manufacturer to produce goods. The end user often purchases machines from multiple manufacturers due to the complexity of production lines. The purpose of the integrator is to ensure that the machines are properly coupled so that production can run automatically and without human intervention at each stage. Integrators play an essential role in the commissioning process because the end user buys the integration from them. It is not uncommon for a single company to play the function of both machine manufacturer and integrator, or to alternate between the two in various partnerships. As a last point, the role of platform provider is entailed because digital twins are digital platforms. Digital twins are created using the platform and software that the platform provider supplies. Transferring data across the many parties involved is essential for digital twins to achieve their optimization goals. However, digital twins have not yet been put to good use, and the difficulties of working together are a major roadblock; further education and experience in this area are required.

4.3.4 Overcoming Difficulties in Collaboration on Digital Platforms

Most of the existing literature on digital platform cooperation focuses on collaborations from the perspective of the unknown economic consequences. Various writers have pointed out difficulties with collaboration when it is examined through the lens of actor networks, which is particularly relevant to industrial platforms. Platforms in the industrial sector are frequently governed by contractual agreements, which make the network approach inappropriate for researching these platforms. This is in contrast to supply-chain platforms. Therefore, problems with cooperation, as seen through a network perspective, do not exist.

Few studies have focused on the difficulties of digital platform collaboration. However, those that have, such as Constantinides and Henfridsson, have identified data management as a significant obstacle. Problems with clearly defining who owns data produced by tangible goods are at the root of this obstacle [24]. The bulk of the prior work on digital twins focuses solely on problems associated with technological advancement. When it comes to systems, for instance, integration and interoperability pose significant challenges. The use of digital twins is complicated by the complexity and high cost of these technologies. In a similar vein, authors note that digital twins' cost-effectiveness suffers due to the high development costs caused by technological complexity.

Sharing information across the value chain is a boon to digital manufacturing, but it can also be a major headache due to different business regulations and attitudes regarding data ownership. Concerning digital twins, authors note that there is uncertainty about the ideal quantity of partners and the length of the partnership. In addition to platform-specific difficulties, the nature of the connections involved in the collaboration should be carefully examined when planning a platform-based effort. One problem with alliances is opportunism. This occurs when one side uses deceit, confusion, and misleading tactics to further their own interests at the expense of the other. Authors addresses how coalitions might lead to one side becoming too reliant on the other. Since the corporation will have less say over future decisions and developments, this is bad for the long run.

Methods for overseeing the necessary partnerships on these platforms need to be investigated, since new technology necessitates their adoption by industrial companies. Management of the networks that encircle industry platforms has always been the primary emphasis of platform literature. Platform providers, according to authors [25]–[26], must build their products in a way that allows for the type of engagement that strengthens the network in order to manage the network effectively, which is fundamental to success. Management approaches that focus on industrial platform networks will not work when researching other digital platforms, just as they will not work when analyzing cooperation issues. Since there is a dearth of research on the management of alternative platform types in the platform literature, researchers can look to works that examine collaboration across non-platform businesses for direction.

4.3.5 Digital Twin Challenges

The primary impetus for establishing the digital twin is the need to integrate several simulation models into a comprehensive system model and to establish a continuous data interchange between physical and virtual objects. As a result, digital twins make it possible to model an object's creation, use, and eventual destruction. In addition, you may substitute virtual tests for the actual ones, adjust the operating mode to your liking, review the results, and then draw conclusions about how the plant should run [27].

4.3.6 Digital Twin Uses

In the digital twin utilization phase, companies work to improve their competitiveness by making better use of digital twins. Since digital twins are applied to business choices or opportunities, they should be employed on demand. In many circumstances, this means they are used continually. Maintaining a steady stream of input for digital twin enhancement is also essential. Digital twin users might be anyone from individuals to physical objects to computer programs. For both software and hardware systems, digital twin functionality should be accessible through an API, even though customers are primarily interested in digital twin services (such visualization). A good interface and access to it are necessities for all system users, whether they are proactive or reactive [28].

4.3.7 Digital Twins in Electric Vehicle Applications

In the past, automotive and aerospace systems were developed utilizing empirical engineering practices. However, in light of increasing performance demands, the need for operational "self-awareness," and the indispensability of external assistance, novel engineering methodologies are now imperative. The advent of the digital twin has facilitated the availability of novel testing and development modeling tools to satisfy evolving needs. Consequently, there has been a consistent surge in research interest in these technologies [29].

When it comes to product-service systems (PSS), authors state that in order to prevent negative customer behavior, one can either apply trust or control and governance monitoring. Collaboration on digital platforms in production can also benefit from this, due to the parallels in the connection structure. In their discussions on control, the authors highlight the importance of having well-defined roles and responsibilities in contracts in order to minimize conflicts and opportunism. While contractual control can reduce disputes, relying solely on legal enforcement might lead to an increase in both conflicts and opportunism. While social links and dynamics directly impact the efficacy of contractual agreements, authors suggest that a social perspective should also be addressed while managing collaborative relationships. Researchers expand on this by stating that reciprocal trust is essential, as studying contracts without considering trust leads to an incomplete picture of reality. A long-term customer connection relies heavily on trust between the people involved in a collaboration. However, there is a lack of literature on the topic of digital platform cooperation management.

4.4 METHOD

The research aimed to improve understanding of digital platform collaboration; thus, it used an exploratory research approach to get there. Our comprehension of the researched area was steadily enhanced through the use of a qualitative technique. In addition, only one case study was conducted because the primary objective of the research was to learn more about effective digital platform collaboration. Because comprehending collaboration within the context of digital platforms necessitates a deeper familiarity with the platform and its surrounding actors, which would have been challenging with another research methodology, this was an ideal fit.

An example of a common digital platform in manufacturing a digital twin was used for this study. This platform provider is a massive multinational technology corporation that serves many different markets with digital and technological solutions. Customers, integrators, and machine builders are all part of its collaborative actors, which are typically smaller enterprises. These people work with digital twins sometimes or every day. According to experts in the field of digital twins, the platform provider was leading the pack when the survey was conducted.

In order to gain a comprehensive grasp of the investigated area, interviews were carried out. With the ability to ask follow-up questions in a semi-structured fashion, we were able to cover all the necessary ground at once and without missing any important details. The interviews were conducted using a pre-made interview guide, which covered topics such as the interviewees' experiences with digital platform collaboration, the difficulties encountered, and strategies for overcoming them. The literature was used as a basis to create the questions. The interview guide was revised to remain current as the research continued and respondents provided fresh perspectives. All eight participating companies had a total of 21 recorded and transcribed semi-structured interviews with study participants. Companies were chosen for their digital twin development progress and to guarantee a balanced representation of the four sorts of actors. In addition, recommendations made during data collection and respondents' prior knowledge of the researched area were both factors in their selection. The last two interviews were conducted to confirm and validate the preliminary outcome. A proper interpretation of the data and an evaluation of the findings' practical applicability were both addressed during the interviews, which included the presentation and discussion of the preliminary findings. Additional interviews were carried out until saturation was achieved, and the respondents were chosen based on suggestions from the previous respondents.

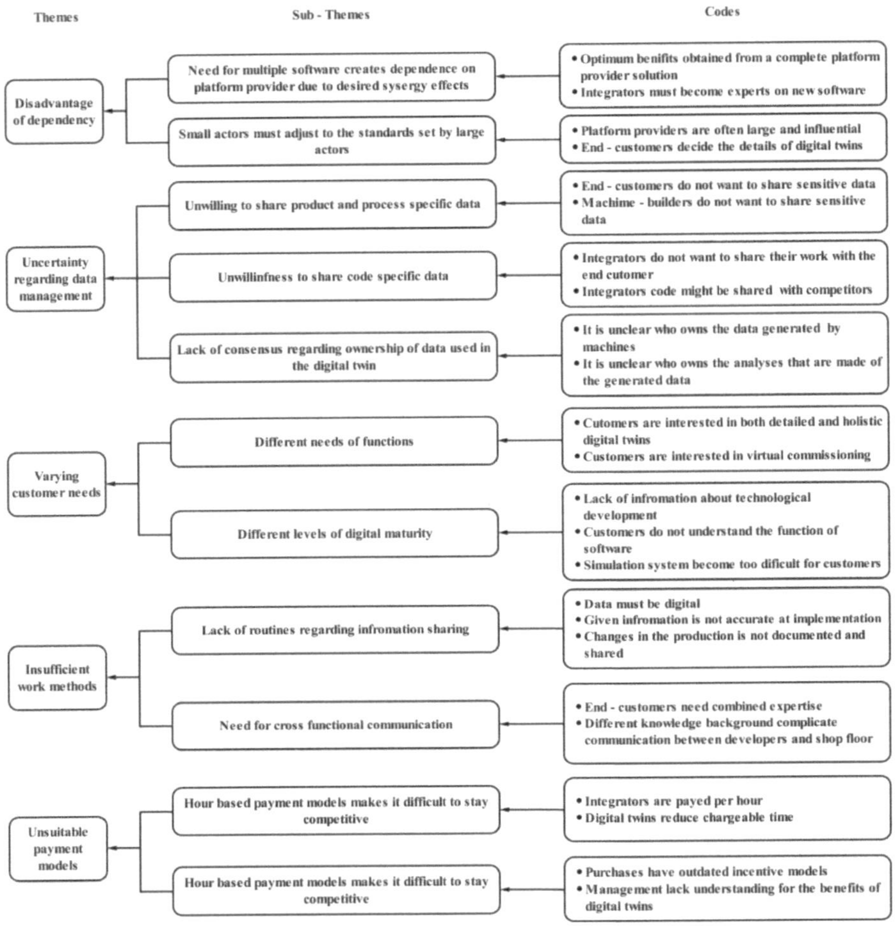

FIGURE 4.3 Thematic Map of Difficulties in Teamwork.

Researchers offered a theme analysis that was used to assess the collected data. The method aims to find patterns in the data and then evaluate and label them. This technique enabled us to expand the current literature by flexibly identifying important similarities of an extensive data set, making it ideal for our investigation. Our data analysis procedure was based on the five-step theme analysis proposed by authors: (1) Getting to know the data, (2) creating preliminary codes, (3) looking for themes and subthemes, (4) reviewing these themes and subthemes, and (5) defining these themes and subthemes. There was no strict progression from one stage of the analysis to the next. Instead, a more recursive approach was used, wherein we iterated between the steps as necessary. Also, the study was refined throughout time, so it did not miss anything crucial. Figures 4.3 and 4.4 display the completed theme maps.

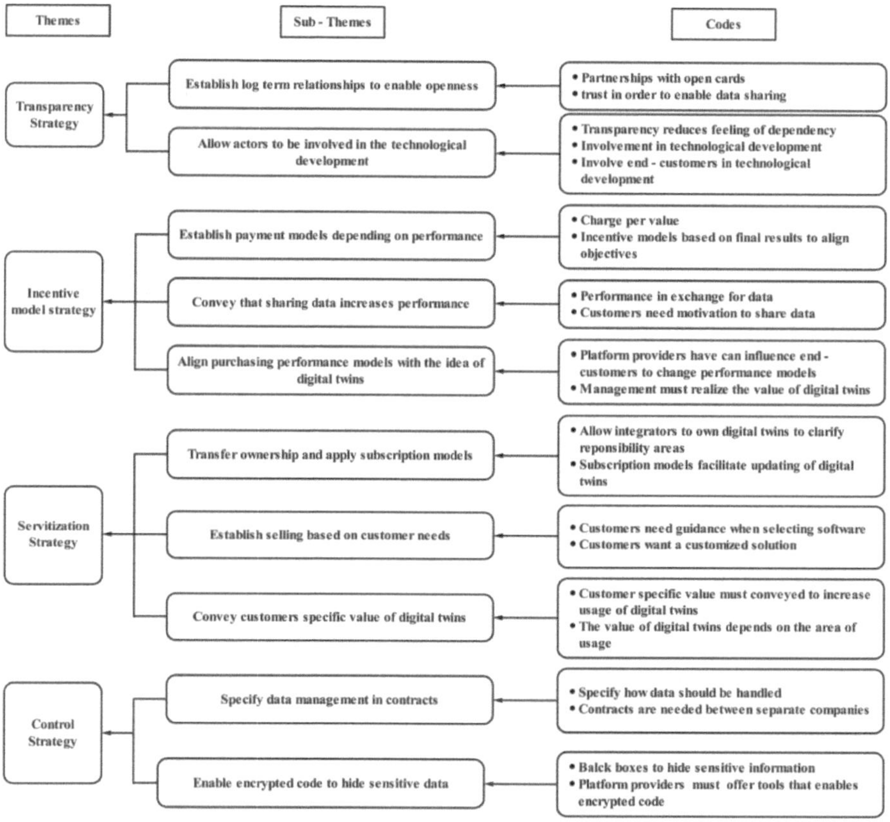

FIGURE 4.4 Thematic Map of Tactics for Cooperation.

4.5 ANALYSIS AND CONCLUSIONS

The results are organized in two parts: the difficulties of collaboration and the methods used to overcome them. Problems that develop as a result of collaborating and obstructing work with digital twins are known as collaboration issues. In order to overcome these obstacles, management can employ collaboration initiatives. The following sections address each group in turn. In addition, a contingency framework is used to match the challenges with the strategies, showing which types of challenges can be handled by particular strategies according to the specific requirements of each partnership.

4.5.1 Challenges

Empirical research into digital twin collaboration in manufacturing reveals five distinct sorts of difficulties: dependency drawbacks, data management ambiguity, inconsistent customer needs, inadequate work techniques, and inappropriate payment structures. Figure 4.3 displays a thematic map that summarizes the results. What follows is a more in-depth analysis of each of these difficulties.

4.5.1.1 Disadvantages of Dependency

The respondents were unsure about the potential drawbacks of being overly reliant on outside parties. This type of challenge is all about how digital twin collaboration can make individual actors feel like their predicament is not as big of a deal. To be more precise, there are two main points to consider while thinking about the problem of dependency's negative aspects: (1) the necessity for various software applications makes one dependent on the platform provider in order to achieve the required synergy, and (2) smaller players are compelled to conform to the standards established by larger ones.

Actors need to use various pieces of software with certain functions in order to work with digital twins. These programs can be offered by many platform firms, but for digital twins to work, they need to be able to integrate with each other without any problems. Because the smoothness depends on each party's chosen platform provider, this complicates the possibilities of collaborating with multiple actors. The question of who is responsible in the event of a malfunctioning interface becomes even more murky when many platform suppliers are involved. As a result, achieving the intended synergy effects requires many software applications, which poses a barrier in terms of the actors' reliance on the platform provider. A respondent clarified that using software from a single provider creates the most optimized solution: "one complete Alpha solution will give optimized benefits, but simultaneously one has backed the customer into a corner that they can never get out of, and they do not like that".

Another difficulty that respondents mentioned as leading to undesirable dependence in the collaboration is the requirement that smaller players conform to the standards established by larger actors. It is impossible to fully utilize digital twins unless all parties involved do their part. As one responder put it: "If Zeta is a strong end customer and Alpha a strong supplier of software, then the integrators become most dependent and have no choice but to follow a certain standard".

When it comes to integrators and builders, the answer mentioned that platform providers and end customers are typically significant companies. Because of their size disparity, larger corporations can dictate the terms of the partnership, forcing smaller ones to conform. So, the smaller corporations have no choice but to agree to any changes proposed by the larger parties to the agreement, no matter how significant those changes may be.

4.5.1.2 Uncertainty about Data Handling

According to the results, there is a lack of clarity about data handling in cross-company collaborations. Careful data handling is essential since data is the building block of digital twins. Digital twin work is complicated and hampered by uncertainty, based on the results, about how businesses share data and who owns particular data. Three main aspects of data management were determined to be extremely difficult: (1) there is a lack of agreement on who owns the data utilized in the digital twin, (2) there is a lack of desire to share data particular to the code, and (3) there is an unwillingness to share data specific to products and processes.

Data on products and processes is being generated at an ever-increasing rate by the industrial sector, in keeping with digitalization. In addition to the data required to construct digital twins, additional data is produced throughout the implementation of digital twins in the form of code development and analysis. Nevertheless, it is evident from the responses that there is a reluctance to provide distinctive data not only to processes and products, but also to codes. Machine manufacturers and end users alike express concern that competitors may gain access to sensitive information about their products and processes and use it to gain an unfair advantage. As digital twins rely on comprehensive product and process data for their operation, they do not alleviate these worries.

As the use of digital twins grows, so does the amount of code-specific data. Because they would rather not have their code stolen, integrators who write code are wary about sharing it with anyone, including end users. The answer went on to explain: what if we share our digital twin with a client, and then they borrow it and use it for their own business? In that case, we lose out. Would you be able to proceed similarly? The consumer will not be able to see through the code entirely.

According to the respondent, if the integrator makes their code publicly available, then the client may share it with a competing supplier, who may theoretically steal their ideas.

Thirdly, there is no agreement on who owns the data utilized in the digital twin, which adds to the data management ambiguity. The issue develops when the parties involved in the collaboration handle data more publicly. The question of ownership becomes murky when, for instance, the machine builder gives the end customer access to some sensitive data, and the end customer uses that data to make money. On the one hand, the machine maker's consent could be seen as giving up their rights to the data; on the other hand, the end customer could be obligated to share the benefit with the builder. "One can get value from the data, but whose data is it?" is how one respondent put it.

4.5.1.3 Changing Demands of the Client

The investigation also showed that clients have quite different requirements for digital twins, which is a third kind of issue. Due to these variations, it becomes more difficult for platform providers to comprehend each customer's predicament and, as a result, to supply an ideal solution. This is an issue because many platform suppliers have a history of delivering generic products and are not accustomed to tailoring their offerings to meet each customer's unique requirements. Two main elements emerged from the analysis of the problem: (1) varying functional requirements and (2) varying degrees of digital maturity.

The platform provider's software portfolio for building digital twins is extensive and ever-evolving, making it impossible for any individual to become proficient in every feature and application. As a result, staff members are oblivious to the unique requirements of each client and are instead intent on making sales on things with which they are familiar. In addition, several respondents felt that providers are trying to market the newest programs to customers regardless of their needs or what programs they already have, probably since digital twins and the techniques used to implement them are still relatively new.

Another finding that adds complexity to the collaboration is that the collaborative players have varying degrees of digital maturity. Because of these variances, items that some consumers find helpful may be too complex for others. According to the responses, the platform provider practically never shares any information with the other parties involved in constructing digital twins. Consequently, a small number of respondents voiced their belief that the level of digital maturity disparities is being exacerbated by the closed approach.

4.5.1.4 Inadequate Working Techniques

The research proved that current procedures are inadequate for using digital twins. The nature of the challenge is based on the fact that the end user requires certain work procedures to update the digital twin in accordance with the present status of production. To be more specific, there are two aspects of work practices that must be taken into account: (1) the absence of regular procedures for recording and sharing information on physical changes to production, and (2) the necessity of communication across different departments.

Maintaining the usefulness of a digital twin requires keeping its data up-to-date with the physical version. The related issue is the absence of procedures for recording and disseminating information on physical changes to the output. Unfortunately, there is currently no developed method for digital twins that automatically updates the virtual environment in response to changes in the physical environment and vice versa. Respondents identified a big issue in modern manufacturing: the need to record production changes for the digital twin to remain relevant. One respondent identified the problem situation as follows: "CAD models are often not updated due to physical changes, and suddenly, the digital twin is not a twin anymore, since it does not match the reality".

One respondent said, "When we work with our customers, the first thing we have to do is to see that everything is accurate, which is a bit humiliating", illustrating how inadequate documentation hinders the intended use of digital twins and necessitates extra effort. However, it is not an easy scenario to handle, because it involves more than just the players mentioned in the research; other players, such as electricians and mechanics, contribute to the problem by making improvements.

There will be a greater demand for communication between departments if the industrial industry is to be digitalized. Both outward communication between the collaborating actors and internal communication between functions at each actor are essential. As an example, according to one respondent: "We want to hire individuals with IT knowledge; but if one does not understand the production and have no sense for what type of data one is optimizing, then you get to conclusions with extremely bizarre solutions that are not appropriate in this context". Given that the desired results from each actor are missing in the absence of effective internal cross-functional collaboration, it stands to reason that external collaboration will also suffer.

4.5.1.5 Inappropriate Methods of Payment

In addition, the results highlight an issue with the applicability of existing payment structures when dealing with digital twins. One kind of problem is the fact that existing pricing methods discourage some actors from collaborating with digital twins, which in turn slows down their adoption. To be more precise, there are two main categories into which the task falls: (1) the value of digital twins is diminished by the purchasing department's performance measurement and (2), staying competitive is made more difficult by hourly payment models when employing digital twins.

There is currently no appropriate payment model available for digital twins, since the concept is still in its early stages and has not yet achieved widespread market adoption. Using digital twins to remain competitive is challenging in today's hourly-based payment paradigm. Problems develop because digital twins are more costly than other methods; nevertheless, in order to maintain competitiveness, the price cannot be raised, regardless of how much more efficient the method is, which shortens the payment term.

Digital twins lose some of their worth due to the current methods of performance monitoring used by purchasing departments, which is another reason why payment models are not acceptable. Many payment systems incentivize shoppers to haggle for the best available price. Nevertheless, according to one respondent: "... they drive down prices as much as possible and thereby, are lowering the given value which could be that one can start a production much earlier and can produce much more than planned". To put it simply, firms' ability to work together effectively around digital twins is hindered by the payment structures used by purchasing departments.

4.6 METHODS OF COOPERATION

Findings from the empirical study point to four management strategies—transparency, incentive models, servitization, and control—as potential solutions to the problems associated with digital twin collaboration in the industrial sector. Figure 4.4 shows a thematic map that summarizes the results. Following this introduction, we will go over each of the methods of cooperation in turn.

4.6.1 Transparency Strategy

According to the results, open communication is key to productive teamwork, including digital twins. Since digital twins necessitate more external information sharing than before, multiple responders stress that parties must use clearer work techniques toward one another to establish crucial trust.

When implemented correctly, the transparency technique shows great promise in meeting specific difficulties. Part one of the plan is to build trusting relationships over time so that everyone involved in the collaboration feels comfortable enough to be honest with one another. However, since these relationships rely on trust and take time to grow, it is crucial to establish them one after the other.

Enabling actors to participate in the creation of technology is another component of the transparency strategy, alongside the establishment of long-term connections.

If the platform supplier takes an open-source approach to development, then they can help the actor to close the digital maturity gap. Customers are more likely to feel that they have some control over the problem when they are involved and transparent throughout the partnership, which can alleviate the unpleasant sensation of dependency. One responder even said, "If we are let in more, the agreement will become even better, and they will get input from us, an end customer, which will help them to create better products". This highlights the significant advantages that the platform provider can gain from customer feedback.

4.6.2 Strategy for Incentive Models

One approach that has shown promise for effective collaboration around digital twins is the implementation of incentive structures. A digital twin collaboration strategy is an approach to system design that aims to encourage actors to act in a way that benefits the coöperation. It is recommended to use three distinct incentive models: (1) Link buying performance models with the concept of digital twins, (2) communicate that data-sharing boosts performance, and (3) set up payment models based on performance.

To solve the problem of inappropriate payment models, a reasonable payment for digital twins would be to set up payment based on performance. For instance, it could take the form of an incentive scheme where everyone is on the same page about a given target, and everyone gets a cut of the profits if we achieve it. However, if things do not work out as planned, then everyone involved will feel the pinch. As an additional illustration, a portion of the improvement might be allocated if the delivered result exceeds expectations.

It is critical to set up a model that shows how sharing data improves performance in addition to an incentive-based payment approach. To get customers to willingly provide their data, you need to provide them with the right incentives. Problems with people's openness to exchanging data on products, processes, and codes can be mitigated in this way. "We strive to collaborate with performance guarantees; we state a specific level of performance if the client is transparent about providing their data", one responder noted. Companies can encourage customers to act in a way that benefits digital twin collaboration by doing this.

Finally, the digital twins' value is impacted by the fact that purchasers' performance is typically measured by their ability to negotiate lower costs. Thus, to fully benefit from digital twins, we need to move beyond the problems caused by inappropriate payment models for purchasing performance metrics. However, this can only happen if upper management supports digital twins and sees their potential. In addition, working together is essential because certain players have more clout with upper management. However, Alpha's assistance is required because they possess superior connections within higher-ranking companies and boards compared to our own. That is why more prominent players, such as platform providers, have a better chance of convincing end users to alter their purchasing performance models than more minor players, such as integrators.

4.6.3 Servitization Strategy

According to the results, a servitization plan is the way to go for a fruitful partnership centered on digital twins. The approach states that digital twins should be viewed as services, rather than products, and that digital twin solutions should be tailored to meet the unique requirements of each customer.

A number of people who took the survey brought up the idea that digital twins should be used to provide services instead of physical products. This would enable businesses to implement subscription models and transfer ownership more easily. An example was given by one responder where the integrator takes ownership of the digital twin and delivers it as a service to the end customer. Both the ownership status and an appropriate payment model can be established in this manner. Also, someone with the knowledge to update the digital twin will be the one to possess it. I think it would be a great addition to your offerings if you offered this as a service, as another responder mentioned. Going in that direction will ensure that we both care about the digital twin's survival.

The likelihood that the end user may divulge information regarding alterations to the physical version grows when there is a shared desire to maintain the digital twin current. The likelihood of a fruitful partnership is thereby enhanced, as this resolves issues stemming from inadequate work practices.

The servitization plan also includes establishing needs-based selling, which, as its name suggests, is an excellent solution to problems caused by customers' diverse wants and needs. One responder said: "I believe that Alpha has to be clearer about the intended customer of their various tools and be better at leading their customers rather than introducing them to a jungle of things that they have to look through themselves". As a platform provider, this strategy is mainly concerned with their offerings. When clients believe that the platform provider's software meets all of their requirements, cooperation between the two parties improves. The likelihood of demonstrating value grows when efforts are made from two distinct directions.

4.6.4 Control Approach

As a last point, one possible tactic to deal with problems that arise in digital twin collaboration is control. When we talk about a control plan, we mean the measures taken to limit or control how a cooperating party acts. Two subthemes comprise the control strategy: (1) enabling encrypted code to conceal sensitive data and (2) specifying data management in contracts.

In order to solve issues linked to inadequate work techniques and confusion around data management, respondents stated that a contract outlining data management is necessary, regardless of the quality of the relationship between the parties involved in the collaboration. "As part of any relationship, there should be contracts outlining who is ready to divulge information and what they can do with it", one responder said.

Having mutual trust is great, but at the outset of any relationship, it is essential to have a written agreement outlining the parameters of the work and the repercussions for failing to adhere to them.

Part two of the cooperation plan is to enable encrypted code, which is now used when making simulation models and should be maintained, to conceal sensitive data. Encrypted code could enable teams to work together in secrecy, according to one responder. Consequently, this approach can help with data management uncertainty to a certain degree. On the other hand, making a simulation model requires a lower level of detail than making a digital twin. As a result, the value of the digital twin will be compromised if all sensitive information is concealed using encrypted code. A control strategy fails miserably at resolving any of the problems. The leading cause of this is that contracts are ineffective at managing unanticipated events. After all, digital twins are still in the development stage, which raises the possibility of such events.

4.6.5 In-house Logistics for Digital Twins

Since in-house logistics systems are complicated due to the large number of entities involved, the ever-changing environment, and the need for decision support, digital twins play an essential role in these processes. Data is an integral part of this intricate system. Preliminary perfect planning of resources is essential for achieving efficiency. Nevertheless, it is preferable to make real-time adjustments as deviations from the original plan occur [30].

4.7 A BACKUP PLAN FOR PRODUCTIVE COOPERATION IN THE MANUFACTURING SECTOR ON DIGITAL PLATFORMS

Given that the digital twin is just a standard digital platform, the results shed light on the difficulties of digital platform collaboration and, more importantly, how to overcome these difficulties. The results do show, though, that not every problem has a perfect solution. The argument in favor of a contingency assessment is that different tactics are better suited to different types of problems. Various techniques' applicability depends largely on the particular actors' current position in relation to the digital platform development life cycle. Figure 4.5 displays a framework that was created to illustrate these findings by matching each obstacle with one of the options. How well a strategy handles each difficulty is indicated by the number of stars (*).

The players usually confront more than one issue when interacting on digital platforms, thus, it is necessary to combine different techniques to reach the greatest conclusion. According to the results, it is possible to generate synergy effects in terms of the selected methods' ability to handle a particular problem or collection of problems by combining them. On the other hand, a servitization plan that defines who owns what data might help with data management uncertainty. When combined, these approaches can overcome the problem of data management uncertainty, even though none of them are ideal on their own. In order to tackle data management uncertainty,

Collaboration Strategies

Collaboration Challenges

	S1: Transparency Strategy	S2: Incentive Model Strategy	S3: Servitization Strategy	S4: Control Strategy
C1: Disadvantages of dependency	High	Low	Medium	Low
C2: Uncertainty regarding data management	Medium	Medium	Medium	Medium
C3: Varyinh customer needs	Medium	Low	High	Low
C4: Insufficient work methods	Low	Low	Medium	Medium
C5: Unsuitable payment models	Low	High	High	Low

→ Digital Platform Collaborattion

FIGURE 4.5 A Backup Plan for Productive Teamwork on Digital Platforms in the Manufacturing Sector.

a synergy effect is generated by combining the incentive model method with the servitization strategy.

It should be taken into account that implementing a plan to solve one problem can actually make other problems more likely. One example is the servitization plan, which involves transferring ownership of a digital platform to another one. This method increases the dependency between players and also makes external data exchange more necessary. The servitization plan can be enhanced with a transparency strategy that encourages long-term connections. This will not only raise the likelihood of data-sharing but also lessen the possibility of unpleasant feelings of dependency. Consequently, to overcome their individual obstacles, each group of stakeholders involved with a digital platform must devise a distinct approach to collaboration. To aid managers in the manufacturing sector with digital platform collaboration, the offered framework can be utilized as a roadmap.

4.8 CONCLUSION

This research has added to our knowledge of online teamwork by outlining potential pitfalls and offering solutions as a contingency framework. With the rise of digital platforms as a survival mechanism in the digitalization era, when effective collaboration is crucial to success, this information is valuable for manufacturing organizations. Consequently, our study adds to the body of knowledge on digital platforms and has significant ramifications for manufacturing industry management.

Three areas are theoretically advanced by our work. First, digital platforms have generally received a lot of attention from academics, but their use in manufacturing has received far less. One of the most significant digital platforms in the manufacturing business is digital twins, which we examine in order to provide insights into this platform's value. The limited literature on digital platforms in manufacturing has chiefly focused on their technological features. By adopting a management viewpoint, we have broadened the platform conversation and brought attention to critical non-technical factors. Research such as this helps to build digital platforms correctly from the beginning, which is particularly important because the use of digital platforms by manufacturing organizations is expected to increase in the following years.

Secondly, while there are a few studies that touch on the topic of digital platform collaboration in manufacturing, we have significantly expanded upon this body of work by naming five obstacles: (1) the drawback of dependency, (2) uncertainty about data management, (3) differing customer needs, (4) inadequate work methods, and (5) inappropriate payment models. We have uncovered a number of previously undiscussed difficulties in collaboration through our mapping efforts. Thus, we have added to the existing body of information about platform collaboration and given a thorough summary of the difficulties in the existing literature.

Lastly, previous studies have failed to address how the manufacturing industry should handle collaboration on digital platforms. Our research has uncovered four distinct approaches that businesses may use to oversee teamwork in digital settings: transparency, reward models, servitization, and control. Furthermore, we added to the literature by developing a contingency framework that links certain difficulties to the collaboration techniques; this framework offers valuable insights into managing collaborations on digital platforms.

Managers in the industrial sector can benefit from our study's findings because digital platforms are an integral aspect of digitalization, which is critical for the future competitiveness of industrial enterprises. We have presented three key points that managers should consider. To start, we have proven that getting everyone on the same page is key to making the most of digital platforms, so managers should not limit themselves to thinking about their own businesses when making the switch. It is essential for businesses to plan for each partnership while keeping their collaborative partners in mind.

A second takeaway is that managers can be proactive while collaborating on digital platforms because we have highlighted potential issues that may develop during the collaboration. Managers can lessen the likelihood of failure, unpleasant surprises financially, and ineffective teamwork by taking the initiative. Finally, we have given managers the resources they need to foster productive collaboration, and we have also shed light on why it is important to evaluate the conditions of each partnership before enabling it. Doing so has given management the resources they need to embrace digital platforms and remain competitive.

There are some caveats to the study and some openings for further investigation. First, while our results are based on data collected from eight separate companies, our study only looks at one platform type in one unique setting. Future research might look at collaboration around different kinds of platforms in the same setting as this study to confirm the results and make them more generalizable. As an alternative, studies in the future can look at different settings where people work together on the same platform. Future research should ideally take into account both recommendations.

Second, our case study reflects the early stages of digital platform establishment in the manufacturing business. This implies that some of the raw data is speculative, since the respondents have used their present knowledge and experience to reason about digital twins. We, therefore, anticipate that future studies will be able to collect empirical information as digital platforms mature, specifically looking at the difficulties that businesses have encountered and how they overcame them. When used in digital platforms, blockchain technology has the potential to greatly facilitate the elimination of uncertainty.

Thirdly, no companies outside of Sweden were involved in the research. This indicates that the requirements of the Swedish industrial sector formed the basis of the results. The prevalence of collaboration challenges and the efficacy of collaboration techniques may differ among nations due to inherent environmental differences. Thus, it is our firm belief that subsequent researchers should conduct studies in other regions. That way, we can bolster the findings' generalizability even further.

REFERENCES

[1] S. P. Kovalyov, "Distributed Energy Resources Management: From Digital Twin to Digital Platform," in *IFAC-PapersOnLine*, Elsevier, 2022, pp. 460–465.

[2] H. Yan, X. Xu, M. Bilal, X. Xia, W. Dou, and H. Wang, "Customer Centric Service Caching for Intelligent Cyber-Physical Transportation Systems with Cloud-Edge Computing Leveraging Digital Twins," *IEEE Transactions on Consumer Electronics*, vol. 70, no. 1, p. 1, 2023.

[3] J. A. Valan, M. Mythily, and J. Anand, "Performance Analysis of Mesh based Multicast Routing Protocol for Adhoc Wireless Networks", " *International Engineering and Technology Publications Journal of Communication Techniques*, vol. 2, no. 2, pp. 99–103, 2008.

[4] V. V. Raja, R. R. Hemamalini, and A. J. Anand, "Performance Evaluation of Agent-based Upstream Congestion Control Approach in WSN," *International Journal of Computer Technology and Applications*, vol. 2, no. 6, 2011.

[5] S. Suruthi, S. S. KR, B. Varsha, S. Saranya, and B. Babu, "Ultrasound Image Denoising Using Cascaded Median Filter and Autoencoder," in *2023 4th International Conference on Smart Electronics and Communication (ICOSEC)*, IEEE, 2023, pp. 296–302.

[6] V. V. Raja, R. R. Hemamalini, and A. J. Anand, "Study of Traditional Upstream Congestion Control Algorithm with Various Metrics in Realistic WSN Scenarios," *International Journal of Computer Science and Information Technologies*, vol. 2, no. 6, pp. 2473–2478, 2011.

[7] V. V. Raja, R. R. Hemamalini, and A. J. Anand, "Study of Traditional Fairness Congestion Control Algorithm in Realistic WSN Scenarios," *International Journal of Engineering Research and Applications (IJERA), ISSN*, vol. 1, pp. 2248–2251, 2011.

[8] A. M.-A. Qadir and A. O. Fatah, "Platformization and the Metaverse: Opportunities and Challenges for Urban Sustainability and Economic Development," *EAI Endorsed Transactions on Energy Web*, vol. 10, pp. 1–29, 2023.

[9] S. He, H. Yang, H. Wang, J. Shen, and W. Li, "Review and Prospect of Software Digital Design for Electric Vehicle Power Control Unit," *Diangong Jishu Xuebao/Transactions of China Electrotechnical Society*, vol. 36, no. 24, pp. 5101–5114, 2021.

[10] V. I. Tinyakova, N. I. Morozova, and T. V Karyagina, "Sustainable Development of the Digital Economy: Global Challenges to Russia's National Security and Prospects for Its Innovative Breakthrough," in *Advances in Science, Technology and Innovation*, Springer, 2022, pp. 471–475.
[11] B. L. MacCarthy and D. Ivanov, "The Digital Supply Chain—Emergence, Concepts, Definitions, and Technologies," in *The Digital Supply Chain*, Elsevier, 2022, pp. 3–24.
[12] E. Blomqvist, M. Lindecrantz, F. Blomsma, P. Lambrix, and B. De Meester, "Decentralized Digital Twins of Circular Value Networks – A Position Paper," in *CEUR Workshop Proceedings*, CEUR-WS, 2022.
[13] S. Seo, B. Seok, and C. Lee, "Digital Forensic Investigation Framework for the Metaverse," *Journal of Supercomputing*, vol. 79, no. 9, pp. 9467–9485, 2023.
[14] V. V. Raja, R. R. Hemamalini, and A. J. Anand, "Multi Agent System Based Upstream Congestion Control in Wireless Sensor Networks," *European Journal of Scientific Research*, vol. 59, no. 2, pp. 241–248, 2011.
[15] M. Yseboodt, "The Digital Transformation is Here to Stay: Infinite Opportunities from Infinite Data," *Cement International*, vol. 19, no. 6, pp. 42–47, 2021.
[16] L. Scime, A. Singh, and V. Paquit, "A Scalable Digital Platform for the Use of Digital Twins in Additive Manufacturing," *Manufacture Letters*, vol. 31, pp. 28–32, 2022.
[17] A. A. A. S. Alsaeedi *et al.*, "Long Term Production Strategy – Application of a Dynamically Integrated Reservoir and Production Model to Identify Compression Requirements and to Address Production Deferral in a Giant Gas Field," in *SPE Middle East Oil and Gas Show and Conference, MEOS, Proceedings*, Onepetro, 2021.
[18] T. Y. Fujii *et al.*, "A Digital Twin Architecture Model Applied with MLOps Techniques to Improve Short-Term Energy Consumption Prediction," *Machines*, vol. 10, no. 1, pp. 1–26, 2022.
[19] S. Honjo, S. Matsumoto, and T. Sano, "Maximizing Production with Real-Time Integrity Operating Windows," in *Proceedings of the Annual Offshore Technology Conference*, (p. D021S021R005), OTC, 2021.
[20] J. A. M.-A. Carriazo and A. Crespo Márquez, "Dynamic Maintenance Management Approach Based on Real Time Monitoring and Artificial Intelligence Using Digital Twins," in *Lecture Notes in Mechanical Engineering*, Springer, 2023, pp. 145–153.
[21] P. Barcik, A. Coufalikova, P. Frantis, and J. Vavra, "The Future Possibilities and Security Challenges of City Digitalization," *Smart Cities*, vol. 6, no. 1, pp. 137–155, 2023.
[22] N. Swaminathan, B. P. Gautham, R. Shukla, C. Malhotra, and T. Gaduparthi, "Digital Engineering Platform For Synergistic Decision-Making in Manufacturing Plant Operations: Resaerch Question," in *Proceedings of the ASME Design Engineering Technical Conference*, vol. 86236, (p. V03BT03A014), 2022.
[23] Q. Qi *et al.*, "Enabling Technologies and Tools for Digital Twin," *Journal of Manufacture Systems*, vol. 58, pp. 3–21, 2021.
[24] S. Suuronen, J. Ukko, R. Eskola, R. S. Semken, and H. Rantanen, "A Systematic Literature Review for Digital Business Ecosystems in the Manufacturing Industry: Prerequisites, Challenges, and Benefits," *CIRP Journal of Manufacturing Science and Technology*, vol. 37, pp. 414–426, 2022.
[25] W. Reim, E. Andersson, and K. Eckerwall, "Enabling Collaboration on Digital Platforms: A Study of Digital Twins," *International Journal of Production Research*, vol. 61, no. 12, pp. 3926–3942, 2023.
[26] F. Naseri *et al.*, "Digital Twin of Electric Vehicle Battery Systems: Comprehensive Review of the Use Cases, Requirements, and Platforms," *Renewable and Sustainable Energy Reviews*, vol. 179, pp. 113280, 2023.

[27] D. Adamenko, S. Kunnen, and A. Nagarajah, "Comparative Analysis of Platforms for Designing a Digital Twin," in *Lecture Notes in Mechanical Engineering*, Springer, 2020, pp. 3–12.

[28] L. Stojanovic *et al.*, "Methodology and Tools for Digital Twin Management—The FA3ST Approach," *IoT*, vol. 2, no. 4, pp. 717–740, 2021.

[29] M. Ibrahim, V. Rjabtšikov, and R. Gilbert, "Overview of Digital Twin Platforms for EV Applications," *Sensors*, vol. 23, no. 3, pp. 1414, 2023.

[30] F. Coelho, S. Relvas, and A. P. Barbosa-Póvoa, "Simulation-Based Decision Support Tool for In-House Logistics: The Basis for a Digital Twin," *Computers & Industrial Engineering*, vol. 153, pp. 107094, 2021.

Part 2

Technologies

5 Digital Twins and Artificial Intelligence

Priya Singh and Sankata Tiwari

5.1 INTRODUCTION

Artificial intelligence (AI) and digital twins are more than simply catchphrases; they are powerful tools for influencing how humans engage with the real and virtual worlds in the future. The two revolutionary technologies have surfaced as game-changers, pushing the limits of industry, innovation, and problem-solving capabilities within the core of the digital revolution. Consider the existence of a virtual equivalent of a system, procedure, or actual object made feasible by digital twins. A digital twin is an intricately realistic virtual model of a physical object, such as a building, product, or whole city. Beyond simply mirroring its physical counterpart's appearance, this digital doppelganger gathers real-time data from sensors to allow for a dynamic recreation of the object's activities. This real-time feedback loop revolutionizes the design, monitoring, and maintenance of complex systems in industries by enabling analysis, prediction, and optimization. Digital twins are transforming efficiency, enabling predictive maintenance, and stimulating innovation in various sectors, including manufacturing, healthcare, and urban planning.

5.1.1 Defining Digital Twins: Bridging the Physical and Digital Worlds

Within contemporary technology, a digital twin is a fantastic idea that connects the material and physical world with its virtual digital twin. Consider a digital replica that flawlessly captures every detail and behavior of a real-world system, process, or object. This digital twin is a dynamic, real-time simulation that changes in tandem with its physical counterpart, not merely a static representation. Fundamentally, a digital twin functions as a link between the digital and physical domains. To recreate the object or system in a virtual setting, it gathers and combines enormous volumes of real-world data from sensors, devices, and other sources. This digital replica is more than just a visual; it includes every stage of the actual entity's existence, from conception and production to operation and maintenance. This mutually beneficial partnership between digital and physical media is revolutionary. Figure 5.1 illustrates the relationship between physical and digital objects. In the digital sphere, engineers and designers can experiment with various situations and adjustments, forecasting the effects of these changes on the real-world equivalent. In the manufacturing industry,

DOI: 10.1201/9781003465904-7

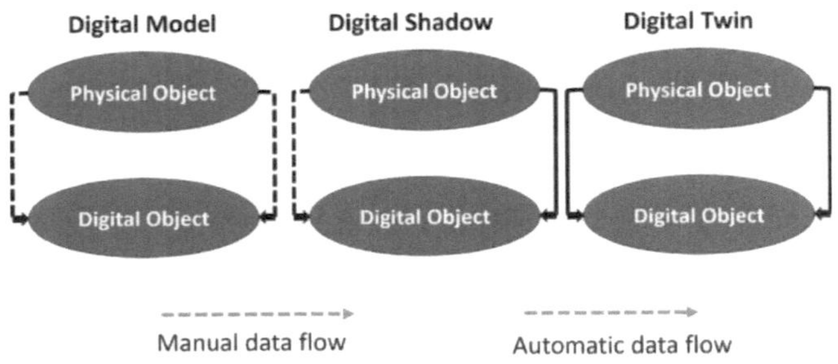

FIGURE 5.1 Digital Model, Shadow, and Twin [2].

for example, a manufacturing line's digital twin can be examined and adjusted in real time to maximize productivity and reduce downtime [1].

This dynamic synchronization makes precise monitoring and analysis possible, promoting proactive problem-solving and wiser decisions. Essentially, the idea of digital twins crosses both digital and physical boundaries, bringing about a new era in which the real and virtual worlds merge. This dynamic synchronization makes precise monitoring and analysis possible, promoting proactive problem-solving and wiser decisions [3]. Essentially, the idea of digital twins crosses both digital and physical boundaries, bringing about a new era in which the real and virtual worlds merge. Industries may reach previously unheard-of levels of productivity, creativity, and comprehension through this mutually beneficial partnership, completely changing how we create, run, and engage with the world.

5.1.2 Unraveling AI: From Machine Learning to Deep Learning

AI is the simulation of human intellect in computers that have been designed with human-like thought and learning processes. It includes various methods and strategies intended to develop intelligent agents capable of carrying out tasks that usually demand human intellect, such as comprehending natural language, identifying patterns, resolving issues, and making choices. Machine learning (ML) is a branch of AI that focuses on creating statistical models and algorithms that let computers learn from data and make predictions or conclusions based on that data. ML algorithms are not expressly built to do a specific task; instead, they analyze data to find patterns and gradually improve [4]. Deep learning (DL) is a subfield of ML that works with neural networks and algorithms modeled after the human brain's architecture and operation. These neural networks comprise layers of interconnected nodes or neurons.

Due to its capacity to handle massive volumes of data and derive insightful conclusions, DL techniques have experienced tremendous success and growth in recent years [5]. Figure 5.2 depicts the functionality of DL and ML, respectively. In summary, DL is a particular method within ML that works with complicated neural networks, enabling machines to learn intricate patterns and representations. At the

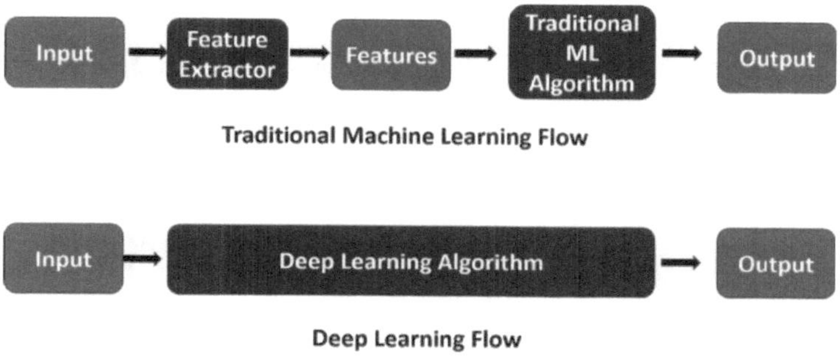

FIGURE 5.2 Functionality of DL and Machine Learning.

same time, AI aims to create intelligent computers. ML is a methodology that enables devices to learn from data. Together, these technologies push the limits of machines, leading to essential advancements in several industries, including healthcare, finance, autonomous cars, and more.

5.2 BUILDING THE FOUNDATION: UNDERSTANDING DIGITAL TWIN TECHNOLOGY

A digital twin is a virtual representation that is more advanced than a static 3D model. It incorporates a multitude of data, such as historical records, sensory inputs, and real-time updates that represent the status and actions of its physical counterpart. Fundamentally, digital twin technology provides a comprehensive perspective of the item or system that it represents by linking the digital and physical domains. Data from the physical world informs the digital twin, and insights from the virtual model can impact decisions and actions in the actual world. This connection is bidirectional. The capacity of digital twins to mimic and predict is one of their main features. Digital twins can replicate real-world situations using diverse technologies, such as the Internet of Things (IoT) sensors, data analytics, and computational modeling. Engineers, scientists, and companies may test theories, carry out tests, and investigate possible outcomes thanks to its simulation feature, without immediately affecting the actual equivalent. For instance, a digital twin of a production line may be used in manufacturing to streamline procedures, spot bottlenecks, and boost overall efficiency—all in a risk-free virtual environment. In addition, digital twins are dynamic entities that undergo continuous evolution and learning [7]. A digital twin can anticipate when maintenance is required, averting expensive malfunctions and downtime, by real-time data analysis from a machine's sensors. Numerous industries use digital twins, including manufacturing, healthcare, urban planning, and space exploration. They help manufacturers to achieve agile production, which makes it possible for companies to quickly adjust to shifting customer needs. Digital twins of human organs are used in healthcare to support customized medicine, which enables physicians to customize therapy according to each patient's specific physiology.

5.2.1 Historical Evolution of Digital Twins: From Concept to Reality

The history of digital twins is intriguing, with theoretical concepts giving way to functional, real-world implementations. A brief synopsis of its journey is as follows:

Origins in the 2000s: Dr. Michael Grieves of the University of Michigan coined "digital twin" in the early 2000s. Virtual representations of those things were first employed in product life cycle management concepts to facilitate better understanding and administration of physical items throughout their life cycle.

Early applications (2010s): With the emergence of the IoT in the early 2010s, real-time data from sensors and other sources was incorporated into digital twins. Digital twins were first used in simulations by the aerospace and automotive industries, enabling engineers to test and improve ideas in a virtual setting before manufacturing.

IoT integration (mid-2010s): Digital twins and IoT technologies had a considerable merger in the middle of the 2010s. This synergy made real-time monitoring and data collecting from physical systems and objects possible. In sectors such as healthcare, digital twins began to gain popularity when they were used to mimic and evaluate patient data.

AI developments in the late 2010s: Predictive analytics was made possible when these technologies were coupled with digital twins. Digital twins evolved from being static representations to being able to model events, predict outcomes, and anticipate behavior.

As of right now (the 2020s): Digital twins are essential to several industries in the 2020s, such as intelligent infrastructure, manufacturing, healthcare, and urban planning. They may be used for anything from supply chain optimization, equipment failure prediction, urban development planning, and building virtual versions of entire cities [3].

Upcoming prospects: In the future, digital twins will continue to develop. Future developments in edge computing, 5G, and augmented reality (AR) may further expand the potential of digital twins.

5.2.2 Core Components of Digital Twins: Data Integration, IoT, and Sensor Networks

Digital twins are intricate virtual representations of real-world systems or items. A few fundamental elements are required to produce precise and useful digital twins: sensor networks, IoT, and data integration. Data integration is the foundation of all digital twins. It entails gathering and merging data into a single platform from several sources, including databases, external systems, and sensors. This information may be presented in text, graphics, numbers, or any other appropriate manner. By guaranteeing the digital twin's access to a wide range of information, data integration makes it possible to accurately and thoroughly replicate the physical system or item that it is modeling. The IoT is essential to the operation of digital twins. IoT devices are embedded in real-world scenarios or objects, and come with sensors and networking capabilities. These gadgets gather real-time data and send it to the digital counterpart for processing and display [8]. The IoT broadens the use of digital twins by offering constant data streams on the device's environment, performance, and state. The foundation of data collecting in the real world comprises sensor networks. These networks comprise strategically positioned, networked sensors on the target or in the monitored surroundings. Numerous characteristics, such as temperature, humidity, motion, and chemical composition, can be detected using sensors. The data collection of these sensors provides insightful information on the behavior of the device and its surroundings. Regarding digital twins, sensor networks provide the raw data necessary for creating an accurate virtual representation [7]. In conclusion, the essential elements of sensor networks, IoT, and data integration are the foundation for creating digital twins. These virtual models are empowered to accurately reflect the world, opening up previously unachievable avenues for industries to innovate, streamline operations, and make well-informed decisions. Digital twins become potent instruments by skillfully combining these elements, revolutionizing our perception of and ability to engage with the complexity of our physical surroundings [9]. There are several varieties of digital twins, each with a distinct function in various applications. In layperson's words, the three primary varieties are physical twins, virtual twins, and hybrid twins. The physical counterpart of a digital twin is the actual, observable thing or system that it represents. For example, the factory's physical machine is the twin of digitally twinned industrial equipment. The genuine stuff being copied and seen are called biological twins, which exist in the physical world. A digital twin of a physical item is created for simulations, testing, and optimization purposes without directly engaging with the original object, improving the real-world counterpart's comprehension, analysis, and management [10]. A virtual twin is a digital copy of a natural system or thing. Unlike a physical twin, a virtual twin only exists digitally. Computer-aided design (CAD) models, simulations, and further digital methods are used in its creation. Virtual twins are especially helpful in developing a product, as they let designers and engineers test features, run simulations, and test concepts without being limited by the real world. They offer an affordable, risk-free atmosphere conducive to experimentation and creativity. The traits of both physical and virtual twins are combined in hybrid twins. A physical item is outfitted with various sensors and IoT devices in a hybrid twin structure to gather real-time data about its surroundings, performance, and state [3]. The digital model is then updated with this real-time data, creating a hybrid twin.

5.2.3 Applications and Use Cases: Real-world Implementations of Digital Twins

Digital twins are transforming several sectors by providing novel approaches to visualize, analyze, and optimize real-world systems. In manufacturing, digital twins are often employed to simulate physical production processes. Manufacturers can model their plants and machinery virtually. They forecast maintenance requirements, track machine performance, and streamline manufacturing processes using real-time IoT sensor data [7]. In line with the tenets of Industry 4.0, this guarantees effective operations, lower downtime, and boosts overall production. Digital twins imitate whole metropolitan environments within the framework of smart cities. City planners use digital twins to simulate public services, energy use, trash management, and traffic flow. By modeling different situations, urban planners may optimize city layouts, improve infrastructure, and improve public services. Cities may become more efficient, sustainable, and receptive to the demands of their citizens with the help of digital twins [11]. In healthcare, digital twins of human organs or whole individuals are utilized for medical simulations and individualized therapy. These virtual models assist clinicians by simulating medical operations, seeing intricate anatomy, and forecasting patient reactions to treatments. By accounting for individual differences in physiology, digital twins provide accurate diagnosis, customized therapy, and better patient outcomes. Digital twins are used in the construction sector to recreate infrastructure projects and structures. They are used in simulations by engineers and architects to test designs, evaluate structural integrity, and examine energy use. Digital twins help with facilities management, utility monitoring, occupancy trends, and general building performance when a facility is serviced. This data-driven strategy guarantees energy efficiency, lowers expenses, and improves the safety and comfort of inhabitants. In the aerospace industry, engineers can mimic flight conditions and forecast performance thanks to digital twins of aircraft and spacecraft. This aids fuel economy and design optimization [12]. Digital twins are used in defense to help with training simulations and mission planning by simulating military hardware and circumstances. Digital twins improve military effectiveness and safety because they faithfully replicate real-world settings. In environmental research, digital twins simulate climatic trends, natural resource availability, and ecosystems. Researchers use these models to forecast environmental changes, evaluate the effects of human activity, and create sustainable behaviors. Digital twins make data-driven decision-making possible, which supports climate change mitigation, natural catastrophe forecasting, and conservation initiatives [13]. Retailers use digital twins to improve customer experience, assess consumer behavior, and optimize store layouts. Digital twins assist with demand forecasting, distribution network optimization, inventory monitoring, and logistics optimization in supply chain management.

5.3 THE AI REVOLUTION: EXPLORING ARTIFICIAL INTELLIGENCE ALGORITHMS

With robots becoming able to learn and make judgments independently and being programmed to do tasks, the AI revolution heralds a revolutionary new era in technology. AI algorithms, which act as the brains of intelligent robots, are at the center of this transformation. Sophisticated mathematical structures called AI algorithms are made to interpret data, identify patterns, and make judgments or predictions. Unlike classical algorithms, AI algorithms can learn from massive data sets. They use DL, ML, and neural networks to analyze enormous volumes of data and spot intricate patterns in it [14]. In conclusion, AI algorithms are the foundation of intelligent systems, enabling machines to mimic human cognitive functions, such as data analysis, pattern recognition, and decision-making.

TABLE 5.1
Difference between Supervised Learning, Unsupervised Learning, and RL

Criteria	Supervised learning	Unsupervised leaning	RL
Definition	ML by labeled data	ML through unlabeled data without any guidance	ML through environment interaction also forms learning and rewords
Type of problems	Regression and Classifications Labeled data	Association & Clustering unlabeled data	Reward based
Type of data	Labeled data	Unlabeled data	No predefined data
Training	External supervision	No supervision	No supervision
Approach	Maps the labeled inputs to the known outputs	Understand patterns & discover outputs	Follow the trial-and-error method

5.3.1 Machine Learning Algorithms: Supervised, Unsupervised, and Reinforcement Learning

An instructor guiding a pupil is what supervised learning is like. This approach pairs input data points with matching output labels to train the algorithm on labeled data. By identifying patterns and connections among the labeled samples, the algorithm gains the ability to translate the input data into the appropriate output. After training, the system can accurately identify new, unknown data or make predictions [16]. Consider it a predictive modeling method for language translation, image recognition, and spam email detection applications. The algorithm uses what it has learned from well-known examples to forecast results for fresh, comparable data. Conversely, unsupervised learning is similar to discovering a new city unaccompanied by a map or guide. In this case, the algorithm operates on unlabeled data, which indicates that the output labels are not predetermined. Instead, the algorithm finds the input data's relationships, structures, or patterns. Unsupervised learning is frequently applied to association and clustering tasks. Clustering combines similar data points to help find naturally occurring clusters within the data [4]. We can better understand the difference between supervised, unsupervised, and reinforcement learning (RL) in Table 5.1.

5.3.2 Deep Learning Architectures: Neural Networks, CNNs, RNNs, and GANs

The architecture of the human brain fundamentally inspires a neural network. Layers of interconnected nodes, or "neurons" make up this structure. Transformations occur as information moves through these layers. Convolutional neural networks (CNNs) are specialized neural networks that handle data resembling a grid, such as pictures

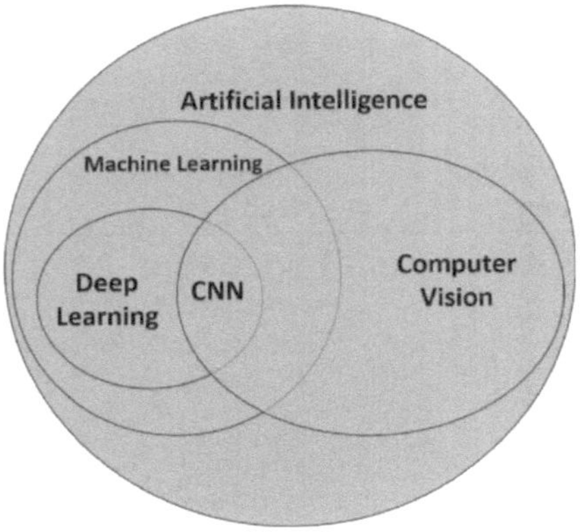

FIGURE 5.3 Relation between ML, DL, CNN, and Computer Vision.

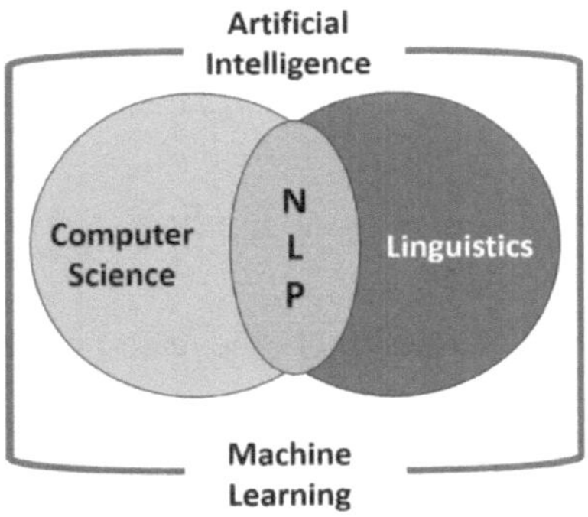

FIGURE 5.4 The intersection of Computer Science and Linguistics is NLP.

and videos. They employ convolutional layers to learn the spatial hierarchies of features from the input images automatically and adaptively. In essence, CNNs are very good at recognizing image patterns [17]. Recurrent neural networks (RNNs) are helpful for tasks such as speech recognition, language translation, and sequence prediction because of their ability to store context and dependencies in sequential data. A generator and a discriminator are two neural networks that compete with one another to form a generative adversarial network (GAN). GANs acquire the ability to produce synthetic data identical to accurate data through this adversarial process [18]. Figure 5.3 better illustrates the relation between ML, DL, and CNN.

5.3.3 Natural Language Processing (NLP) and AI: Enabling Human-like Communication

AI and NLP transform human-machine interactions. These technologies work together to give computers the ability to comprehend, interpret, produce, and react to human language in a way that is strikingly similar to that of humans. Figure 5.4 clearly illustrates that the intersection of computer science and linguistics is NLP.

Fundamentally, NLP enables machines to understand the subtleties of human language, such as context, semantics, and intent [19]. NLP systems can learn from enormous volumes of linguistic data with the aid of AI, mainly ML and DL algorithms, which enables them to identify patterns, comprehend context, and produce coherent and contextually relevant responses [20]. AI-powered NLP systems can carry on natural dialogues and comprehend linguistic nuances such as sarcasm, metaphors, and colloquialisms, just like a human speaker would [14].

The use of AI and NLP in facilitating human-like communication is extensive and diverse. For example, virtual assistants and chatbots are getting more advanced. They can understand user inquiries, respond in depth and appropriately to the context, and learn from each interaction to enhance subsequent responses [20]. In summary, the synergy between AI and NLP drives a paradigm shift in human-computer interactions. It involves more than just word processing; it consists of comprehending human language's complexities and simulating the depth of human interaction [21].

5.3.4 Computer Vision and AI: Enhancing Visual Perception and Interpretation

Computer vision and AI are revolutionizing how machines perceive and interpret visual information. This revolutionary technology, which combines AI and computer vision, is changing several industries, including manufacturing, healthcare, autonomous cars, and entertainment. Computer vision enables machines to comprehend and interpret visual data, including pictures and videos [15].

AI and computer vision improve the interpretation of medical imaging in healthcare. Healthcare professionals can receive assistance in diagnosing diseases and conditions from machines that can analyze MRIs, CT scans, and X-rays [22]. Figure 5.5 shows the CT scan image of a heart.

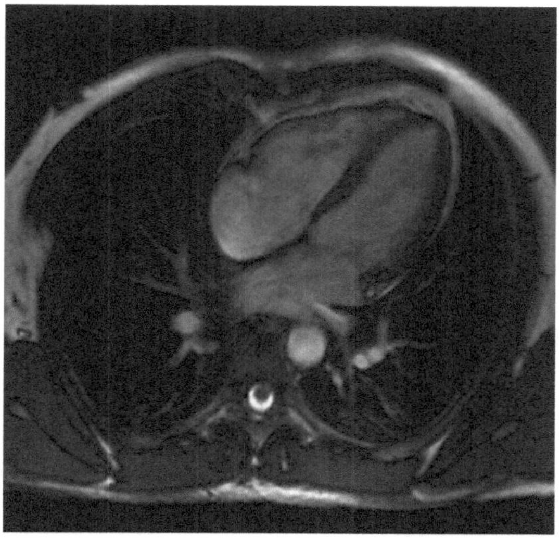

FIGURE 5.5 CT scan of a Heart.

5.4 CONVERGENCE OF DIGITAL TWINS AND AI: TRANSFORMING INDUSTRIES

Global industry transformation is occurring due to the revolutionary synergy between digital twins and AI. As virtual replicas of actual systems or objects, digital twins comprehensively understand their physical counterparts. On the other hand, ML and predictive analytics, two AI tools, are very good at processing large amounts of data, identifying patterns, and making insightful predictions. Personalized medicine in healthcare is made possible by patient digital twins, which are produced by combining AI algorithms with medical data. These virtual models consider each patient's unique characteristics, enabling medical professionals to customize therapies and treatments. Digital twin data is processed by AI, enabling predictive analysis for early diagnosis, effective treatment plans, and disease prevention, all of which improve patient outcomes. AI and digital twins are revolutionizing intelligent city projects and urban planning. Urban planners use AI algorithms to analyze data from sensors, traffic systems, and public services by building digital replicas of entire cities. This integration makes real-time monitoring, proactive traffic management, energy optimization, and practical resource allocation possible, improving cities' sustainability, livability, and responsiveness to citizens' needs.

5.4.1 Industry 4.0 and Digital Twins: Revolutionizing Manufacturing and Supply Chains

AI algorithms combined with digital twins of power grids allow for predictive analysis and real-time energy consumption monitoring. This integration facilitates the shift to sustainable energy sources by balancing supply and demand, improving energy efficiency, and optimizing grid performance. In conclusion, data-driven decision-making, predictive analysis, and creative solutions are made possible by the convergence of digital twins and AI, which is transforming industries. In addition to improving operational effectiveness and cutting costs, this revolutionary synergy stimulates innovation and increases industry adaptability, efficiency, and responsiveness to the ever-evolving demands of the contemporary world: efficiency and lower running expenses [23]. Predictive maintenance is one of the primary uses of digital twins in manufacturing. Manufacturers can anticipate machine failure by continuously using IoT sensors to monitor equipment and integrating the data into digital twins [7]. Digital twins and Industry 4.0 both rely significantly on data. The massive volumes of data gathered from digital twins can provide manufacturers with important insights through advanced analytics and ML algorithms. These insights are the foundation for data-driven decision-making procedures that direct process optimization, product innovation, and strategic planning. By relying on accurate, real-time data, businesses can make decisions that increase productivity, reduce expenses, and boost overall performance. A paradigm shift has occurred with the incorporation of digital twins and Industry 4.0 principles into supply chains and manufacturing processes. Businesses benefit from its unparalleled visibility, swiftness, and effectiveness.

5.4.2 Smart Cities and IoT: Enhancing Urban Planning with AI-driven Digital Twins

AI-driven digital twins are at the forefront of this revolutionary shift in urban planning, brought about by the convergence of smart cities and the IoT technologies. Imagine living in a town where every element is connected and optimized for efficiency, from waste management and public services to energy and transportation use. Incorporating AI and IoT devices into digital twins of urban environments is turning this vision into a reality. IoT devices, including sensors and cameras, are positioned thoughtfully throughout the city to gather data in real time. These devices track a number of different parameters, including energy consumption, traffic flow, and air quality. Figure 5.6 better depicts the idea of having the IoT in the city. Urban planners can obtain a precise and all-encompassing understanding of the dynamics of a town by incorporating this data into digital twins. Advanced virtual models that replicate the whole urban landscape are called digital twins of cities. AI algorithms built into these digital copies handle enormous amounts of data from the IoT devices. AI-powered digital twins are capable of trend identification, complex pattern analysis, and scenario prediction. This predictive capability is invaluable when predicting energy demands, traffic congestion, or environmental issues. Public services and infrastructure can be optimized by urban planners using AI-driven insights from digital twins. For example, traffic signals can be dynamically changed to lessen congestion based

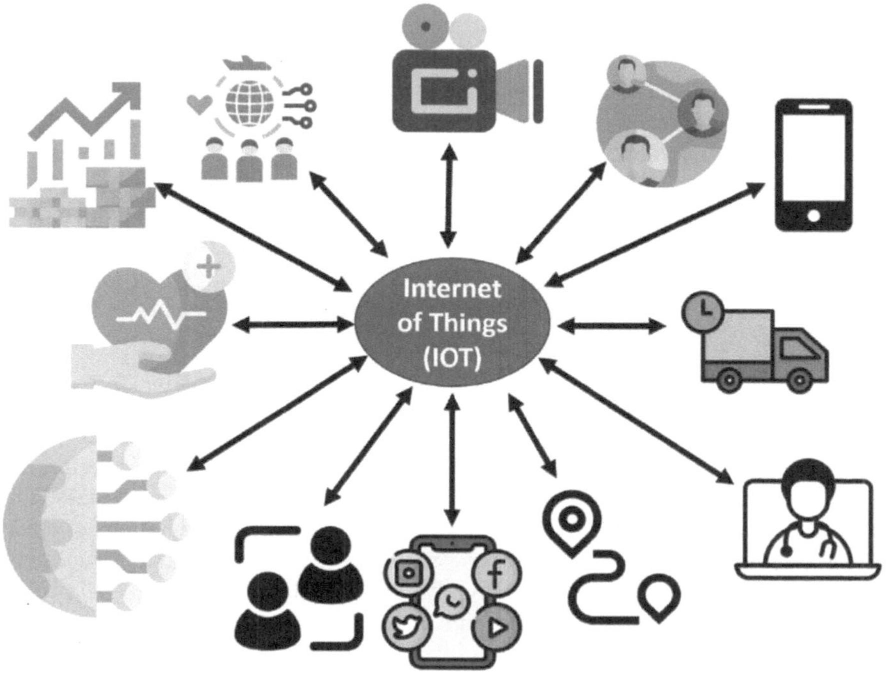

FIGURE 5.6 The IoT.

on real-time traffic data. Energy distribution can be optimized to reduce waste, and collection routes can be optimized by alerting waste management systems when trash cans are full. These improvements result in lower expenses, increased effectiveness, and a more sustainable city environment. Thanks to AI-driven digital twins, cities can now simulate various disaster scenarios, including large-scale accidents and natural disasters. Evacuation plans can be optimized and emergency response strategies can be improved by examining these simulations. Furthermore, AI algorithms can anticipate possible weaknesses in the city's infrastructure, enabling proactive steps to improve disaster resilience. Smart cities use AI-driven digital twins for internal efficiencies and to raise public participation.

AI-driven digital twins contribute significantly to sustainability by maximizing resource utilization, cutting emissions, and raising general efficiency. Cities can use the insights from digital twins powered by the IoT and AI technologies to set and accomplish environmental goals, such as cutting carbon emissions or increasing green spaces. AI-assisted diagnostics and personalized medicine have emerged due to the quick advancement of healthcare innovations. These developments are changing how doctors identify illnesses, create treatment programs, and deliver care, adjusting interventions to each patient's requirements and features. Precision medicine, another name for personalized medicine, is a cutting-edge strategy that tailors medical care and decisions to each patient's unique characteristics. This method considers an individual's genetic composition, way of life, surroundings, and particular biomarkers. Healthcare practitioners can determine genetic variants linked to diseases by examining a patient's genetic and molecular profile. This knowledge makes it possible to choose medication, dosages, and targeted therapies, which improves treatment efficacy and lowers side effects. Oncologists can identify specific tumor mutations through genetic profiling, another application of personalized medicine in cancer treatment. With this knowledge, oncologists can recommend therapies targeting congenital abnormalities, resulting in more individualized and successful cancer care.

Furthermore, predicting disease susceptibility through personalized medicine is essential for enabling preventive measures and lifestyle modifications to reduce risks. Healthcare diagnostics have greatly benefited from AI. A subset of AI called ML algorithms examine enormous volumes of medical data, such as pictures, patient histories, and study results, to spot trends and produce new knowledge [24]–[25].

5.4.3 Environmental Sustainability: Monitoring and Mitigating Climate Change through AI-Infused Digital Twins

Innovative solutions are needed to address the urgent problem of climate change, and digital twins with AI integration show promise as potent tools for tracking and lessening its effects. When paired with AI, these sophisticated digital representations of environmental systems offer a comprehensive method for comprehending, forecasting, and resolving climate-related issues. AI-enhanced digital twins collect real-time data from satellites, IoT devices, and environmental sensors. These sensors monitor sea levels, temperature, air quality, and deforestation rates. By integrating this data into digital twins, scientists and environmentalists can obtain a comprehensive and current understanding of environmental changes. After processing this enormous volume of data, AI algorithms find patterns and trends essential for determining the condition of the environment at the moment. AI-enhanced digital twins promote sustainable practices and increase awareness among communities and individuals. These digital models provide an environmental impact on lifestyle choices and can encourage changes in behavior. Digital twins with AI-infused into them are, in short, vital in the fight against climate change. These digital replicas enable informed decision-making, facilitate proactive measures, and inspire a global movement toward a more sustainable future by fusing real-time data from diverse sources with AI's predictive power.

5.5 CHALLENGES AND ETHICAL CONSIDERATIONS IN THE DIGITAL TWIN AND AI ECOSYSTEM

- Addressing Bias: AI programs trained on skewed data can potentially reinforce and amplify pre-existing prejudices. Therefore, attempts to develop unbiased and fair algorithms must be ongoing.
- Transparency: Concerns are raised when AI decision-making systems lack openness, which makes it necessary to compromise between protecting proprietary technology and offering consumers and other impacted parties enough transparency.
- Ethical Considerations: Ethical issues should be considered when developing AI technology from conception and development to deployment and ongoing evaluation.
- Collaboration for Ethical Implications: Legislators, ethicists, technologists, and the general public must work together to address the ethical implications of AI.
- Strategies for Conscientious AI Development: To guarantee the responsible development and application of AI technology, stronger laws, increased transparency requirements, inclusive and varied teams, and continuous ethical audits are crucial.

5.5.1 Data Privacy and Security: Safeguarding Sensitive Information in the Digital Age

- Fundamental Right to Manage Their Personal Data: To stop fraud, identity theft, and illegal access, people should have the fundamental right to control their personal data.
- Developing and Maintaining Trust with Customers: To foster and sustain customer trust and promote the usage of digital services and online transactions, businesses must prioritize data protection.
- Legal Compliance: Laws (such as the California Consumer Privacy Act (CCPA) and General Data Protection Regulation (GDPR)) dictating how businesses must gather, keep, and manage personal data are enforced by governments and regulatory bodies, and disobedience carries harsh penalties.
- Impact on Reputation and Finances: Data privacy violations can seriously harm a company's reputation, which can have adverse financial and legal effects.

- Data Security Measures: To safeguard data from cyberattacks and illegal access, it is essential to use encryption, multi-factor authentication, frequent software updates, and employee training.
- Business Continuity: Ensuring resilience in the face of data-related incidents, limiting operational disruptions, and preserving business continuity all depend on data security.
- Ethical Responsibility: Companies and organizations have an ethical duty to safeguard individuals' personal information and uphold their right to privacy.
- Assessment of Moral Behavior: Companies are being held increasingly accountable for how they manage consumer data, highlighting the significance of moral principles in addition to legal obligations.

5.5.2 Regulatory Landscape: Navigating Legal and Compliance Challenges

Dealing with intricate legal and compliance issues is necessary when navigating the regulatory environment in AI and emerging technologies. Governments and international organizations are trying to create frameworks that strike a balance between innovation, ethical issues, and the welfare of society as these technologies continue to develop. When tackling compliance and legal matters in the AI space, keep the following points in mind: Strict rules on the collection, use, and storage of personal data are required by data privacy laws, such as the GDPR in the European Union (EU) and comparable legislation in other jurisdictions. Since AI systems frequently use large data sets, it is crucial to make sure these laws are followed. Businesses must use robust data anonymization techniques and get informed consent when handling personal data. Transparency about the decision-making process of AI algorithms is becoming increasingly in demand. Regulations requiring businesses to reveal the inner workings of their AI systems are being considered or implemented by certain jurisdictions, particularly in the financial and healthcare sectors. To meet these transparency requirements, explainable AI (XAI) techniques—which improve the interpretability of AI algorithms for humans—are becoming increasingly important. The biases in AI systems are causing regulators more and more anxiety, primarily when those systems are employed in crucial domains, such as hiring, lending, and law enforcement. To ensure fairness, addressing these issues calls for constant oversight, audits, and, occasionally, regulatory mandates. International standards and guidelines for AI development and application are being worked on by organizations such as the Organization for Economic Cooperation and Development (OECD) and the United Nations (UN). Businesses that operate internationally must abide by all applicable agreements and treaties, and these international standards. Liability for harm or incorrect decisions by AI systems is a complicated legal matter. Legal frameworks must change to define liability precisely, mainly when AI systems function independently. Ensuring responsible deployment of AI requires the establishment of guidelines and accountability mechanisms for mishaps or errors involving the technology. In conclusion, a diversified strategy is needed to effectively navigate the legal and compliance challenges the AI landscape presents. It entails putting ethical frameworks into place, encouraging transparency, staying current on changing regulations, and cultivating global cooperation.

5.6 FUTURE OUTLOOK: SYNERGISTIC GROWTH OF DIGITAL TWINS AND AI

The combination of AI and digital twins is a promising development in the rapidly changing field of technology. It holds the promise of a time when the virtual and real worlds will coexist peacefully. Businesses in the manufacturing, healthcare, and urban planning sectors stand to gain a great deal from this combination. Large-scale data sets produced by digital twins are analyzed by AI algorithms, which help companies to make better decisions, streamline operations, and increase productivity [30]. Predictive analytics is enhanced when digital twins and AI are combined. Digital twins provide historical and current data that AI algorithms can use to predict future events accurately. This ability to predict the future is beneficial for planning equipment breakdowns and supply chain optimization [8].

5.6.1 Emerging Trends: Quantum Computing, Edge AI, and Swarm Intelligence

A few trends in the rapidly evolving field of technology have the potential to completely change how we view and engage with the digital world. Among these, the disruptive technologies that are reshaping industries and pushing the limits of what we previously believed was possible are quantum computing, Edge AI, and swarm intelligence. Computing power has undergone a paradigm shift with the advent of quantum computing. Quantum computers use quantum bits, also known as qubits, which are multistate entities that can exist simultaneously, in contrast to classical computers that use bits (0s and 1s). This inherent parallelism enables quantum computers to solve complicated problems, such as cryptography and optimization, exponentially faster than their classical counterparts. With further research, quantum computing may revolutionize industries such as AI, drug development, and climate modeling, launching a new phase of scientific and technological discoveries. AI algorithms can now be used closer to the data source with Edge AI, eliminating the need to transfer massive volumes of data to centralized servers for processing. Edge AI reduces bandwidth usage, improves real-time decision-making, and minimizes latency by processing data locally on devices or Edge servers. This trend is significant for applications that require snap decisions, such as AR, IoT devices, and autonomous cars [31]. Swarm intelligence takes its cues from how essential individuals work together to solve complicated issues in natural systems such as ant colonies and bird flocks. In the context of technology, swarm intelligence is the cooperative behavior of networks of independent entities, frequently with little to no central authority. Applications for this method can be found in data analysis, robotics, and optimization. Swarm intelligence algorithms open the door to novel solutions in manufacturing, logistics, and disaster relief industries by enabling systems to adapt, learn, and evolve based on their interactions with the environment and other agents [32]. The convergence of these movements reveals an exciting future. Imagine real-time processing at the edge using quantum-powered algorithms that improve swarm intelligence systems' capacity for learning [33]. This combination may result in unprecedented effectiveness, flexibility, and problem-solving abilities. With the further development of quantum computing, Edge AI, and swarm intelligence, these technologies stand alone as revolutionary breakthroughs and work in unexpected ways in concert, pointing to a time when technology will closely resemble the complexity and effectiveness of the natural world.

5.6.2 Collaborative Innovation: Research, Academia, and Industry Partnerships

In the current innovation landscape, cooperation between academic institutions, corporate partners, and research institutions have emerged as a critical component for advancing real change and promoting growth. The future is being shaped in the following ways by this successful collaboration between industry, university, and research: Academic institutions, businesses, and research facilities all contribute various resources and areas of expertise. Together, they produce a potent mashup of creativity, skill, and technology. Academic institutions foster talent and skill development; research institutes bring state-of-the-art information, and companies present real-world problems and valuable applications [34]. This combination of resources creates an atmosphere where creative ideas can be explored, tested, and refined. Working together promotes interdisciplinary research, in which specialists from different domains come together to tackle challenging issues. For example, combining computer science with healthcare knowledge has resulted in breakthroughs in medical technology and innovative healthcare solutions. Collaborative efforts have been made to expedite technology transfer from research papers and labs to valuable, commercial goods. Working together, the academic and business communities expose students to the real world. Students can gain practical insights into industrial processes and difficulties through cooperative research projects, internships, and industry-academia partnerships. This exposure helps them to develop their talents and match industry demands with their education, resulting in a more skilled and prepared workforce for the crew. Addressing global food security, healthcare crises, and climate change requires collaborative creativity. Collaborative efforts use industry resources, academic knowledge, and research discoveries to create sustainable solutions [35]. For instance, collaborations between educational institutions, agricultural businesses, and agricultural researchers have produced creative farming methods that guarantee food security while reducing environmental effects [36]. The cooperative synergy of research institutions, academia, and industrial partners sparks transformative innovation.

5.6.3 Predictions and Speculations: Envisioning the Next Frontier of Technological Advancements

Some fascinating technical developments are about to revolutionize how we work, live, and interact as we approach the dawn of a new era. It is predicted that quantum computers will reach a point where they can solve puzzles that traditional computers cannot. This can transform industries, including materials science, medicine development, cryptography, and more, opening doors to previously unthinkable discoveries. The human brain and computers may work together seamlessly in the future thanks to results in neurotechnology. Brain-computer interfaces, or BCIs, have the potential to change healthcare, communication, and even entertainment by enabling direct contact with gadgets [37]. Imagine using brain signals to communicate thoughts or to operate devices [38]. Personalized medicine and treating hereditary illnesses may benefit significantly from developing genetic engineering tools such as clustered regularly interspaced short palindromic repeats (CRISPR). Editing genes could transform healthcare by improving human longevity and treating previously incurable ailments. VR, AR and AI can potentially build intelligent, immersive worlds [39]. The amalgamation of virtual classrooms, augmented workplaces, and AI-powered simulations for many industries has the potential to revolutionize how we learn, work together, and see the world. Given the urgency of addressing climate change, novel technologies are probably on the horizon. Advanced renewable energy systems, carbon capture technology, sustainable agricultural methods, and environmentally friendly materials are a few examples. These developments may have a significant impact on creating a more sustainable future. Technological developments in propulsion, life support, and resource management in space may open doors to establishing long-term lunar bases, Mars colonies, and interstellar travel, extending humankind's reach beyond Earth. Applications for blockchain technology are not limited to cryptocurrency. Supply chain management, decentralized social networks, and safe voting systems are just a few of the industries that blockchain's transparent, unchangeable protocols have the potential to transform [40]. Innovative AI-powered learning systems may provide individualized instruction. Adaptive learning systems, interactive simulations, and customized educational content have the potential to completely change how people learn and develop new abilities, increasing both the effectiveness and accessibility of education. The cooperation of scientists, legislators, ethicists, and society will be crucial in building a future where technology improves the human experience and promotes a more sustainable, inclusive, and enlightened world as humanity explores this new frontier.

5.7 CONCLUSION

AI and digital twins interact dynamically to transform industries, technology, and problem-solving. Digital twins are complex virtual copies of physical systems that close the gap between digital and physical worlds by enabling two-way communication and real-time information sharing. Industries use these virtual equivalents to revolutionize design, monitoring, and maintenance procedures in conjunction with real-world sensor data. AI, including DL and ML, enables computers to simulate human cognitive processes. DL advances AI capabilities by drawing inspiration from neural networks in the human brain. Digital twins and AI work together to improve prediction abilities and encourage proactive problem-solving. The development of digital twins from concept to actual use reflects advances in ML, AI, and IoT. Digital twins are made possible by the fundamental components of the IoT, sensor networks, and data integration, which allow for accurate representations of real-world systems. Digital twins can be physical, virtual, or hybrid, and they can be used for anything from safe space testing to operation replication in the real world. Applications improve productivity, preventive maintenance, and decision-making in various industries, including urban planning, manufacturing, healthcare, and environmental research. Intelligent systems, where sophisticated algorithms enable machines to learn, evaluate data, spot patterns, and make decisions independently, are the product of the AI revolution. The coming together of digital twins and AI heralds a new era in which the virtual and physical worlds coexist. Because of this synergistic relationship, industries today have never-before-seen prospects for production, innovation, and understanding. AI and digital twins have a profoundly transformational effect on how we create, work, and interact with the world, indicating a never-ending quest to push the boundaries of technology.

BIBLIOGRAPHY

[1] H. Boyes and T. Watson, "Digital twins: An analysis framework and open issues," *Comput. Ind.*, vol. 143, no. February, p. 103763, 2022, doi: 10.1016/j.compind.2022.103763.

[2] A. Fuller, Z. Fan, C. Day, and C. Barlow, "Digital Twin: Enabling technologies, challenges and open research," *IEEE Access*, vol. 8, pp. 108952–108971, 2020, doi: 10.1109/ACCESS.2020.2998358.

[3] A. Sharma, E. Kosasih, J. Zhang, A. Brintrup, and A. Calinescu, "Digital Twins: State of the art theory and practice, challenges, and open research questions," *J. Ind. Inf. Integr.*, vol. 30, no. November 2020, p. 100383, 2022, doi: 10.1016/j.jii.2022.100383.

[4] I. H. Sarker, "Machine learning: Algorithms, real-world applications and research directions," *SN Comput. Sci.*, vol. 2, no. 3, pp. 1–21, 2021, doi: 10.1007/s42979-021-00592-x.

[5] I. H. Sarker, "Deep learning: A comprehensive overview on techniques, taxonomy, applications and research directions," *SN Comput. Sci.*, vol. 2, no. 6, pp. 1–20, 2021, doi: 10.1007/s42979-021-00815-1.

[6] U. I. Musa and S. Ghosh, "Advancing digital twin through the integration of new AI algorithms," *Int. Res. J. Eng. Technol.*, vol. 10, no. 03, pp. 922–932, April, 2023.

[7] L. Gopal, H. Singh, P. Mounica, N. Mohankumar, N. P. Challa, and P. Jayaraman, "Digital twin and IOT technology for secure manufacturing systems," *Meas. Sensors*, vol. 25, no. August 2022, p. 100661, 2023, doi: 10.1016/j.measen.2022.100661.

[8] S. M. Zayed, G. M. Attiya, A. El-Sayed, and E. E. D. Hemdan, "A review study on digital twins with artificial intelligence and internet of things: Concepts, opportunities, challenges, tools and future scope," *Multimed. Tools Appl.*, no. 0123456789, 2023, doi: 10.1007/s11042-023-15611-7.

[9] M. Javaid, A. Haleem, and R. Suman, "Digital Twin applications toward Industry 4.0: A review," *Cogn. Robot.*, vol. 3, no. April, pp. 71–92, 2023, doi: 10.1016/j.cogr.2023.04.003.

[10] C. Czwick and R. Anderl, "Cyber-physical twins – Definition, conception and benefit," *Procedia CIRP*, vol. 90, no. March, pp. 584–588, 2020, doi: 10.1016/j.procir.2020.01.070.

[11] Z. Wang, "Digital twin technology," *Ind. 4.0 – Impact Intell. Logist. Manuf.*, 2020, doi: 10.5772/intechopen.80974.

[12] J. Guo and Z. Lv, "Application of Digital Twins in multiple fields," *Multimed. Tools Appl.*, vol. 81, no. 19, pp. 26941–26967, 2022, doi: 10.1007/s11042-022-12536-5.

[13] R. da S. Mendonça, S. de O. Lins, I. V. de Bessa, F. A. de Carvalho Ayres, R. L. P. de Medeiros, and V. F. de Lucena, "Digital twin applications: A survey of recent advances and challenges," *Processes*, vol. 10, no. 4, pp. 1–12, 2022, doi: 10.3390/pr10040744.

[14] D. Khurana, A. Koli, K. Khatter, and S. Singh, "Natural language processing: State of the art, current trends and challenges," *Multimed. Tools Appl.*, vol. 82, no. 3, pp. 3713–3744, 2023, doi: 10.1007/s11042-022-13428-4.

[15] C. O. Jaynes, "Computer vision and artificial intelligence," *XRDS Crossroads, ACM Mag. Students*, vol. 3, no. 1, pp. 7–10, 1996, doi: 10.1145/332148.332152.

[16] S. Naeem, A. Ali, S. Anam, and M. M. Ahmed, "An unsupervised machine learning algorithms: Comprehensive review," *Int. J. Comput. Digit. Syst.*, vol. 13, no. 1, pp. 911–921, 2023, doi: 10.12785/ijcds/130172.

[17] L. Alzubaidi *et al.*, *Review of deep learning: Concepts, CNN architectures, challenges, applications, future directions*, vol. 8, no. 1. Springer International Publishing, 2021. doi: 10.1186/s40537-021-00444-8.

[18] I. Goodfellow *et al.*, "Generative adversarial networks," *Commun. ACM*, vol. 63, no. 11, pp. 139–144, 2020, doi: 10.1145/3422622.

[19] D. Khurana, A. Koli, and K. Khatter *et al.*, "Natural language processing: State of the art, current trends and challenges." *Multimed Tools,* vol. 82, pp. 3713–3744, 2023. doi: 10.1007/s11042-022-13428-4

[20] S. Kumar Behera and M. M. Nayak, "Natural language processing for text and speech processing: A review paper," *Int. J. Adv. Res. Eng. Technol.*, vol. 11, no. 11, pp. 1947–1952, 2020, [Online]. Available: www.iaeme.com/Home/journal/IJARET1947editor@iaeme.comhttp://iaeme.comhttp://iaeme.com.

[21] S. Dridi, "Unsupervised learning- A systematic literature review," *A Comp. Anal. Linear Regres. Support Vector Regres.*, no. 2015, pp. 1–8, 2015.

[22] V. Wiley and T. Lucas, "Computer vision and image processing: A paper review," *Int. J. Artif. Intell. Res.*, vol. 2, no. 1, p. 22, 2018, doi: 10.29099/ijair.v2i1.42.

[23] M. Javaid, A. Haleem, R. P. Singh, R. Suman, and E. S. Gonzalez, "Understanding the adoption of Industry 4.0 technologies in improving environmental sustainability," *Sustain. Oper. Comput.*, vol. 3, no. September 2021, pp. 203–217, 2022, doi: 10.1016/j.susoc.2022.01.008.

[24] A. Golande and T. Pavan Kumar, "Heart disease prediction using effective machine learning techniques," *Int. J. Recent Technol. Eng.*, vol. 8 , no. 1 Special Issue 4, pp. 944–950, 2019.

[25] N. Kumar, N. Narayan Das, D. Gupta, K. Gupta, and J. Bindra, "Efficient automated disease diagnosis using machine learning models," *J. Healthc. Eng.*, vol. 2021, 2021, doi: 10.1155/2021/9983652.

[26] Y. K. Dwivedi *et al.*, "Climate change and COP26: Are digital technologies and information management part of the problem or the solution? An editorial reflection and call to action," *Int. J. Inf. Manage.*, vol. 63, no. November 2021, 2022, doi: 10.1016/j.ijinfomgt.2021.102456.

[27] S. Srinivasan, "Privacy and data protection in Japan," *Gov. Inf. Q.*, vol. 9, no. 2, pp. 121–133, 1992, doi: 10.1016/0740-624X(92)90046-O.

[28] E. Narayanan and B. Muthukumar, "A machine learning framework for providing data integrity and confidentiality for sensitive data cloud applications," *Int. J. Syst. Assur. Eng. Manag.*, no. 0123456789, 2022, doi: 10.1007/s13198-022-01741-y.

[29] R. P. Romansky and I. S. Noninska, "Challenges of the digital age for privacy and personal data protection," *Math. Biosci. Eng.*, vol. 17, no. 5, pp. 5288–5303, 2020, doi: 10.3934/MBE.2020286.

[30] K. K. Alnowaiser and M. A. Ahmed, "Digital Twin: Current research trends and future directions," *Arab. J. Sci. Eng.*, vol. 48, no. 2, pp. 1075–1095, 2023, doi: 10.1007/s13369-022-07459-0.

[31] R. Singh and S. S. Gill, "Edge AI: A survey," *Internet Things Cyber-Physical Syst.*, vol. 3, no. March, pp. 71–92, 2023, doi: 10.1016/j.iotcps.2023.02.004.

[32] D. Chopra and P. Arora, "Swarm intelligence in data science: Challenges, opportunities, and applications," *Procedia Comput. Sci.*, vol. 215, no. 2022, pp. 104–111, 2022, doi: 10.1016/j.procs.2022.12.012.

[33] H. Ahmed and J. Glasgow, "Swarm intelligence: Concepts, models and applications technical report 2012-585," *Tech. Rep.*, vol. 585, no. February, pp. 1–50, 2012, doi: 10.13140/2.1.1320.2568.

[34] S. Ankrah and O. AL-Tabbaa, "Universities-industry collaboration: A systematic review," *Scand. J. Manag.*, vol. 31, no. 3, pp. 387–408, 2015, doi: 10.1016/j.scaman.2015.02.003.

[35] M. Englund and J. Gabrielsson, "Barriers and outcomes of the collaboration between industry and academia in a new approach: The Living Labs Authors: Supervisor." Thesis of "Master of Science in Management of Innovation and Business Development", Halmstad University School of Business and Engineering, pp. 1–35, 2019. https://urn.kb.se/resolve?urn=urn:nbn:se:hh:diva-14000

[36] P. Kettunen, J. Järvinen, T. Mikkonen, and T. Männistö, "Energizing collaborative industry-academia learning: A present case and future visions," *Eur. J. Futur. Res.*, vol. 10, no. 1, 2022, doi: 10.1186/s40309-022-00196-5.

[37] I. Olaronke, I. Rhoda, I. Gambo, O. Oluwaseun, and O. Janet, "Prospects and problems of brain computer interface in healthcare," *Curr. J. Appl. Sci. Technol.*, vol. 29, no. 6, pp. 1–17, 2018, doi: 10.9734/cjast/2018/44358.

[38] Y. Singh, "A review paper on brain computer interface," *International Journal of Engineering Research & Technology (IJERT)*, vol. 3, no. 10, pp. 1–6, 2015.

[39] F. A. Ran, P. D. Hsu, J. Wright, V. Agarwala, D. A. Scott, and F. Zhang, "Genome engineering using the CRISPR-Cas9 system," *Nat. Protoc.*, vol. 8, no. 11, pp. 2281–2308, 2013, doi: 10.1038/nprot.2013.143.

[40] Z. Zheng, S. Xie, H. Dai, X. Chen, and H. Wang, "An overview of blockchain technology: Architecture, consensus, and future trends," *Proc. – 2017 IEEE 6th Int. Congr. Big Data, BigData Congr. 2017*, no. June, pp. 557–564, 2017, doi: 10.1109/BigDataCongress.2017.85.

6 Blockchain Empowered Digital Twins

A Comprehensive Review of Integration and Innovation

Naveenkumar Jayakumar, Narayanan Prasanth, T. Joshva Devadas, Saravanan Krishnan, Saurabh Saoji and Shrikant Jadhav

6.1 INTRODUCTION

In an era defined by rapid technological evolution, two groundbreaking concepts have emerged as the driving forces behind transformative change: digital twins and blockchain technology. The integration of digital twins into various sectors holds the promise of unprecedented efficiency, accuracy, and innovation. The fusion of these innovations promises to revolutionize industries, redefining how to perceive, interact with, and secure the digital and physical realms. This chapter embarks on a journey that unveils the significance of digital twins across industries and explores the transformative potential that blockchain integration brings to these dynamic ecosystems. Digital twins are virtual replicas of physical objects, systems, or processes. They emulate their real-world equivalents in both form and behavior, allowing for real-time monitoring, analysis, and optimization. By leveraging cutting-edge technologies [1], such as the Internet of Things (IoT), sensors, artificial intelligence (AI), and data analytics, digital twins create a bridge between the physical and virtual realms. As industries continue to embrace this concept, the transformative power of digital twins is poised to redefine how to design, operate, and interact with the world around them. Some of the use cases are discussed below to highlight the growing significance of digital twins [2]–[3].

6.1.1 Growing Significance

a. Manufacturing and Industry 4.0: In the manufacturing sector, digital twins enable predictive maintenance, production optimization, and quality control. By simulating various scenarios, manufacturers can anticipate machinery failures, reduce downtime, and enhance overall productivity. This has given rise to the Industry 4.0 revolution, where factories become smart and interconnected entities that adapt to changing conditions [4].

DOI: 10.1201/9781003465904-8

b. Healthcare and Biomedical Engineering: Digital twins have found a unique application in the healthcare domain. They enable personalized medicine by replicating an individual's physiology and medical history. These virtual models aid in diagnosing diseases, simulating treatment outcomes, and guiding surgical procedures, ultimately improving patient care and treatment efficacy [5]–[6].
c. Aerospace and Engineering: In the aerospace and engineering sectors, digital twins simulate the behavior of complex systems, such as aircraft, spacecraft, and infrastructure. These simulations facilitate design optimization, testing, and risk assessment, leading to safer and more efficient solutions. The aviation industry, for instance, employs digital twins to monitor aircraft health, predict maintenance needs, and enhance overall performance [7]–[9].
d. Urban Planning and Smart Cities: The concept of digital twins extends to urban planning, where cities are replicated in virtual environments. By integrating real-time data from sensors and IoT devices, urban planners can predict traffic patterns, optimize resource allocation, and create more sustainable urban landscapes. The result is the emergence of smart cities that enhance residents' quality of life [9].
e. Energy and Sustainability: Digital twins also have a crucial function in the energy sector, which is to monitor and optimize energy consumption. They aid in managing energy grids, identifying inefficiencies, and facilitating the integration of renewable energy sources. This contributes to sustainable energy management and reduces environmental impact [2]–[7].
f. Supply Chain Management: Digital twins offer enhanced transparency and traceability in supply chain management. By replicating the movement of goods, materials, and products, businesses can identify bottlenecks, streamline operations, and ensure compliance with regulations. This leads to improved inventory management and optimized logistics [10].

In the realm of technological innovation, the convergence of blockchain technology and digital twin ecosystems has ushered in a new era of enhanced capabilities and unparalleled trustworthiness. The chapter also highlights the potential of integrating blockchain technology to elevate the capabilities and instill unwavering trust within digital twin ecosystems, thereby reshaping industries and revolutionizing business processes. Some of the benefits and challenges based on the integration are discussed in further sections [8], [11]–[13].

6.1.2 The Power of Blockchain Integration

a. Immutability and Data Integrity: The hallmark feature of blockchains is their capacity to ensure data immutability. By recording transactions in an unalterable and transparent ledger, blockchain technology fortifies the integrity of digital twin data. Every interaction, modification, or update is securely timestamped and cryptographically sealed, establishing an incontrovertible audit trail. In the realm of digital twins, this translates to trusted historical records of asset behavior, maintenance activities, and overall performance.

b. Decentralization and Data Security: Blockchain's decentralized architecture eliminates the need for intermediaries, fostering a peer-to-peer network where transactions are verified by consensus. This self-governing model bolsters data security within digital twin ecosystems. Unauthorized access and malicious tampering become significantly more challenging, as alterations must be approved by consensus across the network.
c. Seamless Data Sharing and Collaboration: Digital twin ecosystems often involve multiple stakeholders, necessitating secure yet transparent data sharing. Blockchain's encrypted and permissioned data-sharing mechanisms enable seamless collaboration across entities. Manufacturers, suppliers, regulators, and consumers can access relevant data while maintaining authority regarding their data confidentiality [14]–[15].
d. Smart Contracts and Automation: Blockchains introduce the notion of smart contracts, which are self-executing codes that automate predetermined actions that are triggered when particular conditions are met. Within the domain of digital twins, smart contracts can trigger automatic maintenance requests, updates, or even payments based on real-time data. This streamlines processes, reduces administrative overhead, and accelerates response times [16].
e. Trustworthy Supply Chain Management: Blockchain's traceability and transparency enhance supply chain management within digital twin ecosystems. Every stage of the supply chain, from the acquisition of raw materials to the delivery of the final product, may be recorded on an immutable blockchain. These combat counterfeiting, ensure product authenticity, and bolster consumer trust.
f. Enhancing Interoperability: Integrating blockchains into digital twin ecosystems can bridge data silos and enhance interoperability. The standardized and secure data exchange facilitated by blockchains ensures seamless communication between diverse components of the ecosystem. This leads to a comprehensive view of asset behavior and performance.

6.2 BACKGROUND

6.2.1 Navigating the Blockchain Landscape

Fundamentally, a blockchain refers to a distributed and decentralized digital record-keeping system that has ushered in a novel era of security, transparency, and trust. Blockchains, often referred to as a "chain of blocks," are a revolutionary approach to data management. In contrast to conventional centralized databases, in which a single entity governs data, blockchains function as a decentralized and widely distributed ledger. Data is stored across a network of interconnected computers, or nodes, ensuring a high level of security, resilience, and transparency. The core principle of blockchain technology is decentralization, which means that it does not rely on a central authority to validate and authorize transactions. Instead, blockchains employ a consensus mechanism where a network of nodes verifies transactions, and once consensus is reached, the transaction is added to the blockchain. This decentralized approach eliminates the need for intermediaries, reduces vulnerabilities, and empowers network participants. Immutability is a fundamental concept in blockchain technology. When data is integrated into a blockchain, it is impossible to modify or delete. Each block in the chain has a unique cryptographic hash derived from the block's data and the previous block's hash. Modifying any data in a block would require changing the entire subsequent chain—a task that would demand an impractical amount of computational power. This unchangeable quality ensures the dependability and honesty of the data retained on the blockchain. Blockchain technology is closely associated with transparency, offering an enhanced level of visibility into transactions and data. Each participant within the network can view the entire transaction history stored on the blockchain. This transparency fosters accountability, as any discrepancies or fraudulent activities are easily detectable and traceable. In manufacturing organizations, such as supply chain management or financial services, lucidity is paramount, and blockchains offer a transformative solution. Blockchains employ advanced cryptographic techniques to secure data and transactions. Every transaction is secured with a digital signature that ensures the sender and receiver are authentic. Furthermore, transactions are organized into blocks, with each block being connected to the preceding one through a cryptographic hash. This not only prevents unauthorized modifications, but it also guards against unauthorized access to sensitive information [17]–[19].

Understanding the nuances of different blockchain types and consensus mechanisms is crucial to comprehending the versatility and applicability of this technology. Various types of blockchains exist, encompassing public, private, and consortium blockchains.

a. Public Blockchains: Public blockchains are the bedrock of decentralization, characterized by open access and transparency. Anyone can join the network, participate in transactions, and validate blocks. The most iconic example is Bitcoin's blockchain, where miners engage in a competition to solve intricate mathematical puzzles, thereby verifying transactions and appending them to the blockchain. The decentralized nature of public blockchains ensures that no single entity maintains control, which ensures security and trust. However, the openness comes with scalability and privacy challenges, making public blockchains more suitable for applications where transparency is paramount.
b. Private Blockchains: In contrast, private blockchains restrict access to a selected group of participants. These participants are often known entities with established relationships. Privacy and control are key features of private blockchains. Use cases range from supply chain management to internal record-keeping within organizations. Unlike public blockchains, where anyone can participate in validation, private blockchains typically utilize a permissioned consensus mechanism where a central authority validates transactions. This approach boosts efficiency and scalability, but it sacrifices some degree of decentralization and transparency.

c. Consortium Blockchains: Consortium blockchains bridge the gap between public and private blockchains. In this model, a consortium of organizations collaborates to operate the blockchain network. The participating entities share governance responsibilities and participate in consensus mechanisms. Consortium blockchains often find applications in industries requiring collaboration and data sharing among multiple stakeholders, such as financial services and supply chain management [20]–[22].

In addition to the three main types of blockchains, there are also a number of different consensus mechanisms that can be used to secure the network. Consensus mechanisms are the algorithms that blockchain networks use to attain a consensus regarding the ledger's status. They guarantee that all network participants reach an agreement on the legitimacy of transactions and the sequence in which they are registered.

a. Proof of Work (PoW): This system, utilized in Bitcoin, mandates that participants (miners) solve intricate mathematical problems to authenticate transactions. PoW demands a substantial amount of energy but provides strong security.
b. Proof of Stake (PoS): PoS assigns the right to validate transactions based on the stake participants hold in the network. The more tokens held, the higher the chance of being selected. PoS is more energy-efficient than PoW.
c. Delegated Proof of Stake (DPoS): DPoS involves token holders electing delegates who validate transactions. It combines decentralization with efficiency, allowing for faster transaction confirmation.
d. Proof of Authority (PoA): PoA relies on designated validators who are known and trusted. It is often used in private and consortium blockchains for efficiency and control.
e. Practical Byzantine Fault Tolerance (PBFT): PBFT ensures consensus by having nodes agree on the next block. It is faster than PoW but is typically used in private or consortium blockchains [23]–[24].

6.2.2 Unveiling the Digital Twins System

These systems provide a virtual replica of real-world assets, as shown in Figure 6.1, enabling real-time monitoring, analysis, and optimization. This subsection delves into the architecture, components, and key features that define digital twin systems, illuminating their significance across diverse industries. The architecture of a digital twin system consists of interconnected components that synergize to create a dynamic virtual counterpart of a physical entity. The architecture can be conceptualized in three layers:

a. Physical Layer: This layer encompasses the real-world asset, system, or process that the digital twin replicates. Sensors, actuators, and IoT devices collect data from the physical entity, forming the foundation of the digital twin's functionality.

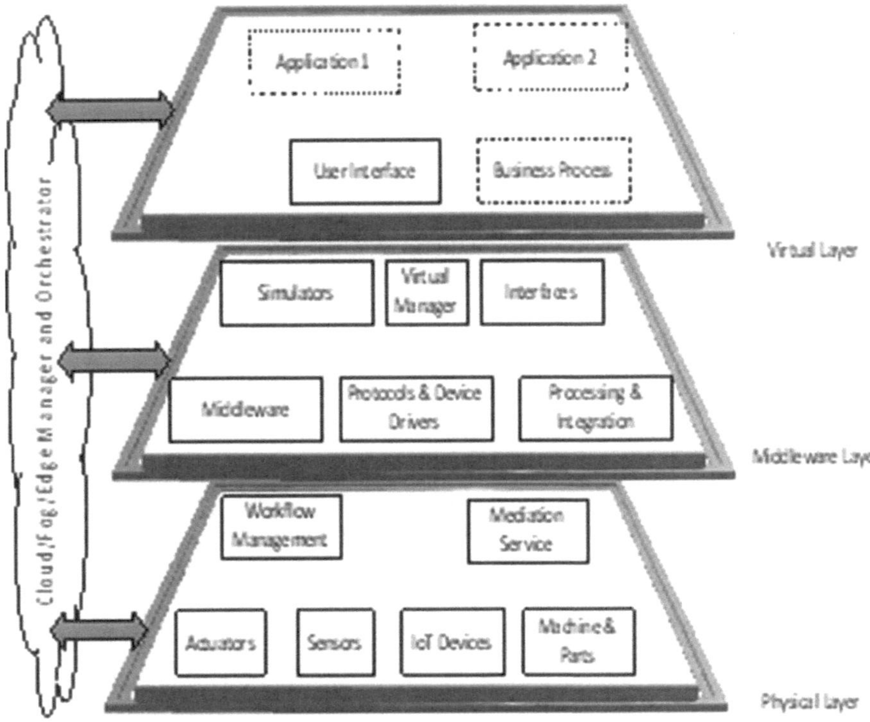

FIGURE 6.1 Three layers of Digital Twins.

b. Virtual Layer: The virtual layer comprises the digital representation of the physical entity. This digital twin is constructed using data collected from sensors and various sources, creating a detailed and accurate virtual model.
c. Analytics and AI Layer: The analytics and AI layer leverages data from the physical and virtual layers to enable real-time analysis, simulation, and optimization. AI algorithms process data to generate insights, predict outcomes, and facilitate decision-making [13], [25].

Digital twin systems consist of several components that collectively empower their functionality:

a. Sensors and Data Sources: Sensors embedded in the physical entity collect real-time data on parameters, such as temperature, pressure, and motion. This data serves as the foundation for creating an accurate digital replica.
b. Data Processing and Integration: The data collected from sensors and various sources undergo processing and integration. This step involves cleaning, aggregating, and transforming raw data into meaningful insights.

c. Virtual Model: The virtual model is a digital portrayal of the physical entity, capturing its geometry, behavior, and attributes. This model is continuously updated to reflect changes in the real-world counterpart.
d. Connectivity and Communication: Seamless communication between the physical and virtual layers is facilitated through IoT networks, enabling real-time data exchange and synchronization.
e. Analytics and Simulation: AI algorithms analyze data to generate insights, predict future behavior, and simulate scenarios. This capability empowers decision-makers to optimize processes and enhance performance.

The key features of digital twin systems include:

a. Real-time data collection: Digital twin systems collect data from the physical object in real time. This enables the digital twin to be kept up-to-date with the latest information about the physical object.
b. Data processing: Digital twin systems use data processing algorithms to process the data that is collected from the physical object. This enables the digital twin to generate insights about the physical object's performance.
c. Simulation: Digital twin systems can be used to simulate the performance of the physical object. This enables users to test different scenarios and to optimize the physical object's design and operation.
d. Prediction: Digital twin systems can be used to predict the future behavior of the physical object. This enables users to take corrective action before problems occur.
e. Optimization: Digital twin systems can be used to optimize the physical object's design and operation. This enables users to improve the performance of the physical object and to reduce costs.

6.3 RELATED WORK

The paper discusses the study of supply chain traceability (SCT) in light of developing efficient traceability solutions powered by blockchain technology, as elaborated in reference [10]. The study explores how traceability is tackled in both academic and practical literature across diverse sectors. It incorporates case studies from various sectors and the facilitation of an executive workshop to address research questions. The outcomes of the research have led to the creation of a framework that outlines the potential scope of SCT and the levels of granularity implemented throughout the supply chain. The paper makes three key contributions. Firstly, it uncovers the main drivers for adopting blockchains for SCT based on empirical data. Secondly, it highlights the crucial contextual factors that cause variations in the scope of SCT solutions adopted by different organizations within their supply chains. Finally, it presents a new framework that delineates the potential scope of SCT applicable to a standard supply chain, offering new perspectives on two essential dimensions of scope: the range of SCT in terms of start and end points across a supply chain, and the depth in terms of the granularity of information captured [10].

In reference [26], the authors describe the creation of a digital twin prototype that is connected to a blockchain based on Ethereum. The goal of this project is to create a method to maintain the authenticity of unique tokens that represent physical assets, using digital twin technology as a safeguard against tampering. This method is demonstrated through a case study focused on matrix-structured manufacturing systems for small and medium-sized enterprises, which consists of three main layers: a physical layer, a connection layer that facilitates interaction between ERC721 tokens and digital twins, and a foundational blockchain layer. The study primarily concentrates on the connection layer, which ensures that non-fungible tokens are synchronized with their respective physical assets by using data from a tamper-proof digital twin. The implementation of this concept in the case study and its comparison with a situation involving a start-up company reveals several distinctions. Although the proposed conceptual framework provides many benefits, its effectiveness is contingent upon certain contextual factors.

In the research paper [13], the authors propose an innovative digital twin system that leverages blockchain technology to improve predictive maintenance. This system uses the power of blockchains to securely store a wealth of data related to asset performance, including sensor data and historical maintenance records. This rich database of information is crucial in forecasting potential asset failures, enabling the timely execution of maintenance tasks and preventing expensive operational interruptions.

In the research paper [27], the authors have carried out an exhaustive review of the literature published in the past five years, with a focus on the intersection of blockchains, digital twin technology, and related advancements in the construction industry. The main catalysts for this integration are identified as AI and the IoT. The study employs a bibliometric analysis to investigate patterns of co-authorship, relationships between keywords, trends in citations, and bibliometric links among various sources. The results highlight that the fusion of these technologies is still in its nascent stages, necessitating continued investigation by academic researchers and input from industry experts. To facilitate an understanding of the unique roles and advantages that each technology contributes to this integration, the study presents a conceptual model. This research also emphasizes the need for interdisciplinary collaboration among scholars and practitioners, not just within the construction sector, but also across various technological domains. This collective effort is crucial for propelling progress in this emerging and promising field.

In the study [5], the authors propose an innovative collaborative digital twin framework, underpinned by blockchain technology. The main goal is to create a decentralized epidemic alert system that can tackle not only existing issues, such as Covid-19, but also future pandemics. By merging cutting-edge technologies, such as blockchains, digital twins, and AI, this framework provides a solid solution for decentralized epidemic alerts. The suggested approach focuses on leveraging data-driven collaboration through digital twins, strengthened by blockchain technology, to effectively handle pandemic situations, such as Covid-19 outbreaks. The framework consists of four interconnected layers, each contributing to the operational effectiveness of the blockchain-integrated digital twin collaboration: the physical layer, the blockchain-integrated digital twin layer, the data analytics layer, and the decision-making layer. This all-encompassing framework underscores the potential of using advanced technologies for a collaborative and decentralized approach to managing epidemics.

Cyber-physical systems (CPS) are sophisticated embedded systems that interact dynamically and continuously with their environment. They are integral to a variety of fields, such as smart grids, traffic networks, intelligent buildings, cooperative robotics, automotive, and avionics systems. CPS seamlessly blend computational processes with physical activities. In the realm of smart home applications, CPS often involve sensor nodes with low capacity that sense the physical environment and transmit data through the network. Due to resource limitations, these systems typically require lightweight security mechanisms. On the other hand, a blockchain is a decentralized technology that provides a secure, public ledger. It is the underlying technology for cryptocurrencies such as Bitcoin that has found applications in various sectors, including IoT, manufacturing, transportation, and supply chain. Blockchains combine consensus algorithms, distributed data storage, and secure protocols to build robust and reliable systems. They ensure that transactions cannot be altered, providing a high level of security and trust. The key difference between CPS and blockchains lies in their primary functions and applications. CPS are designed to bridge the physical and digital worlds, enabling real-time monitoring and control of physical processes. On the other hand, blockchains are designed to provide a secure, transparent, and tamper-proof system for recording transactions. While both technologies can be used independently, they can also be integrated to enhance the security and efficiency of systems. For example, blockchains can be used to enhance the security of CPS by providing a tamper-proof record of sensor data, thereby preventing fraudulent activities and enhancing trust in the system [28]–[31].

The study [32] presents an innovative approach for data transfer in manufacturing systems based on digital twins, utilizing blockchain technology. The problem being addressed is the bottleneck and latency in cloud-based data transfer during the production of varied, small batches. The suggested solution is an architecture termed "cloud-edge interplay workshop" that enables peer-to-peer data transfer via a blockchain. This strategy lightens the burden on the cloud as edge computing processes local data, guaranteeing secure and efficient data transfer. The advantages encompass improved efficiency of interactions, meeting the data demands of digital twins, and heightened adaptability of the workshop. Nonetheless, the research is mainly concentrated on machining and assembly, suggesting a need to broaden to other procedures. Future research avenues include optimizing system information transmission and job shop scheduling for enhanced efficiency and cost-effectiveness. This pioneering solution offers the potential for efficient data transfer in dynamic manufacturing environments, with scope for further refinement and broader relevance.

The study [33] in question delves into the possibilities of merging digital twins and blockchain technologies within the construction sector. It scrutinizes existing research on the utilization of these technologies across different stages of the building life cycle, emphasizing the advantages of their integration for superior life cycle management. The paper unveils a rapid increase in related research, with a particular emphasis on the construction phase, and underscores the varied applications of digital twins and blockchains. It champions the union of digital twins and blockchains to tackle issues of collaboration and information fragmentation. The paper acts as a crucial resource for academics and industry experts, providing insights into this methodology and setting the stage for future research and practical implementation of combined digital twins and blockchain solutions. Essentially, the paper [33] urges further exploration into the integration of digital twins and blockchains to exploit their potential and revolutionize building life cycle management.

The research paper [34] In question investigates the potential of merging blockchains and digital twins to boost the security and efficiency of the industrial Internet of Things (IioT) network, thereby setting the stage for Industry 5.0. The study tackles the issue of managing a growing number of IoT devices and guaranteeing secure data transmission. The suggested solution is a blockchain-based PoA mechanism for IioT applications, employing authority master nodes for effective network management and secure data transactions. The advantages encompass lower energy consumption, improved data security, real-time service quality, and a decentralized, scalable, and secure structure. Simulation results indicate enhanced trustworthiness compared to conventional IioT networks. The study implies that the integration of blockchains and digital twins could markedly enhance IioT security, efficiency, and scalability, thereby establishing a robust foundation for Industry 5.0.

Research [35] investigates the security of digital twins and threats to user privacy from malicious communication activities. It proposes a sturdy digital twin network (DTN) structure and a model for malware propagation to safeguard distributed machine learning (ML) systems. Blockchains, which are decentralized, transparent, and anonymous mechanisms, guarantee secure data exchange and privacy. A data distribution service (DDS) architecture based on blockchains bolsters data security with the IoT support. The BL-FCL algorithm, which merges blockchains and digital twins, accelerates model training, facilitates incremental learning, and upholds user privacy. The prediction accuracy of BL-FCL competes with federated averaging (FedAvg) schemes but outperforms them by 20%–60%, rectifying training inaccuracies while preserving security. This study is vital for DTN security and the digital economy, offering a guide for the application of blockchains and distributed learning in the digital domain.

6.4 OBSERVATIONS AND PROCEDURE FOR TOTAL FUSION OF DIGITAL TWIN WITH BLOCKCHAINS

The research papers reviewed [1]–[10], [27] mention the incorporation of digital twins with blockchains, to a certain degree, but there are no clear guidelines or processes for integrating these two technologies. The absence of detailed discussions on the process or procedure of integrating digital twins and blockchains in existing papers could be attributed to several factors, such as:

a. Scope and Focus: Many research papers aim to explore specific challenges, benefits, or implications of integrating blockchains and digital twin technologies. Due to limited space and the complexity of both technologies, authors may choose to focus on discussing the theoretical and practical aspects of the integration, rather than providing a step-by-step process.

b. Varied Implementations: The process of integrating digital twins and blockchains can vary significantly based on the industry, use case, and specific technologies being employed. Providing a comprehensive integration process may be challenging due to the diversity of scenarios in which this integration can occur.
c. Technical Complexity: Integrating blockchains and digital twin technologies involves intricate technical considerations. Describing the technical details in a concise and understandable manner can be challenging, and authors may opt to address this topic in more technical or specialized publications.
d. Rapidly Evolving Field: The field of blockchain technology and digital twins is rapidly evolving, with new tools, protocols, and best practices emerging frequently. As a result, the process of integration may change or become more refined over time, making it difficult for authors to provide a one-size-fits-all procedure.
e. Audience Considerations: Some papers may be targeted toward a specific audience, such as researchers, professionals, or decision-makers. The integration process could be considered a more operational or implementation-oriented topic, but it may not lead to the primary focus of the paper.

While the existing papers may not delve into the detailed process of integrating digital twins and blockchains, this presents an opportunity for future research and publications to explore this aspect more comprehensively. As the technology landscape evolves and practical implementations become more common, more in-depth discussions on the procedural aspects of integrating these two transformative technologies can be anticipated and are required. Based on the methodologies proposed by various researchers, a common set of tasks can be carried out that may facilitate the integration of blockchains and digital twins. Integrating blockchains into a digital twin ecosystem is a multifaceted endeavor that requires meticulous planning, technical expertise, and collaboration. By following this detailed process and procedure, organizations can successfully leverage the combined power of blockchains and digital twins to enhance transparency, security, and efficiency across various industries. Here is a detailed process and procedure for integrating blockchains in a digital twin ecosystem, which is also represented in Figure 6.2:

a. Define Objectives and Use Cases: Delineate the precise objectives and applications for incorporating blockchains into the digital twin environment. Determine how the attributes of blockchains (i.e., data immutability, transparency, decentralized control) can enhance their existing functionality.
b. Select the Right Blockchain Platform: Choose the appropriate blockchain platform based on factors such as scalability, consensus mechanism, privacy features, and the needs of the digital twin ecosystem [3]. Consider whether a public, private, or consortium blockchain is most suitable.

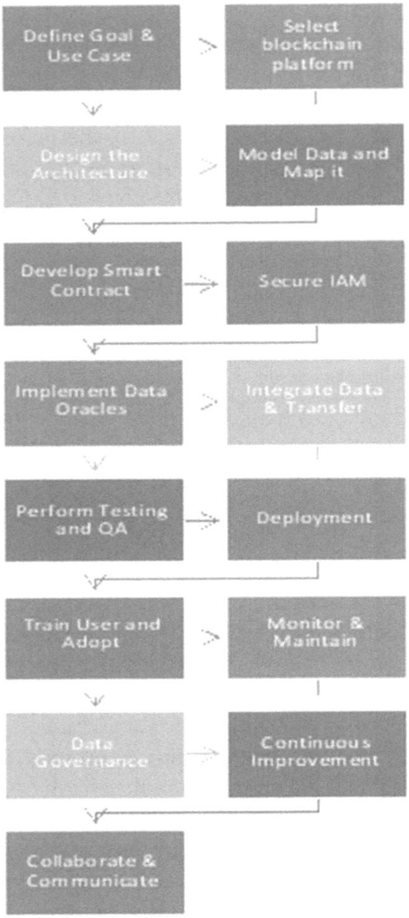

FIGURE 6.2 Integrating Blockchains in a Digital Twin Process.

c. Design the Architecture: Develop a clear architectural design that outlines how the blockchain will interact with the digital twin components, sensors, databases, and external stakeholders. Decide on the data structure, smart contract logic, and integration points.
d. Data Modeling and Mapping: Define how the data from the digital twin will be stored on the blockchain. Map the relevant data fields to blockchain transactions and decide which data will be stored directly on the blockchain and which will be referenced.
e. Develop Smart Contracts: Design and code smart contracts that automate processes within the digital twin ecosystem. Smart contracts can automate asset transactions, trigger events based on real-time data, and enforce rules for interactions among stakeholders.

f. Secure Identity and Access Management: Put in place robust authentication and access management for individuals engaging with the blockchain-infused digital twin. Identity management solutions make certain that solely approved entities can access and adjust the data.
g. Implement Data Oracles: Connect real-world data sources, such as sensors and IoT devices, to the blockchain. Data oracles act as bridges between off-chain data and the blockchain, ensuring that the blockchain's data remains accurate and up-to-date.
h. Data Integration and Migration: Transfer the current data from the digital twin environment to the blockchain. Guarantee the maintenance of data consistency and integrity throughout the migration procedure.
i. Testing and Quality Assurance: Conduct comprehensive evaluations of the fusion of blockchain and digital twin elements. Carry out assessments in terms of functionality, security, and performance to confirm the seamless operation of the combined system.
j. Deployment: Implement the blockchain-incorporated digital twin environment within a regulated setting. Vigilantly oversee the system in its initial phases to detect and address any potential problems.
k. User Training and Adoption: Provide training to relevant stakeholders on how to interact with the new blockchain-integrated system. Ensure that users understand how to use smart contracts, access data, and perform tasks within the ecosystem.
l. Monitoring and Maintenance: Continuously monitor the performance of the blockchain-integrated digital twin ecosystem. Regularly update smart contracts, apply security patches, and address any scalability or performance issues that arise.
m. Data Governance and Compliance: Implement mechanisms to ensure compliance with data protection regulations and industry standards. Establish clear data ownership and governance policies.
n. Continuous Improvement: Collect input from users and stakeholders to pinpoint opportunities for improvement. Consistently refine the system to boost functionality, user satisfaction, and overall effectiveness.
o. Collaboration and Communication: Maintain open lines of communication between blockchain developers, digital twin experts, and other stakeholders. Collaboration ensures that the integration remains aligned with business objectives.

6.5 SYNERGIES BETWEEN BLOCKCHAIN TECHNOLOGY AND DIGITAL TWINS

Blockchains, recognized for their decentralized and secure characteristics, naturally complement digital twins, which generate virtual duplicates of physical assets or processes for real-time supervision and enhancement. This pairing promises to enhance data integrity, security, and collaboration within digital twin ecosystems. The decentralized architecture of blockchains ensures that transactions are verified by consensus across a network of participants. This consensus mechanism aligns with the goals of digital twins, where accurate and real-time data is critical for effective decision-making. Incorporating blockchains into digital twin ecosystems establishes an immutable and tamper-resistant log of each interaction and update, which improves data integrity and credibility [36]. The security features of blockchain technology address one of the central concerns in digital twin systems: data vulnerability. As digital twins generate and exchange sensitive data, the cryptographic security mechanisms of blockchains safeguard against unauthorized access and tampering. This fortification of data security is particularly relevant in industries such as healthcare, manufacturing, and supply chain management, where sensitive information is at stake.

Blockchain's ability to enable secure and transparent data sharing aligns with the multi-stakeholder nature of digital twin ecosystems. Whether it is manufacturers, suppliers, regulators, or customers, blockchains ensure that participants can access and share data while retaining control over their own information. The integration of blockchain technology can also streamline transactions, automate contractual agreements, and enhance audit trails within digital twin ecosystems. These improvements lead to operational efficiency and cost reduction, fostering a more resilient and agile environment. However, challenges remain, such as scalability and interoperability. Blockchain's resource-intensive nature could pose challenges when integrating with real-time systems such as digital twins. Addressing these challenges requires innovative solutions to ensure seamless integration without compromising performance.

Blockchain technology presents a transformative solution to several critical challenges within digital twin ecosystems, including data integrity, provenance, secure data sharing, and privacy. By leveraging its decentralized and cryptographic features, blockchains address these challenges and enhance the trustworthiness and efficiency of digital twin systems. Blockchain's core principle of immutability ensures data integrity within digital twin ecosystems. Every transaction or update is securely documented as a block and connected to the preceding one via a cryptographic hash. This forms an immutable chain of data, preventing unauthorized modifications or tampering. Since digital twins depend on precise and up-to-date data for their virtual representations, the unchangeable nature of blockchains ensures the trustworthiness of the data, fostering trust among stakeholders. Provenance, the capability to track the origin and data's history, is crucial for ensuring the authenticity of information within digital twin ecosystems. Blockchain's transparent and tamper-proof ledger enables the tracking of every transaction and modification made to the digital twin's data. This ensures a clear record of the journey of the data from its source to its current state. In industries such as supply chain management, where tracing the origin of products is essential, blockchains provide an auditable trail of provenance. Secure data sharing is a fundamental requirement within digital twin ecosystems, where multiple stakeholders collaborate and exchange information. Blockchain's cryptographic mechanisms enable secure and controlled data sharing. Participants can grant and revoke access to specific data through encryption and smart contracts, ensuring that sensitive information is only accessible to authorized parties. This empowers secure collaboration while maintaining data privacy and ownership.

Privacy is a critical concern, especially in sectors such as healthcare, where personal information is involved. The design of blockchains allows for pseudonymous data sharing, where participants are identified by their cryptographic keys instead of personal information. This enhances privacy while enabling transparent and traceable data sharing. Moreover, permissioned blockchains, such as those used in consortium environments, provide more nuanced control over who can access and view data, ensuring compliance with privacy regulations.

The decentralized structure of blockchains enables participants to retain control over their data. Smart contracts can be utilized to specify access rights, guaranteeing that only approved entities have the ability to reach particular data points. This controlled data access mitigates the risk of unauthorized usage, manipulation, or leakage of sensitive information within the digital twin ecosystem. Blockchain's transparent and tamper-proof nature enhances auditability and accountability within digital twin ecosystems. Every transaction is recorded in an immutable ledger, creating a comprehensive audit trail that can be easily verified. This helps in identifying discrepancies, pinpointing the source of errors, and holding responsible parties accountable.

6.6 INTEGRATION PROCEDURES IN BLOCKCHAIN FOR DIGITAL TWINS: A REVIEW

A procedural outline of how blockchains ensure data integrity and immutability in digital twin systems is given below:

1. Input:
• Raw data from physical assets and sensors.
• Digital twin models and attributes.
2. Initialization:
• Initialize a blockchain network with a suitable consensus system, such as PoW or PoS.
• Set up cryptographic key pairs for participants.
3. Create Transactions:
• Generate transactions that represent data updates from physical assets and digital twin models.
• Include a timestamp and relevant metadata for each transaction.
4. Hashing:
• Hash the content of each transaction using a cryptographic hash function (e.g., SHA-256).
• The hash value uniquely represents the content of the transaction.

(continued)

5. Linking Transactions:
• Each new transaction includes the hash of the previous transaction, creating a linked chain of hashes.
6. Consensus Mechanism:
• Network participants (nodes) validate and agree upon the validity of new transactions.
• Consensus mechanisms guarantee concurrence among nodes regarding the condition of the blockchain.
7. PoW Example:
• Miners work on intricate mathematical challenges to confirm transactions.
• The initial miner who is to solve the puzzle appends the new block of transactions to the blockchain.
• This procedure demands substantial computational exertion, rendering it challenging to modify already-existing blocks.
8. Immutable Blocks:
• After a block is integrated into the blockchain, it becomes exceedingly challenging to modify.
• Modifying the information in a block would necessitate evaluating the hash for that block and all the blocks that follow it.
9. Security Measures:
• Each participant's data is secured using their private key, ensuring authorized access only.
• The public key verifies the authenticity of transactions and updates.
10. Data Verification:
• To verify the integrity of data within a block, recalculating the hash of that block should match the stored hash.
• Any tampering with the data would change the hash value and be immediately detected.
11. Immutable History:
• The linked chain of hashes creates an immutable history of all transactions.
• Modifying one block would necessitate altering the information in subsequent blocks, a task that is computationally impractical.
12. Data Consistency:
• All participants have access to the same copy of the blockchain, ensuring data consistency.
• Any discrepancies are easily identified, and the source of errors can be traced.
13. Verification and Trust:
• The transparency and immutability of blockchains build trust among stakeholders.
• Participants can independently verify the history of data changes and updates.

6.6.1 Mechanisms of Hashing, Cryptographic Signatures, and Consensus Algorithms

Blockchain technology employs a combination of cryptographic mechanisms and consensus algorithms to ensure secure data recording, integrity, and immutability. These mechanisms play a pivotal role in maintaining the trustworthiness of the data that is preserved within the blockchain. In a digital twin ecosystem, hashing is used to ensure the integrity and immutability of data generated by physical assets and sensors. When data is collected from these sources, it is hashed using a cryptographic hash function. The resulting hash is then stored on the blockchain along with the data itself. This hash becomes a unique identifier for that specific set of data. Any alterations to the data will produce a distinct hash, promptly notifying participants of potential tampering. For example, if a sensor records temperature readings from a manufacturing machine, then the temperature data is hashed before being added to the blockchain. If someone attempts to alter the temperature reading, then the hash will no longer match, indicating that the data has been compromised. Cryptographic signatures have a vital role in verifying the legitimacy and source of transactions and modifications within a digital twin environment. When participants interact with digital twin data, they use their private keys to sign the transactions before adding them to the blockchain. These signatures are cryptographic proofs that the transaction was initiated by an authorized participant. For instance, if a technician performs maintenance on a machine and updates its digital twin model, then they would sign the transaction with their private key. Others can then verify the transaction's authenticity using the technician's public key. This guarantees that modifications to the digital twin data can only be performed by authorized individuals. Consensus algorithms are essential in maintaining a shared and synchronized view of the digital twin ecosystem among all participants. Nodes within the environment employ consensus algorithms to reach a consensus on the authenticity and sequence of transactions and updates [37]. This procedure guarantees that all involved parties possess a uniform and precise depiction of the digital twin. For instance, in a manufacturing scenario where multiple stakeholders collaborate on optimizing production processes using digital twin models, a consensus algorithm such as PBFT may be used. This algorithm enables nodes to collectively agree on changes and optimizations to the digital twin model, ensuring that all parties have a unified understanding of the changes being made.

6.6.2 Integration of Hashing in a Blockchain-Powered Digital Twin Ecosystem: Outline

1. Input:
• Data collected from physical assets or processes.
• Cryptographic hash function (e.g., SHA-256).
• Blockchain network for data storage and immutability.

(continued)

2. Output:	
	• Confirmation of data integrity or detection of potential tampering using blockchain technology.
3. Procedure:	
3.1 Collect Data:	
	• Receive data from sensors, processes, or physical assets in the digital twin ecosystem.
3.2 Apply Hashing:	
	• Apply the chosen cryptographic hash function (e.g., SHA-256) to the collected data.
	• Generate a hash value unique to the input data.
3.3 Create Transaction on the Blockchains:	
	• Create a transaction to store both the original data and its corresponding hash on the blockchain.
	• Include the hash value as part of the transaction data.
3.4 Add Transaction to the Blockchain:	
	• Send the transaction to the blockchain network for validation and consensus.
	• Upon consensus, the transaction is added to the blockchain ledger.
3.5 Update Data:	
	• If the data in the digital twin environment is updated, then collect this new data.
3.6 Hash New Data:	
	• Apply the same hash function to the new data, generating a new hash value.
3.7 Retrieve Stored Hash from the Blockchain:	
	• Retrieve the stored hash associated with the original data from the blockchain.
3.8 Compare Hashes:	
	• Compare the newly generated hash with the stored hash from the blockchain.
3.9 Hash Match Check Using the Blockchain:	
	If the hashes match:
	Data integrity is confirmed through the blockchain's immutability.
	No tampering has occurred.
	If the hashes do not match:
	Data integrity is compromised, or tampering is detected.
	Alert or Notify:
	If the hashes do not match, then raise an alert or notify relevant stakeholders about the potential data tampering.

Example: Suppose that a digital twin represents an IoT-based environmental monitoring system. Here is how the algorithm would apply in a blockchain context:

- The environmental sensor data is collected: Temperature = 25°C.
- The hashing is applied: Hash (Temperature = 25°C) = "8c5a2f..."

- Create a blockchain transaction to store data and hash.
- The transaction is added to the blockchain.
- The environmental sensor data is updated: Temperature = 26°C.
- The hashing is applied to the new data: Hash (Temperature = 26°C) = "f36e9b..."
- Retrieve the stored hash from the blockchain.
- Compare the hashes: If the stored hash ("8c5a2f...") matches the new hash ("f36e9b..."), then data integrity is confirmed.
- No alert is raised, as the hashes match.

6.6.3 Integration of Cryptographic Signatures in a Blockchain-Powered Digital Twin Ecosystem

1.	**Input:**
	• Data to be recorded in the digital twin.
	• Participants' cryptographic key pairs (private and public keys).
	• Blockchain network for data storage and immutability.
2.	**Output:**
	• Securely recorded data with cryptographic signatures on the blockchain.
3.	**Procedure:**
3.1.	Collect Data:
	• Receive data from sensors, processes, or physical assets within the digital twin ecosystem.
3.2.	Generate Cryptographic Signature:
	• Use the sender's private key to digitally sign the collected data.
	• This generates a unique cryptographic signature that authenticates the data.
3.3.	Create Transaction on the Blockchain:
	• Create a transaction to include the original data and its corresponding cryptographic signature.
	• The transaction contains both data and the digital signature.
3.4.	**Add Transaction to the Blockchain:**
	• Send the transaction to the blockchain network for validation and consensus.
	• Upon consensus, the transaction is added to the blockchain ledger.
3.5.	**Verify Transaction Signature:**
	• Any individual with access to the blockchain has the ability to confirm the legitimacy of the transaction.
	• The sender's public key is employed to authenticate the digital signature.
3.6.	**Update Data:**
	• In the event of data modifications within the digital twin environment, gather the updated information.

(*continued*)

3.7. Generate a New Cryptographic Signature:
• Apply the sender's private key to sign the updated data, generating a new digital signature.
3.8. Retrieve Stored Signature from the Blockchain:
• Retrieve the stored cryptographic signature associated with the original data from the blockchain.
3.9. Verify New Signature:
• Use the sender's public key to verify the newly generated signature.
3.10. Signature Verification Check:
• If the new signature is verified against the stored signature, then data authenticity is confirmed.
• If the verification fails, then the data integrity may be compromised.
3.11. Alert or Notify:
• If the signature verification fails, then raise an alert or notify relevant stakeholders about potential data tampering.

Example: Suppose a digital twin represents a supply chain process.

- The supplier records the product details: Product ID = ABC123.
- Generate a cryptographic signature using the supplier's private key.
- Create a blockchain transaction containing the product data and the cryptographic signature.
- The transaction is added to the blockchain.
- The supplier updates the product details: Product ID = ABC124.
- Generate a new cryptographic signature for the updated data.
- Retrieve the stored signatures from the blockchain.
- Verify the new signature using the supplier's public key.
- If the verification succeeds, then data authenticity is confirmed.
- No alert is raised, as the signatures match.

6.6.4 Integration of Consensus Algorithm in a Blockchain-Powered Digital Twin Ecosystem

1. Input:
• Data intended for storing within the digital twin.
• Engagements of individuals with the digital twin.
• Selected consensus mechanism (e.g., PoW, PoS).
• Blockchain network for data storage and consensus.
2. Output:
• Agreement on the order and validity of transactions within the digital twin ecosystem.

3.	**Procedure:**
3.1.	Collect Data and Interactions:
	• Gather data from sensors, processes, or physical assets within the digital twin ecosystem.
	• Record participants' interactions, such as updates or changes to the digital twin data.
3.2.	**Transaction Creation:**
	• Create a transaction for each data update or interaction in the digital twin.
	• Transactions can include new data, changes, or other relevant information.
3.3.	**Propose Transactions:**
	• Participants propose their transactions to be included in the next block of the blockchain.
	• Depending on the consensus algorithm, different participants may have the authority to propose transactions.
3.4.	**Validate Transactions:**
	• Other participants validate the proposed transactions based on the consensus algorithm's rules.
	• Transactions are checked for correctness, authenticity, and adherence to predefined rules.
3.5.	**Reach Consensus:**
	• Based on the chosen consensus algorithm, participants work together to agree on the order and validity of transactions.
	• This process involves communication and collaboration among nodes in the network.
3.6.	**Add Transactions to Block:**
	• After consensus is achieved, the transactions that have been collectively agreed upon are included in a fresh block within the blockchain.
	• The block also includes a link to the preceding block, forming a sequential chain.
3.7.	**Block Validation:**
	• Participants verify the newly added block's validity according to the consensus algorithm.
	• This involves confirming that transactions were properly validated and that the block structure adheres to protocol rules.
3.8.	**Block Addition to the Blockchain:**
	• After the new block is verified, it is incorporated into the blockchain by attaching it to the current chain.
3.9.	**Data Synchronization:**
	• The updated blockchain is shared among all participants in the digital twin ecosystem.
	• All participants have access to the latest validated data and transactions.

(*continued*)

Example: Consider a digital twin ecosystem that simulates traffic management.

- The traffic sensors collect real-time data on road conditions.
- The transactions are created for each sensor's data update.
- The sensor nodes propose transactions based on the newly collected data.
- Other nodes validate the proposed transactions, ensuring accuracy and adherence to traffic rules.
- The nodes collaborate to reach a consensus on the order of transactions.
- Agreement is established, and verified transactions are appended to a fresh block.
- The fresh block's authenticity is validated by the nodes in the network.
- The confirmed block is included in the blockchain, forming an unbroken chain.
- All participants receive the updated blockchain, synchronizing their digital twin data.

6.6.5 Smart Contracts in Automating Processes and Interactions

Smart contracts are essential in automating processes and interactions within digital twin ecosystems that are integrated with blockchains. These are automated contracts in which the agreement's terms between parties are encoded directly within the code. By embedding business regulations and reasoning within the contract, smart contracts enable the mechanization of procedures and engagements, which are subsequently triggered automatically when specific conditions are fulfilled. In digital twin ecosystems integrated with blockchain technology, smart contracts can mechanize diverse procedures and engagements, including data sharing, access control, and asset monitoring. As an example, a smart contract may autonomously provide access to data stored on a digital twin when specific conditions are met, such as when a user has the appropriate permissions or when a particular event occurs. Smart contracts can also automatically track the movement of assets within a digital twin ecosystem, providing real-time visibility and transparency. Smart contracts can enforce rules and regulations within a digital twin ecosystem. For example, they can ensure that data is only shared with authorized parties or that certain actions can only be performed by users with the appropriate permissions, enhancing security and trust within the ecosystem.

6.7 ANALYZING THE SECURITY IMPLICATIONS OF COMBINING BLOCKCHAINS AND DIGITAL TWINS

Combining blockchains and digital twins can offer numerous benefits, but it also introduces several security implications that need to be carefully considered. The following are the security implications of this integration:

6.7.1 Data Integrity

a. Implication: Blockchains ensure data immutability, but if the digital twin's data source is compromised before being added to the blockchain, then it could result in the propagation of incorrect data.

b. Mitigation: Implement strong data validation mechanisms before data is added to the blockchain. Use cryptographic hashing to detect any unauthorized changes to the data.

6.7.2 Identity and Access Management

a. Implication: Access to digital twins must be controlled to prevent unauthorized modifications or unauthorized parties from accessing sensitive data.
b. Mitigation: Implement robust identity and access management solutions. Use cryptographic signatures to authenticate participants and control access permissions.

6.7.3 Privacy Concerns

a. Implication: Sensitive data in digital twins may be exposed on the blockchain, potentially compromising privacy.
b. Mitigation: Implement techniques such as zero-knowledge proofs to maintain privacy while still benefiting from blockchain's transparency.

6.7.4 Smart Contracts

a. Implication: Smart contracts can execute automated actions based on predefined conditions. If they are not properly coded, then they can lead to unintended consequences or vulnerabilities.
b. Mitigation: Conduct thorough, smart contract audits and testing to identify and rectify vulnerabilities. Employ best practices in coding and testing.

6.7.5 Consensus Mechanisms

a. Implication: The consensus algorithm's security and decentralization may be compromised if a single entity or a group of colluding entities gains majority control.
b. Mitigation: Choose consensus mechanisms carefully based on use case requirements. Employ mechanisms such as PoW, PoS, or PBFT with appropriate decentralization levels.

6.7.6 Data Leakage Through Blockchain Transactions

a. Implication: Even though blockchains are secure, metadata within transactions can sometimes reveal sensitive information.
b. Mitigation: Carefully design and structure transactions to minimize exposure of sensitive metadata.

6.7.7 Supply Chain Attacks

a. Implication: If a malicious actor gains control over a part of the supply chain, then it may compromise the integrity of data fed into the digital twin.

b. Mitigation: Implement robust security measures in the supply chain to prevent unauthorized access and tampering.

6.7.8 Scalability and Performance

a. Implication: Blockchain scalability and performance limitations may impact the real-time nature of data updates in dynamic digital twin ecosystems.
b. Mitigation: Choose a blockchain platform that aligns with the performance requirements of the digital twin ecosystem. Explore solutions such as layer two scaling.

6.7.9 Cross-Chain Attacks

a. Implication: If digital twin data is spread across multiple blockchains, then attacks on one blockchain may indirectly affect the accuracy and integrity of the digital twin ecosystem.
b. Mitigation: Implement security measures in cross-chain communication. Ensure that each blockchain maintains its security.

6.8 ANALYZING THE PERFORMANCE IMPLICATIONS

The choice of blockchain platform, consensus mechanism, data storage strategy, and optimization techniques play a vital role in ensuring that the integration results in an efficient and effective digital twin ecosystem.

6.8.1 Data Processing Speed

a. Implication: Blockchain networks, especially those with PoW consensus, can have slower transaction processing times, which may impact real-time updates in dynamic digital twin environments.
b. Mitigation: Select consensus mechanisms that give priority to speed, such as PoS, or consider newer consensus protocols such as DPoS or PBFT.

6.8.2 Scalability

a. Implication: As blockchain networks grow, scalability can become a challenge, leading to congestion and slower transactions.
b. Mitigation: Investigate blockchain platforms that incorporate scalability enhancements, such as sharding, secondary layer scaling solutions (e.g., the Lightning Network), or recently developed consensus algorithms tailored for scalability.

6.8.3 Network Latency

a. Implication: Transactions on blockchain networks may experience latency due to network propagation and verification processes.

b. Mitigation: Opt for blockchain networks with lower block confirmation times and use optimization techniques to reduce network latency.

6.8.4 Storage Costs

a. Implication: Storing data on a blockchain can lead to increased storage costs, especially for large-scale digital twin systems.
b. Mitigation: Leverage off-chain storage solutions for non-critical data while using the blockchain primarily for storing critical metadata or pointers to external data.

6.8.5 Energy Consumption

a. Implication: Some blockchain networks, such as PoW-based ones, consume a significant amount of energy, which can be a concern in energy-efficient applications.
b. Mitigation: Choose energy-efficient consensus mechanisms or consider private blockchains in cases where energy consumption is a concern.

6.8.6 Transaction Costs

a. Implication: Blockchain transactions can involve fees, especially on public networks, which may not be suitable for microtransactions.
b. Mitigation: Consider blockchain platforms with lower transaction fees or explore options such as sidechains for handling smaller transactions.

6.8.7 Smart Contract Complexity

a. Implication: Complex smart contracts can increase the execution time of transactions, affecting the overall performance.
b. Mitigation: Efficiently design and optimize smart contracts to minimize execution complexity. Consider off-chain computation for complex operations.

6.8.8 Blockchain Integration Overhead

a. Implication: Integrating blockchains may introduce additional overhead in terms of development, maintenance, and integration efforts.
b. Mitigation: Carefully assess the benefits versus the overhead of integration. Ensure proper planning and resource allocation.

6.8.9 Consensus Mechanism Choice

a. Implication: Different consensus mechanisms have varying performance characteristics. Choosing an inappropriate one may lead to performance bottlenecks.
b. Mitigation: Evaluate the requirements of the digital twin system and choose a consensus mechanism that aligns with the desired performance levels.

6.8.10 Data Synchronization

a. Implication: Maintaining data consistency across distributed nodes in the blockchain network may introduce synchronization delays.
b. Mitigation: Implement efficient synchronization mechanisms and explore solutions, such as state channels to optimize data synchronization.

6.9 FUTURE RESEARCH DIRECTIONS BASED ON REVIEW

The blockchain technology braced and powered digital twins offer a promising landscape for future research and innovation. Scalability solutions are sought to facilitate real-time updates in large-scale ecosystems, while hybrid consensus mechanisms are explored to harmonize security, scalability, and energy efficiency. Developing interoperability standards becomes vital for seamless communication among diverse blockchain-based digital twin ecosystems. Decentralized identity management systems are poised to enhance privacy, security, and authentication across these ecosystems. Pioneering privacy-preserving techniques will enable secure data sharing and analysis without compromising sensitive information. Energy-efficient blockchain platforms are envisioned as aligning with resource-constrained environments. Dynamic smart contracts hold the potential for adaptive adjustments in response to changing conditions. Cross-chain solutions will facilitate a secure data exchange across blockchain networks. Quantum-resistant algorithms are researched to safeguard against potential quantum computing threats. Integrating real-time analytics and AI into digital twin ecosystems, with blockchains as the backbone, offers unprecedented trust and data integrity. Regulatory compliance frameworks, user-centric interfaces, and AI-driven security measures stand as essential pillars. Moreover, practical implementations and case studies across industries showcase tangible benefits. As industry standards and regulations evolve, they will pave the way for secure and efficient blockchain-based digital twin deployments across various industries and applications, ushering in a fresh era of innovation.

6.10 CONCLUSION

The progress of blockchain-powered and braced digital twin technologies has revealed promising pathways for innovation in a wide range of industries. This comprehensive exploration and analysis of integrating blockchains into digital twin ecosystems reveal a spectrum of opportunities and considerations. While this integration offers heightened security, immutability, and collaborative possibilities, it necessitates addressing intricate challenges, such as scalability, privacy preservation, and consensus efficiency. The proposed solutions, such as multi-factor authentication, dynamic smart contracts, and federated identity management, emphasize the intricate balance required between security and usability. The outlined research directions offer a roadmap for future endeavors, encompassing quantum-resistant algorithms, AI-driven security, and regulatory compliance frameworks. As industries transition toward blockchain-integrated digital twins, our findings highlight the importance of aligning technological innovations with ethical considerations, regulatory requirements, and user-centric interfaces. Collectively, this chapter paves the way for advancements in blockchain-based digital twin ecosystems, empowering industries to embrace transformative potential while ensuring data integrity, security, and trust in a rapidly evolving digital landscape.

REFERENCES

[1] N. Jayakumar and D. P. Joshi, 'Big Data & Disruptive Computing Platforms Braced Internet of Things: Facets & Trends', in *Internet of Things, Smart Computing and Technology: A Roadmap Ahead*, N. Dey, Parikshit. N. Mahalle, P. M. Shafi, V. V. Kimabahune, and A. E. Hassanien, Eds., in Studies in Systems, Decision and Control., Cham: Springer International Publishing, 2020, pp. 119–150. doi: 10.1007/978-3-030-39047-1_6.

[2] A. Hakiri, A. Gokhale, S. Ben Yahia, and N. Mellouli, 'A Comprehensive Survey on Digital Twin for Future Networks and Emerging Iot Industry', SSRN, preprint, 2023. doi: 10.2139/ssrn.4535810.

[3] N. Crespi, A. T. Drobot, and R. Minerva, 'The Digital Twin: What and Why?', in *The Digital Twin*, N. Crespi, A. T. Drobot, and R. Minerva, Eds., Cham: Springer International Publishing, 2023, pp. 3–20. doi: 10.1007/978-3-031-21343-4_1.

[4] N. Jayakumar and S. Choudhari, 'Digital Twins: Towards the Oil and Gas Industry 4.0', in *Digital Twins: The Industry 4.0 Use Cases: The Technologies, Tools, Platforms and Applications*, Nova Science Publishers, Inc, 2023, 979-889113057-9, pp. 145–169. https://doi.org/10.52305/EODP0452

[5] R. Sahal, S. H. Alsamhi, K. N. Brown, D. O'Shea, and B. Alouffi, 'Blockchain-Based Digital Twins Collaboration for Smart Pandemic Alerting: Decentralized COVID-19 Pandemic Alerting Use Case', *Comput. Intell. Neurosci.,* vol. 2022, p. 7786441, 2022, doi: 10.1155/2022/7786441.

[6] D. Lee, S. H. Lee, N. Masoud, M. S. Krishnan, and V. C. Li, 'Integrated Digital Twin and Blockchain Framework to Support Accountable Information Sharing in Construction Projects', *Autom. Constr.,* vol. 127, p. 103688, Jul. 2021, doi: 10.1016/j.autcon.2021.103688.

[7] H. Xu, J. Wu, Q. Pan, X. Guan, and M. Guizani, 'A Survey on Digital Twin for Industrial Internet of Things: Applications, Technologies and Tools', *IEEE Commun. Surv. Tutor.*, pp. 1–1, 2023, doi: 10.1109/COMST.2023.3297395.

[8] M. Li, Z. Li, X. Huang, and T. Qu, 'Blockchain-Based Digital Twin Sharing Platform for Reconfigurable Socialized Manufacturing Resource Integration', *Int. J. Prod. Econ.,* vol. 240, p. 108223, Oct. 2021, doi: 10.1016/j.ijpe.2021.108223.

[9] R. Sahal, S. H. Alsamhi, K. N. Brown, D. O'Shea, C. McCarthy, and M. Guizani, 'Blockchain-Empowered Digital Twins Collaboration: Smart Transportation Use Case', *Machines*, vol. 9, no. 9, Art. no. 9, Sep. 2021, doi: 10.3390/machines9090193.

[10] W. A. H. Ahmed and B. L. MacCarthy, 'Blockchain-Enabled Supply Chain Traceability – How Wide? How Deep?', *Int. J. Prod. Econ.,* vol. 263, no. C, 2023, Accessed: Nov. 03, 2023. [Online]. Available: www.ideas.repec.org//a/eee/proeco/v263y2023ics0925527323001950.html.

[11] A. Hakiri, B. Sellami, S. Ben Yahia, and P. Berthou, 'A Blockchain Architecture for SDN-Enabled Tamper-Resistant IoT Networks', in *2020 Global Information Infrastructure and Networking Symposium (GIIS)*, Oct. 2020, pp. 1–4. doi: 10.1109/GIIS50753.2020.9248492.

[12] H. R. Hasan et al., 'A Blockchain-Based Approach for the Creation of Digital Twins', *IEEE Access*, vol. 8, pp. 34113–34126, 2020, doi: 10.1109/ACCESS.2020.2974810.

[13] B. Hellenborn, O. Eliasson, I. Yitmen, and H. Sadri, 'Asset Information Requirements for Blockchain-Based Digital Twins: A Data-Driven Predictive Analytics Perspective', *Smart Sustain. Built Environ.*, vol. ahead-of-print, no. ahead-of-print, Jan. 2023, doi: 10.1108/SASBE-08-2022-0183.

[14] A. Hakiri, S. B. Yahia, and S. Gokhale Aniruddha, 'Hyper-5G: A Cross-Atlantic Digital Twin Testbed for Next Generation 5G IoT Networks and Beyond', in *2023 IEEE 26th International Symposium on Real-Time Distributed Computing (ISORC)*, Nashville, TN, USA: IEEE, May 2023, pp. 230–235. doi: 10.1109/ISORC58943.2023.00041.

[15] K. Jung, 'Introduction to the Crypto Revolution', in *The Quiet Crypto Revolution: How Blockchain and Cryptocurrency Are Changing Our Lives*, K. Jung, Ed., Berkeley, CA: Apress, 2023, pp. 1–6. doi: 10.1007/978-1-4842-9627-1_1.

[16] P. Raj, 'Chapter Thirteen – Empowering Digital Twins with Blockchain', in *Advances in Computers*, vol. 121, S. Aggarwal, N. Kumar, and P. Raj, Eds., in The Blockchain Technology for Secure and Smart Applications across Industry Verticals, vol. 121., Elsevier, 2021, pp. 267–283. doi: 10.1016/bs.adcom.2020.08.013.

[17] 'Using Blockchain to Drive Supply Chain Transparency and Innovation', Deloitte United States. Accessed: Nov. 03, 2023. [Online]. Available: www2.deloitte.com/us/en/pages/operations/articles/blockchain-supply-chain-innovation.html.

[18] 'A Comparative Analysis of Blockchain Architecture and Its Applications: Problems and Recommendations'. Accessed: Nov. 03, 2023. [Online]. Available: www.researchgate.net/publication/337752303_A_Comparative_Analysis_of_Blockchain_Architecture_and_Its_Applications_Problems_and_Recommendations.

[19] L. C. X. Team, 'Data Security in Blockchain With Cryptography', *LCX*. Accessed: Nov. 03, 2023. [Online]. Available: www.lcx.com/data-security-in-blockchain-with-cryptography/

[20] R. Belchior, A. Vasconcelos, S. Guerreiro, and M. Correia, 'A Survey on Blockchain Interoperability: Past, Present, and Future Trends'. arXiv, Mar. 22, 2021. doi: 10.48550/arXiv.2005.14282.

[21] A. S. Yadav, N. Singh, and D. S. Kushwaha, 'Evolution of Blockchain and Consensus Mechanisms & Its Real-World Applications', *Multimed. Tools Appl.*, vol. 82, no. 22, pp. 34363–34408, Sep. 2023, doi: 10.1007/s11042-023-14624-6.

[22] B. R. Blog, 'Public Versus Private Blockchains', Blockchain Research Lab. Accessed: Nov. 03, 2023. [Online]. Available: www.blockchainresearchlab.org/portfolio/public-versus-private-blockchains/.

[23] W. Yao, F. P. Deek, R. Murimi, and G. Wang, 'SoK: A Taxonomy for Critical Analysis of Consensus Mechanisms in Consortium Blockchain', *IEEE Access*, vol. 11, pp. 79572–79587, 2023, doi: 10.1109/ACCESS.2023.3298675.

[24] D. Mingxiao, M. Xiaofeng, Z. Zhe, W. Xiangwei, and C. Qijun, 'A Review on Consensus Algorithm of Blockchain', in *2017 IEEE International Conference on Systems, Man, and Cybernetics (SMC)*, Oct. 2017, pp. 2567–2572. doi: 10.1109/SMC.2017.8123011.

[25] B. Sellami, A. Hakiri, and S. Ben Yahia, 'Deep Reinforcement Learning for Energy-Aware Task Offloading in Join SDN-Blockchain 5G Massive IoT Edge Network', *Future Gener. Comput. Syst.*, vol. 137, pp. 363–379, Dec. 2022, doi: 10.1016/j.future.2022.07.024.

[26] C. P. Nielsen, E. R. Da Silva, and F. Yu, 'Digital Twins and Blockchain – Proof of Concept', *Procedia CIRP*, vol. 93, pp. 251–255, 2020, doi: 10.1016/j.procir.2020.04.104.

[27] H. Sadri et al., 'Integration of Blockchain and Digital Twins in the Smart Built Environment Adopting Disruptive Technologies—A Systematic Review', *Sustainability*, vol. 15, no. 4, p. 3713, Feb. 2023, doi: 10.3390/su15043713.

[28] H. Rathore, A. Mohamed, and M. Guizani, 'A Survey of Blockchain Enabled Cyber-Physical Systems', *Sensors*, vol. 20, no. 1, Art. no. 1, Jan. 2020, doi: 10.3390/s20010282.

[29] S. Thite and N. Jayakumar, 'A Session Key Based Security Mechanism for Cyber Physical System', 2021. doi: 10.3233/APC210201.

[30] S. Thite and J. NaveenKumar, 'SecureAuthKey: Design and Evaluation of a Lightweight Key Agreement Technique for Smart Home Applications', *Int. J. Inf. Comput. Secur.*, vol. 18, no. 3–4, pp. 406–424, 2022, doi: 10.1504/ijics.2022.125267.

[31] S. Shourya, I. Venkatachalam, H. Patel, and M. Mittal, 'Comparative and Preventive Analysis of Dictionary Attacks', *Lect. Notes Electr. Eng.*, vol. 1071 LNEE, pp. 443–451, 2024, doi: 10.1007/978-981-99-4713-3_42.

[32] S. Liu, Y. Lu, J. Li, X. Shen, X. Sun, and J. Bao, A Blockchain-Based Interactive Approach between Digital Twin-Based Manufacturing Systems. *Computers & Industrial Engineering*, vol. 175, p. 108827, 2023.

[33] N. A. N. Adu-Amankwa, F. P. Rahimian, N. Dawood, and C. Park, Digital Twins and Blockchain Technologies for Building Lifecycle Management. *Automation in Construction*, vol. 155, p. 105064, 2023.

[34] A. Sasikumar, S. Vairavasundaram, K. Kotecha, V. Indragandhi, L. Ravi, G. Selvachandran, and A. Abraham, Blockchain-Based Trust Mechanism for Digital Twin Empowered Industrial Internet of Things. *Future Generation Computer Systems*, vol. 141, pp. 16–27, 2023.

[35] Z. Lv, C. Cheng and H. Lv, "Blockchain-Based Decentralized Learning for Security in Digital Twins," in IEEE Internet of Things Journal, vol. 10, no. 24, pp. 21479-21488, 15 Dec.15, 2023, doi: 10.1109/JIOT.2023.3295499.

[36] A. K. Tyagi, T. T. George, and G. Soni, 'Blockchain-Based Cybersecurity in Internet of Medical Things (IoMT)-Based Assistive Systems', in *AI-Based Digital Health Communication for Securing Assistive Systems*, IGI Global, 2023, pp. 22–53. doi: 10.4018/978-1-6684-8938-3.ch002.

[37] P. R. F. Bin Kaha, S. B. Rahayu, A. A. Wardhana, M.-G. Lee, A. L. A. Lokman, and N. N. Azmi, 'Designing an E-Voting Framework Using Blockchain: A Secure and Transparent Attendance Approach', in *2023 IEEE 8th International Conference On Software Engineering and Computer Systems (ICSECS)*, Aug. 2023, pp. 371–376. doi: 10.1109/ICSECS58457.2023.10256270

7 Big Data and Digital Twins

Sayed Abdulhayan and Senan Ali Abd

7.1 INTRODUCTION

A digital twin is a virtual representation of a physical system or process that exists in the real world. It is created by using real-time data and advanced modeling techniques to simulate and analyze the behavior of the physical entity. Digital twins can be used in various industries, such as manufacturing, construction, oil and gas, healthcare, and retail. They offer several benefits, including real-time monitoring, data analysis, and the simulation of different scenarios. Digital twins can improve operational efficiency, asset management, and decision-making by providing detailed information on the performance and condition of assets. They can also be used to simulate the performance of new assets before they are built, reducing the risk of errors. The technology behind digital twins involves integrating sensors, the Internet of Things (IoT) devices, and machine learning algorithms to continuously optimize the system [1]–[5].

7.1.1 Benefits of Digital Twins

Digital twin technology offers several benefits across various industries. In manufacturing, construction, and oil and gas, digital twins enable real-time monitoring, data analysis, and the simulation of different scenarios, leading to improved efficiency and decision-making [1]. In engineering, digital twins can enhance processes and outcomes by applying engineering analysis and design principles, resulting in improved healthcare services [6]–[7]. In healthcare, human digital twins (HDTs) can transform patient care by modeling patients' tissues, organs, and physiological processes. HDTs can optimize treatment plans, monitor disease progression, and enable precision surgical planning [8]. Additionally, digital twins have the potential to revolutionize cancer patient care by providing unique insights, patient-specific clinical decision-making support, and the potential for virtual clinical trials [9]. Overall, digital twins have immense potential to improve efficiency, enhance patient care and outcomes, reduce costs, and advance medical research and education.

DOI: 10.1201/9781003465904-9

7.2 ASPECTS AND APPLICATIONS

Here are a few key aspects and applications of digital twins:

1. Virtual Replicas: Digital twins are accurate digital counterparts of physical objects or systems. They include all relevant data and characteristics, such as geometry, sensors, materials, and behavior.
2. Real-Time Data: Digital replicas are continuously refreshed with real-time information acquired from sensors, IoT devices, and various other sources, thereby enabling them to accurately portray the present condition and operational efficiency of the tangible entity.
3. Simulation and Prediction: Digital twins enable simulation and predictive analysis. They can be used to model how a physical system will behave under different conditions or how it may respond to changes, helping with optimization and decision-making.
4. Monitoring and Control: Digital twins offer a way to monitor and control physical assets remotely. This is particularly useful in scenarios where physical access may be limited or hazardous.
5. Manufacturing: In manufacturing, digital twins can be used to replicate and analyze the production process. This allows for process optimization, quality control, and the identification of potential issues before they occur.
6. Healthcare: In healthcare, digital twins can represent a patient's physiology or their specific body parts. They are used for personalized medical treatment, surgical planning, and monitoring patient health in real time.
7. Urban Planning: In urban planning, digital twins can replicate cities or specific infrastructures, such as transportation systems. Planners and policymakers can use these models to optimize traffic flow, energy consumption, and disaster preparedness.
8. Aerospace: Digital twins are widely used in aerospace to monitor the health and performance of aircraft, spacecraft, and engines. They aid in predictive maintenance and safety.
9. IoT and Smart Cities: Digital twins play a role in the development of IoT devices and smart cities. They help to manage and optimize complex networks of interconnected devices and services.
10. Energy: In the energy sector, digital twins can simulate and optimize the operation of power plants, grids, and renewable energy systems.
11. Building and Construction: Digital twins are used in the construction industry to replicate building designs and monitor construction progress, enabling better project management.
12. Environmental Modeling: Environmental scientists use digital twins to model and simulate natural systems, such as ecosystems, weather patterns, and ocean currents.

7.3 DIGITAL TWINS AND BIG DATA

The use of digital twins in conjunction with big data offers a powerful combination for enhanced understanding, analysis, and optimization of physical systems and processes. Here is how digital twins and big data work together in their applications:

1. Data Integration and Collection
   - Big Data Sources: Big data analytics collect and process vast amounts of data from various sources, including sensors, IoT devices, social media, and others.
   - Digital Twin Data Integration: Digital twins provide a structured framework for integrating real-time data from these sources, creating a dynamic virtual representation of the physical entity.
2. Real-Time Monitoring and Analysis
   - Big Data Processing: Big data platforms analyze real-time and historical data to identify patterns, anomalies, and trends.
3. Digital Twins Enable Visualization: The digital twin functions as a visual depiction of the tangible structure or operation, thereby facilitating users to oversee and comprehend the data within its pertinent framework.
4. Predictive Modeling
   - Big Data Analytics: Big data tools are capable of producing prognostic models by utilizing past data and present observations.
   - Digital Twins Simulation: Digital twins enable these models to be applied to the virtual representation, thereby facilitating the simulation of the physical system's behavior under diverse circumstances.
5. Optimization
   - Big Data Insights: Big data analytics reveal areas for optimization and improvement within the data.
   - Digital Twins Enable Testing: Digital twins provide a safe environment for testing changes and strategies without affecting the physical system, allowing for experimentation and fine-tuning.
6. Maintenance and Predictive Analysis
   - Big Data for Maintenance: Big data can be used to identify maintenance needs and predict equipment failures.
   - Digital Twins for Simulation: Digital twins can simulate the impact of maintenance actions, thereby helping to plan and execute maintenance more effectively.
7. Supply Chain and Manufacturing
   - Big Data in Supply Chain: Big Data analytics track inventory, demand, and production data across the supply chain.
   - Digital Twins for Manufacturing: Digital Twins can model the entire manufacturing process, providing real-time insights and simulations to optimize production.
8. Urban Planning and Smart Cities
   - Big Data for City Data: Big data analytics process information from traffic sensors, environmental sensors, and social media.

- Digital Twins for Urban Planning: Digital twins of urban areas enable city planners to visualize and simulate the impact of changes in infrastructure, transportation, and services.

9. Healthcare and Personalized Medicine
   - Big Data in Healthcare: Big data in healthcare collects patient data, medical records, and genomic information.
   - Digital Twins for Patients: Digital twins can create personalized models of patients, allowing for simulations to determine the best treatment options and track their health in real time.

The amalgamation of digital twins and big data augments the process of decision-making, mitigates risks, and nurtures the development of novel ideas across diverse sectors. It allows for data-driven insights and simulations, making it easier to understand complex systems, predict outcomes, and optimize processes. Moreover, it enables real-time monitoring and control, helping organizations to react swiftly to changing conditions and minimize disruptions.

Data integration and collection within the framework of digital twin technology and big data encompass the ongoing consolidation of data from diverse origins to generate a dynamic virtual depiction of a tangible system or entity. Here are a few examples of how digital twins and big data work together in this context:

1. **Smart Buildings**
   - **Digital Twin**: A digital twin is created for a smart building to model the structure, heating, ventilation, and air conditioning (HVAC) systems, lighting, and security.
   - **Big Data Integration**: IoT sensors are placed throughout the building to collect real-time data on temperature, occupancy, lighting levels, and security events. This data is integrated into the digital twin.
   - **Use Case Example**: The digital twin enables building managers to monitor energy usage, occupancy patterns, and security in real time. With big data integration, they can optimize HVAC and lighting based on occupancy, reduce energy consumption, and enhance security protocols.
2. **Manufacturing and Industry 4.0**
   - **Digital Twin**: In a manufacturing plant, a digital twin models the entire production process, including machines, sensors, and assembly lines.
   - **Big Data Integration**: The sensors attached to the machines collect data on temperature, pressure, machine performance, and product quality. This real-time data is integrated into the digital twin.
   - **Use Case Example**: When a sensor detects a deviation in product quality or machine performance, the digital twin updates in real time. With big data analytics, manufacturing teams can identify issues promptly, adjust machine settings, and ensure consistent product quality.
3. **Healthcare and Remote Patient Monitoring**
   - **Digital Twin**: A digital twin can model a patient's health status, medical history, and vital signs.

- **Big Data Integration**: Wearable devices and medical sensors continuously collect real-time data on the patient's heart rate, blood pressure, and other vital signs. This data is integrated into the patient's digital twin.
- **Use Case Example**: Healthcare providers can use the digital twin to remotely monitor patients, detect anomalies in vital signs, and make adjustments to treatment plans. Real-time integration of big data enhances the ability to provide timely care and personalized treatment.

4. **Urban Traffic Management**
   - **Digital Twin**: A digital twin of a city's transportation system models road networks, traffic signals, and public transportation infrastructure.
   - **Big Data Integration**: Traffic cameras, GPSs, and social media feeds provide real-time data on traffic conditions, accidents, and public transportation schedules. This data is integrated into the digital twin.
   - **Use Case Example**: Urban planners and traffic management authorities can use the digital twin to monitor real-time traffic patterns and respond to congestion and accidents promptly. With big data integration, they can optimize traffic signal timings, reroute public transportation, and improve overall traffic flow.
5. **Environmental Monitoring**
   - **Digital Twin**: A digital twin represents an environmental system, such as a forest ecosystem.
   - **Big Data Integration**: Sensors, satellite imagery, and weather data continuously provide real-time information on factors such as temperature, humidity, and vegetation health. This data is integrated into the digital twin.
   - **Use Case Example**: Environmental scientists can use the digital twin to monitor the health of the ecosystem, detect potential issues, such as forest fires or invasive species, and implement timely conservation efforts based on real-time data.

In each of these examples, the combination of digital twins and big data integration enables real-time monitoring and decision-making. This allows for the creation of virtual representations of physical systems or entities that are continuously updated with data from various sources. It helps to optimize operations, respond to changes, and improve efficiency and resource utilization.

Digital twin technology has been formulated with the assistance of extensive data sets, specifically in the domains of healthcare, positive energy districts (PED), and intelligent manufacturing. Substantial data is employed to amass invaluable insights from the generated data and enhance the examination of digital replicas [1]–[4]. In the context of PED, digital twin components include virtual models, sensor network integration, data analytics, and stakeholder layers [5]. However, there is a lack of comprehensive tools for digital twin creation, with most tools focusing on industrial applications or data collection and visualization based on building information models (BIM) or geographical information systems (GIS). The incorporation of large-scale information and digital replica technology is of utmost importance in attaining intelligent production, as the amalgamated examination of production large-scale information can yield advantages across all facets of production. In the context of smart urban management, data availability is a major challenge for implementing digital twin cities, and systematic approaches for large-scale urban data collection are needed.

Christina et al. [10] discuss how big data is used to optimize digital twins in healthcare, but it does not specifically mention data integration and collection.

Michelle and Hendra [11] discuss how big data is used to optimize digital twins in healthcare, but it does not specifically mention data integration and collection.

Xingxing et al. [12] mention that most existing tools for digital twins focus on data collection, communication, and visualization based on BIM or GIS. However, there is a lack of knowledge and skill in data analysis and interoperable interaction with different stakeholders. The paper does not specifically mention the integration of big data with digital twin technology.

Qinglin and Fei [13] discuss the concept and applications of big data and digital twins in manufacturing. However, it does not specifically address the integration and collection of data between digital twin technology and big data.

The notable weaknesses or drawbacks in the approach used in these papers are as follows:

- One notable weakness in the approach used in these papers is the limited availability of data for digital twin models of cities. The papers acknowledge that data collection for comprehensive digital twin models is a major challenge, and that there is little knowledge about systematic approaches for large-scale urban data collection.
- Another weakness is the lack of full-function tools for digital twin models in PED. The review of tools and platforms for digital PED twins found that most tools either focus on industrial applications or are designed for data collection, communication, and visualization based on BIM or GIS. There is a need for further development of tools that can fully support digital PED twin models.
- Moreover, the articles fail to offer a thorough examination of the constraints and difficulties involved in integrating big data and digital twin technologies. While they discuss the similarities and differences between big data and digital twins, they do not delve into the specific challenges and opportunities in integrating these technologies for smart manufacturing and healthcare applications.

Future research directions or unanswered questions suggested by these papers are as follows:

Directions and unanswered questions in the field of digital twin cities and smart urban management. These include:

- There needs to be further exploration of systematic approaches for large-scale urban data collection to address the challenge of data availability for comprehensive digital twin models.
- Research is being conducted to examine the fusion of big data and digital twin technologies in the context of intelligent manufacturing. This investigation

encompasses the creation of frameworks and methodologies to effectively combine these technologies and harness their synergistic capabilities.

- Advancement of tools and platforms for digital twin models in PED, with a focus on addressing challenges in data analysis, semantic interoperability, business models, and data security and management.
- There needs to be an exploration of the potential applications of digital replicas and the vast amounts of information in the realm of healthcare, encompassing the creation of customized HDT models and the enhancement of healthcare procedures and infrastructure.
- Further research on the fusion of digital twins, big data, and services in smart manufacturing needs to take place, with a focus on the convergence of these technologies and their significance in optimizing manufacturing processes.
- An investigation of the challenges and opportunities in optimizing the sustainability of digital twin models for PED, including areas such as data analysis, semantic interoperability, and data security and management, needs to be prioritized.
- An exploration of the potential advantages and disadvantages of employing digital replicas and substantial information technologies in the healthcare sphere should be undertaken, encompassing deliberations on data confidentiality, safeguarding, and moral consequences.

7.3.1 Real-Time Monitoring and Analysis

Real-time monitoring and analysis with digital twin technology and big data enable organizations to continuously track, analyze, and respond to changes and events in real time. Here are a few examples of how digital twins and big data are applied in the context of real-time monitoring and analysis:

1. **Smart Grid Management**
    - **Digital Twin**: A digital twin represents an electrical grid, including power generation, distribution infrastructure, and consumer demand.
    - **Big Data Integration**: Sensors, smart meters, and weather data provide real-time information on energy consumption, grid performance, and weather conditions. This data is integrated into the digital twin.
    - **Use Case Example**: Real-time monitoring and analysis within the digital twin can detect grid disturbances, adjust energy distribution, and predict potential issues, enhancing grid stability and reliability.
2. **Healthcare and Patient Monitoring**
    - Digital Twin: A patient's digital twin models their physiological parameters, medical history, and treatment plans.
    - Big Data Integration: Wearable devices and medical sensors continuously collect real-time data on vital signs, medication adherence, and other health metrics. This data is integrated into the patient's digital twin.
    - Use Case Example: Healthcare providers can use the digital twin to monitor a patient's health in real time, detect anomalies, and trigger alerts or interventions, ensuring timely and personalized care.

3. **Aircraft Performance Monitoring**
    - **Digital Twin**: A digital twin represents an aircraft, including its engines, avionics, and structural components.
    - **Big Data Integration**: Sensors and avionics systems on the aircraft continuously provide real-time data on the engine's performance, structural health, fuel consumption, and flight conditions. This data is integrated into the digital twin.
    - **Use Case Example**: Real-time monitoring within the digital twin can assess aircraft health, identify performance deviations, and predict maintenance needs, ensuring safe and efficient flights.
4. **Traffic Management in Smart Cities**
    - **Digital Twin**: A digital twin models a city's transportation system, including roads, traffic signals, and public transportation infrastructure.
    - **Big Data Integration**: Traffic cameras, GPSs, and social media feeds provide real-time data on traffic conditions, accidents, and public transportation schedules. This data is integrated into the digital twin.
    - **Use Case Example**: Real-time analysis within the digital twin can optimize traffic signal timings, reroute public transportation, and provide real-time traffic updates to improve traffic flow and reduce congestion.
5. **Environmental Monitoring**
    - **Digital Twin**: A digital twin represents an environmental system, such as a forest ecosystem.
    - **Big Data Integration**: Sensors, satellite imagery, and weather data continuously provide real-time information on factors such as temperature, humidity, and vegetation health. This data is integrated into the digital twin.
    - **Use Case Example**: Real-time monitoring and analysis within the digital twin can detect changes in environmental conditions, such as the spread of wildfires or invasive species, enabling timely responses and conservation efforts.

In these examples, digital twins combined with big data integration enable real-time monitoring, analysis, and decision-making. This enables organizations to respond to changing conditions, optimize operations, and enhance safety, efficiency, and resource utilization.

Digital twin technology, combined with advancements in big data, has greatly improved real-time monitoring and analysis capabilities [14]–[15]. It allows for the creation of virtual models that can simulate and predict the behavior of complex systems, such as power systems and turbofan engines [16]–[17]. By harnessing real-time data and employing machine learning (ML) algorithms, digital replicas can furnish precise and contemporary information for the purpose of decision-making [18]. This technological advancement has been successfully employed in a myriad of fields, encompassing power systems and the diligent study of natural catastrophes. Nevertheless, the widespread acceptance and integration of digital replicas have been impeded by obstacles such as the absence of an all-encompassing reference framework, apprehensions relating to security, and a reliance on auxiliary technologies.

Despite these formidable hurdles, digital replicas have demonstrated great potential in ameliorating the efficiency and efficacy of contemporaneous monitoring and analysis within diverse industries.

Hao HanYong et al. [14] examine the application of digital twin technology in the real-time observation and evaluation of power systems. However, they do not explicitly address the utilization of big data within this particular framework.

Research by Yaseen et al. [15] discusses the advancements in ML, the IoT, and big data that have contributed to the improvements in digital twin technology, particularly in real-time monitoring and forecasting properties.

In the study conducted by Leila et al. [16], the focus lies on the utilization of data science-based digital twin models in power systems for the purpose of real-time analysis and decision-making. However, there is no explicit mention of the incorporation of big data for real-time monitoring and analysis in this context.

Similarly, Koshimura and Mas [17] explore the application of digital twin technology in enhancing disaster response systems. Nevertheless, there is no specific reference to the utilization of big data for real-time monitoring and analysis in their research.

Lastly, Mei et al. [18] examine the use of digital twin technology for real-time test data analysis and performance monitoring of a turbofan engine. However, the utilization of big data in this context is not explicitly discussed in their study.

The notable weaknesses or drawbacks in the approach used in these papers are as follows:

- The absence of publicly accessible implementation specifics in these papers represents a flaw in the approach employed, thereby posing a challenge in evaluating the efficacy of the digital twin methodology.
- The domain dependence of digital twins can be identified as a further limitation, as it has the potential to impede their extensive implementation across various industries.
- Security concerns regarding the sharing of data in digital twins are also a drawback that needs to be addressed.
- The reliance of digital twins on other technologies is another weakness, as their effectiveness is dependent on the availability and integration of these technologies.
- The lack of a universal reference framework for digital twins is a drawback that hinders their development and comparison with other concepts.
- In terms of the specific applications mentioned in the papers, one weakness is the need for improved data collection frequency and accuracy to enhance the results of load balancing and energy storage optimization.
- The use of ML methods in digital twins for cybersecurity could benefit from the fusion of deep learning and control theory to improve the speed, accuracy, and predictability of attacks.

The future research directions or unanswered questions suggested by these papers are as follows:

- Future research avenues proposed by these papers encompass the establishment of a comprehensive and all-encompassing framework for digital twins, with the objective of streamlining their implementation and enabling their juxtaposition with alternative concepts. The exploration of a digital twin's complete capabilities and targeted resolution of unique challenges in every domain, such as power systems, disaster response systems, and renewable energy systems, necessitates further investigation and inquiry.
- Unanswered questions include the exploration of quantitative metrics for evaluating the effectiveness of digital twins and the identification of novel research questions to advance the theory and practice of digital twins.
- Further research is needed to address the weaknesses and drawbacks of the current approach, such as the lack of publicly available implementation details, domain dependence, security concerns, and reliance on other technologies.
- There is a need for research on improving data collection frequency and accuracy in digital twins to enhance their performance in areas such as load balancing, energy storage optimization, and real-time monitoring.
- The fusion of ML methods, such as deep learning, with control theory in digital twins for cyber security is an area that requires further investigation to improve the speed, accuracy, and predictability of attacks.

7.4 PREDICTIVE MODELING

Predictive modeling with digital twin technology and big data involves using historical and real-time data to create predictive models that simulate and forecast the behavior of physical systems or processes. The following are a few examples illustrating how digital twins and big data are applied to predictive modeling:

1. **Predictive Maintenance in Manufacturing**
   - **Digital Twin**: A digital twin is a manufacturing machine that replicates the machine's physical components, sensors, and historical performance data.
   - **Big Data Integration**: Sensors on the machine continuously collect data on factors such as temperature, vibration, and component wear. This real-time data is integrated into the digital twin.
   - **Use Case Example**: By analyzing historical and real-time data in the digital twin, predictive models can be created to forecast when machine components are likely to fail. Maintenance can be scheduled proactively, minimizing downtime and reducing maintenance costs.
2. **Energy Efficiency in Smart Buildings**
   - **Digital Twin**: A digital twin of a smart building models the building's structure, HVAC systems, and lighting.
   - **Big Data Integration**: Sensors throughout the building collect data on factors such as temperature, occupancy, and energy consumption. This data is integrated into the digital twin.

- **Use Case Example**: A possible output could be the optimization of a building's energy consumption, which can be achieved by constructing predictive models within the digital twin. These models employ both historical and real-time data to anticipate the appropriate timing for adjusting HVAC settings, lighting, and other systems. The objective is to minimize energy usage while simultaneously ensuring comfort levels are maintained.

3. **Traffic Management in Smart Cities**
   - **Digital Twin**: A digital twin of a city's transportation system replicates road networks, traffic signals, and public transportation infrastructure.
   - **Big Data Integration**: Data sources, such as traffic cameras, GPSs, and social media feeds provide real-time data on traffic conditions, accidents, and public transportation schedules. This data is integrated into the digital twin.
   - **Use Case Example**: Predictive models within the digital twin use historical and real-time data to forecast traffic congestion, identify accident-prone areas, and predict public transportation delays. These insights enable proactive measures, such as adjusting traffic signal timings or planning for infrastructure improvements.
4. **Healthcare and Disease Management**
   - **Digital Twin**: A patient's health condition, complete medical records, and strategies for treatment are represented by their digital twin.
   - **Big Data Integration**: Wearable sensors, electronic health records, and medical data continuously provide real-time information on the patient's vital signs, medication adherence, and relevant medical data. This data is integrated into the patient's digital twin.
   - **Use Case Example**: Predictive models within the patient's digital twin can analyze historical data and real-time health information to forecast disease progression or detect early warning signs of health issues. Healthcare providers can use these predictions to adjust treatment plans or interventions proactively.

In the given instances, the utilization of digital twins exploits the integration of vast quantities of data to construct prospective models that provide discernment into forthcoming occurrences or circumstances. These models facilitate anticipatory decision-making, culminating in heightened effectiveness, diminished expenditures, augmented security, and superior results across diverse sectors and implementations.

Digital twin technology, when combined with big data, enables predictive modeling for various engineering and industrial domains. The use of data-driven models in digital twins allows for real-time updates and predictions by leveraging data and computational models [19]. However, the fidelity of available data and the scarcity of accurate sensor data can hinder the efficient learning of surrogate models in digital twins [20]. To address this challenge, a novel framework called hybrid polynomial correlated function expansion (H-PCFE) integrates polynomial correlated function expansion (PCFE) with the Gaussian process (GP) to create a robust multi-fidelity surrogate model [21]. This framework, along with the introduction of deep-HPCFE, is a cascading arrangement of models with different fidelities, and effectively addresses the issue of erroneous predictions from low-fidelity models [22]. Additionally, the combination of digital twin technology and big data platforms allows for real-time monitoring, analysis, and management of various fields, such as user management, equipment health management, and product quality management [23].

The role of predictive engineering analytics in the digital twin is discussed by Agus et al. [19], but there is no specific mention of the use of big data in predictive modeling.

The challenges of developing accurate surrogate models for digital twin systems due to the lack of detailed application-specific information and accurate sensor data are discussed by Desai et al. [20]. A multi-fidelity modeling framework is proposed to address these challenges, but there is no specific mention of the use of big data in predictive modeling for digital twin technology.

The use of data-driven models in digital twin technology is discussed by Et.al [21], but there is no specific mention of the use of big data or predictive modeling.

The use of digital twin modeling based on big data platforms for numerical analysis is discussed by Sujing [22], but there is no specific mention of predictive modeling.

The use of data-driven methods, such as neural networks, for performance evaluation in digital twin networks is discussed by Marco et al. [23]. However, there is no specific mention of the use of big data or predictive modeling in relation to digital twin technology.

The notable weaknesses or drawbacks in the approach used in these papers are as follows:

- One notable weakness in the approach used in these papers is the lack of detailed application-specific information, which poses challenges to the seamless implementation of digital twin technology in practical systems.
- Another weakness is the fidelity of available data and the scarcity of accurate sensor data, which hinder the efficient learning of surrogate models that connect physical systems and digital twin models.
- The fixed forms of classical data-driven techniques limit their flexibility in modeling complicated network mechanisms, leading to limited applications.
- The models used in the papers may not be flexible enough to handle rapidly evolving network changes, such as topology, configurations, or mechanisms, which can affect the accuracy of performance metrics generated by the models.
- The lack of data or the inability to measure certain types of data can limit the effectiveness of digital twin models. Predictive engineering analytics, such as science-based simulations, can help to generate missing data, but there are still scenarios where engineers require data that cannot be measured.

Future research directions or unanswered questions suggested by these papers are as follows:

- Future research directions suggested by these papers include exploring data collection methods, model construction techniques, and application prospects for digital twin networks.

- There is a need for further research on how to handle diversified inputs and limited data collected from production networks to improve the capacities of models used in digital twin networks.
- The papers highlight the opportunities for using data-driven methods in digital twin systems, but they also point out the challenges and obstacles that need to be addressed to fully realize the potential of these methods.
- The papers suggest that there is a need for research on enhancing multi-fidelity modeling for digital twin systems and uncertainty quantification, particularly in engineering and industrial domains, such as aerospace, infrastructure, and automotive industries.
- There is a need for further research in developing robust multi-fidelity surrogate models that can effectively connect physical systems and digital twin models, and address the issue of erroneous predictions from low-fidelity models.
- The papers suggest that future research should focus on the integration of predictive engineering analytics with operating data to improve the performance, life, and integrity of equipment, systems, and facilities in digital twin systems.
- Further research is needed to explore the potential of combining different levels of science-based modeling approaches, such as finite element analysis (FEA) and computational fluid dynamics (CFD), with field data to provide real-time data and insights for operational decision-making in digital twin systems.

7.5 OPTIMIZATION

Optimization with digital twin technology and big data involves using data-driven insights to enhance the performance, efficiency, and decision-making of physical systems or processes. Here are a few examples of how digital twins and big data are applied to optimization:

1. **Supply Chain Optimization**
   - **Digital Twin**: A digital twin models the entire supply chain, including suppliers, warehouses, transportation, and inventory levels.
   - **Big Data Integration**: Data from IoT devices, radio frequency identification (RFID) tags, and inventory management systems continuously provide real-time data on supply chain activities. This data is integrated into the digital twin.
   - **Use Case Example**: By analyzing historical and real-time data in the digital twin, predictive models can optimize inventory levels, transportation routes, and order fulfillment. This leads to reduced costs, minimized stockouts, and improved customer satisfaction.
2. **Energy Grid Optimization**
   - **Digital Twin**: A digital twin replicates an energy grid, including power generation sources, distribution infrastructure, and consumer demand.
   - **Big Data Integration**: Data from smart meters, weather forecasts, and energy generation sensors continuously provide real-time data on

energy consumption and production. This data is integrated into the digital twin.
- **Use Case Example**: Predictive models within the digital twin can optimize energy distribution and storage, leveraging real-time data and forecasts to balance supply and demand efficiently. This reduces energy waste and enhances the stability of the grid.

3. **Manufacturing Process Optimization**
   - **Digital Twin**: A digital twin models a manufacturing process, including machinery, assembly lines, and quality control.
   - **Big Data Integration**: Sensors on machines collect data on production rates, quality metrics, and equipment performance. This real-time data is integrated into the digital twin.
   - **Use Case Example**: By analyzing historical and real-time data in the digital twin, optimization models can adjust production parameters and quality control processes to maximize output and minimize defects. This improves production efficiency and product quality.
4. **Aircraft Performance Optimization**
   - **Digital Twin**: A digital twin represents an aircraft, including its engines, avionics, and structural components.
   - **Big Data Integration**: Sensors and avionics systems on the aircraft continuously provide real-time data on engine performance, structural health, fuel consumption, and flight conditions. This data is integrated into the digital twin.
   - **Use Case Example**: Real-time optimization models within the digital twin can adjust engine parameters, route planning, and fuel consumption based on real-time data and weather forecasts. This optimizes aircraft performance, reduces fuel costs, and enhances safety.
5. **Traffic Management in Smart Cities**
   - **Digital Twin**: A digital twin models a city's transportation system, including roads, traffic signals, and public transportation infrastructure.
   - **Big Data Integration**: Traffic cameras, GPSs, and social media feeds provide real-time data on traffic conditions, accidents, and public transportation schedules. This data is integrated into the digital twin.
   - **Use Case Example**: Optimization models within the digital twin can adjust traffic signal timings and public transportation schedules in real time based on current traffic conditions and anticipated congestion. This reduces traffic congestion and improves the flow of vehicles and public transport.

In the provided examples, the amalgamation of digital twins and the integration of big data facilitates the formulation of optimization models that employ both historical and real-time data to arrive at informed decisions and make necessary adjustments. This results in a heightened level of efficiency, decreased expenses, enhanced quality, and improved performance in a wide range of industries and applications.

Digital twin technology, when combined with big data, presents prospects for optimization across various domains. It enables the ability to predict, optimize, and make decisions in real time by generating virtual representations of physical assets [24]. The proliferation of data and advancements in big data analysis techniques, along with the advent of IoT, 5G, and cloud computing, have greatly contributed to the advancement of digital twin technology [25]. Decision-making mechanisms based on digital twin technology can optimize production processes, reduce costs, and enhance operational efficiency [26]. Data plays a pivotal role in constructing virtual models, establishing cyber-physical connections, and executing intelligent operations within digital twin systems [27]. The integration of digital twin technology with big data facilitates real-time analysis, monitoring, and optimization across various domains, such as power systems and integrated energy systems [28]. By harnessing the potential of digital twin technology and big data, researchers and practitioners can enhance the efficiency and effectiveness of decision-making processes within diverse industries.

Safiye et al. [24] delve into the examination of how the integration of digital twin technology with the analysis of large-scale data can enhance the efficiency of the production process, minimize expenses, and augment operational effectiveness within a flexible manufacturing environment.

Zhang et al. [25] explore the utilization of digital twin technology in conjunction with the Internet of Things, cloud computing, and big data for the real-time assessment and monitoring of electromagnetic and thermal fields in power systems. Nevertheless, it fails to explicitly address the optimization aspect of big data in relation to digital twin technology.

Meng et al. [26] emphasize the significance of data in digital twin technology, albeit without specifically mentioning the application of big data in terms of optimization.

Ferranti et al. [27] scrutinize the adoption of a data-driven digital twin approach to optimize long-term production planning in a chemical production site. Nevertheless, it does not explicitly address the utilization of big data in this context.

Shumin et al. [28] explore the utilization of digital twin technology to enhance the energy efficiency of an integrated energy system. However, they do not explicitly mention the utilization of big data in conjunction with digital twin technology for optimization.

The notable weaknesses or drawbacks in the approach used in these papers are as follows:

- The existing technology for system load forecasting is not accurate enough, resulting in a relatively low energy efficiency coefficient even after system optimization.
- The digital twin approach relies on simple regression models to represent the consumption of raw materials and utilities, which may not capture the complexity of the production processes accurately.
- The research paper explores the subject of digital twin data, delving into the emerging requirements for data pertaining to digital twins. However, it fails to explicitly address any shortcomings or limitations associated with this approach.

- The scholarly article on the digital twin model of energy equipment emphasizes the transformative role that digital twins play in this context. Nevertheless, it neglects to discuss any weaknesses or drawbacks that may be inherent to this approach.
- The scholarly document on the digital twin-based flexible manufacturing system modeling, which utilizes a fuzzy approach, sets out to optimize the production process and enhance operational efficiency. Regrettably, it does not acknowledge any weaknesses or drawbacks that may be associated with this approach.

Future research directions or unanswered questions suggested by these papers are as follows:

- Future investigations can direct their attention toward the enhancement of the precision of system load forecasting, aiming to further amplify the optimization of energy efficiency in integrated energy systems. There exists a possibility for additional research in the development of more sophisticated models, capable of grasping the intricacies of production processes in a chemical production facility, surpassing the limitations of simple regression models. The notion of digital twin data (DTD) that is introduced in one of the scholarly articles provides opportunities for future studies in domains such as data collection, extraction, fusion, interaction, and iterative refinement for digital twins.
- The paper on digital twin models of energy equipment suggests future research directions in real-time analysis and monitoring, as well as the establishment method of digital twin models for different components in power systems.
- The paper on digital twin-based flexible manufacturing system modeling with a fuzzy approach suggests future research in optimizing production processes, reducing costs, and increasing operational efficiency using digital twin-based integrated smart production models and fuzzy decision-making approaches.

7.6 MAINTENANCE AND PREDICTIVE ANALYSIS

Maintenance and predictive analysis with digital twin technology and big data involve using historical and real-time data to forecast equipment or system maintenance needs and to plan maintenance activities efficiently. Here are examples of how digital twins and big data are applied to maintenance and predictive analysis:

1. **Predictive Maintenance in Manufacturing**
   - Digital Twins: A digital twin models a manufacturing machine, including its physical components and historical performance data.
   - Big Data Integration: Sensors on the machine continuously collect data on factors, such as temperature, vibration, and component wear. This real-time data is integrated into the digital twin.
   - Use Case Example: Predictive maintenance models within the digital twin analyze historical and real-time data to predict when machine components

are likely to fail. Maintenance can be scheduled proactively to minimize downtime and reduce maintenance costs.

2. **Aircraft Engine Health Monitoring**
   - Digital Twins: A digital twin represents an aircraft engine, including its physical components and historical performance data.
   - Big Data Integration: Sensors and avionics systems on the aircraft engine continuously provide real-time data on factors such as temperature, pressure, and vibration. This data is integrated into the digital twin.
   - Use Case Example: Predictive analysis within the digital twin can monitor the engine's health in real time and predict when maintenance is needed based on deviations from historical data. Maintenance crews can schedule repairs or replacements to prevent in-flight issues.
3. **Energy Grid Infrastructure Maintenance**
   - **Digital Twins**: A digital twin models components of an energy grid, including transformers, power lines, and substations.
   - **Big Data Integration**: Sensors on grid infrastructure continuously collect data on factors such as temperature, load, and electrical characteristics. This real-time data is integrated into the digital twin.
   - **Use Case Example**: Predictive analysis within the digital twin can monitor the condition of grid components and predict when maintenance is needed. Maintenance teams can schedule inspections and repairs to prevent power outages and failures.
4. **Fleet Vehicle Maintenance**
   - **Digital Twins**: A digital twin models a fleet of vehicles, including historical maintenance records and real-time sensor data.
   - **Big Data Integration**: Sensors in each vehicle continuously collect data on engine performance, fuel efficiency, and wear and tear. This real-time data is integrated into the digital twin.
   - **Use Case Example**: Predictive maintenance models within the digital twin can analyze real-time and historical data to predict when individual vehicles in the fleet require maintenance. Fleet managers can schedule maintenance to optimize vehicle uptime and reduce operating costs.
5. **Healthcare Equipment Maintenance**
   - **Digital Twins**: A digital twin models healthcare equipment, such as medical devices and diagnostic tools.
   - **Big Data Integration**: Sensors on the equipment continuously collect data on usage, error rates, and wear and tear. This real-time data is integrated into the digital twin.
   - **Use Case Example**: Predictive maintenance analysis within the digital twin can forecast when medical equipment will require servicing or calibration based on real-time data and historical performance. Healthcare providers can schedule maintenance to ensure equipment reliability.

In these examples, Digital twins, in conjunction with big data integration, enable predictive maintenance and analysis. They use historical and real-time data to predict maintenance needs, enabling organizations to schedule maintenance proactively, minimize downtime, and optimize asset performance and longevity.

Digital twin technology, in conjunction with the utilization of big data, assumes a pivotal role in the realm of maintenance and predictive analysis [29]–[33]. The domain of digital twin engineering fabricates a dynamic data model of industrial assets, which is adaptable to alterations in the surrounding environment and operations through the application of real-time sensory data. Functioning as an entity, the digital twin is capable of proactively detecting potential issues pertaining to its tangible counterpart while simultaneously prognosticating the remaining lifespan of the asset by capitalizing on models rooted in the principles of physics, as well as data-driven analytics. By virtue of digital twins and sophisticated big data, a sophisticated framework is established for the purpose of smart predictive maintenance, thereby furnishing comprehensive and efficacious tools for diagnosing faults and monitoring the health of assets. Employing data fusion, ML, artificial intelligence (AI), and the analysis of big data, digital twin simulation, analysis, and prediction are rendered actionable within the context of predictive maintenance. Consequently, these advanced technologies facilitate the exploration and amalgamation of data originating from a multitude of sensors, thereby enabling the construction of predictive maintenance models tailored to the diverse manufacturing industries.

Liu et al. [29] explore the application of digital twin technology in the context of predictive maintenance. However, the authors do not explicitly address the utilization of big data in relation to maintenance and predictive analysis.

Pulcini et al. [30] propose a smart predictive maintenance framework that leverages digital twins and smart big data for the purpose of predictive maintenance and health monitoring.

Guc et al. [31] put forward a smart predictive maintenance framework that employs digital twins and smart big data for the purpose of predictive maintenance and health monitoring.

Pulcini et al. [32] discuss the utilization of data fusion and ML algorithms for the predictive maintenance of industrial assets through digital twin simulation and analysis. However, the authors do not specifically mention the integration of digital twin technology with big data for maintenance and predictive analysis.

Avornu et al. [33] examine the use of data fusion and ML algorithms for predictive maintenance of industrial assets through digital twin simulation and analysis. However, the authors do not specifically mention the integration of digital twin technology with big data for maintenance and predictive analysis.

The notable weaknesses or drawbacks in the approach used in these papers are as follows:

- The papers do not mention any notable weaknesses or drawbacks in the approach used in their research.

Future research directions or unanswered questions suggested by these papers are as follows:

The papers put forth potential avenues for future research in data fusion pertaining to the field of predictive maintenance, with a specific focus on the utilization of digital twin technology. These papers emphasize the necessity of a comprehensive and efficient predictive maintenance framework that integrates smartness and leverages industrial AI technologies. Additionally, the papers suggest the adoption of smart big data and the creation of digital twin models for the purpose of predictive maintenance. Furthermore, the application of digital twin technology in proactively identifying potential issues and predicting the remaining useful life of industrial assets is also discussed.

7.6.1 Supply Chain and Manufacturing

Supply chain and manufacturing processes can benefit significantly from the combination of digital twin technology and big data. These technologies enable organizations to optimize their operations, enhance efficiency, and make data-driven decisions. Here are examples of how digital twins and big data are applied to supply chain and manufacturing:

1. **Supply Chain Optimization**
   - **Digital Twins**: Digital twin models the entire supply chain, including suppliers, manufacturing facilities, warehouses, transportation, and distribution networks.
   - **Big Data Integration**: Data from IoT devices, RFID tags, and inventory management systems continuously provide real-time data on supply chain activities. This data is integrated into the digital twin.
   - **Use Case Example**: By analyzing historical and real-time data in the Digital Twin, predictive models can optimize inventory levels, transportation routes, and order fulfillment. This leads to reduced costs, minimized stockouts, and improved customer satisfaction.
2. **Manufacturing Process Optimization**
   - **Digital Twins**: Digital twin models a manufacturing process, including machinery, assembly lines, and quality control.
   - **Big Data Integration**: Sensors on machines collect data on production rates, quality metrics, and equipment performance. This real-time data is integrated into the digital twin.
   - **Use Case Example**: By analyzing historical and real-time data in the digital twin, optimization models can adjust production parameters and quality control processes to maximize output and minimize defects. This improves production efficiency and product quality.
3. **Warehouse Management and Inventory Control**
   - **Digital Twins**: A digital twin models a warehouse, including layout, shelving, and storage areas.
   - **Big Data Integration**: IoT sensors, barcode scanners, and RFID technology continuously provide real-time data on inventory levels, order picking times, and storage conditions. This data is integrated into the digital twin.
   - **Use Case Example**: Predictive models within the digital twin can optimize warehouse layout and inventory placement based on historical and real-time data. This reduces order fulfillment times, minimizes errors, and saves storage space.

4. **Quality Control in Manufacturing**
   - **Digital Twins**: A digital twin models a manufacturing facility and the production process.
   - **Big Data Integration**: Sensors and cameras on production lines continuously collect data on product dimensions, visual defects, and performance metrics. This data is integrated into the digital twin.
   - **Use Case Example**: Real-time analysis within digital twin can monitor product quality and deviations from standards. Predictive models can detect quality issues and trigger immediate adjustments to the manufacturing process to reduce defects and improve product quality.
5. **Supplier Network Optimization**
   - **Digital Twins**: A digital twin represents the entire supplier network, including suppliers, lead times, and inventory levels.
   - **Big Data Integration**: Data from supplier performance metrics, lead times, and real-time inventory data are integrated into the digital twin.
   - **Use Case Example**: By conducting an analysis of both historical and contemporaneous data within the digital twin, prognostic models can enhance the process of supplier identification and estimation of delivery periods. This capacity enables organizations to formulate judicious determinations regarding supplier affiliations, consequently diminishing expenses and guaranteeing a steadfast supply chain.

Digital twin technology, when combined with the integration of big data, holds significant implications for the domains of supply chain and manufacturing. The utilization of digital twins enables the creation of virtual models that accurately replicate physical assets and processes, thereby facilitating real-time monitoring, analysis, and optimization. Through the utilization of big data, digital twins have the capacity to capture and analyze extensive amounts of information from various sources, including sensors, machines, and systems, thereby providing valuable insights and predictive capabilities. This amalgamation of digital twin technology and big data integration not only enables the enhancement of supply chain planning, resource utilization, and operational performance, but also facilitates internal integration and collaboration within the manufacturing value chain, thereby leading to an improved creation of value and enhanced efficiency. Moreover, the implementation of digital twin technology offers a multitude of benefits in terms of reliability, efficiency, and productivity within the realm of global production, rendering it an invaluable tool in the contexts of supply chain planning and asset management. Ultimately, the combination of digital twin technology, big data, and supply chain management possesses the potential to drive innovation, enhance decision-making processes, and optimize performance across the manufacturing and supply chain domains.

Wang et al. [34] explore the utilization of manufacturing big data for the analysis of the endogenous driving mechanism of the supply chain and industrial chain derived from the 4.0 value chain of manufacturing enterprises. They propose a framework for the enterprise digital endogenous integration tower three-dimensional (3D) process and analyze the nature and derivation mechanism of the supply chain based on the 4.0 value chain drive. However, the authors do not specifically address the integration of digital twin technology with big data within the context of supply chain and manufacturing.

Li and Li [35] examine the application of digital twin technology in the analysis of big data pertaining to enterprise supply chains under the green financial system. Their focus lies on the impact of digital twin technology on corporate finance, market, and operational performance.

Khan et al. [36] discuss the utilization of digital twin technology in supply chain planning, albeit without explicit mention of the utilization of big data in conjunction with digital twins.

Ouahabi et al. [37] discuss the utilization of digital twin technology in supply chain planning, albeit without explicit mention of the utilization of big data in conjunction with digital twins.

The notable weaknesses or drawbacks in the approach used in these papers are as follows:

- The approach used in these papers is focused on bibliometric analysis and mapping of scientific publications in the field of supply chain digital twins.
- However, there are no specific mentions of weaknesses or drawbacks in the approach used in these papers in the given abstracts and contexts.

Future research directions or unanswered questions suggested by these papers are as follows:

The papers put forward potential avenues for future research and areas that have not yet been explored within the realm of the supply chain digital twin. One possibility for future research involves combining the study of IoT for Sustainability in Agriculture with the digital twin concept. Another potential direction for research could involve investigating the utilization of digital twin technology in the examination of large data sets within enterprise supply chains, particularly within the context of the green financial system. Moreover, the papers emphasize the necessity for further research into the internal driving mechanism of the supply chain and industrial chain that stems from the 4.0 value chain of manufacturing enterprises. In addition, there is a recommendation for future research on the digital twin framework for the Distribution Centre (DC) and its advantages and capabilities compared to existing planning methods. All in all, these papers offer valuable insights into possible future research avenues and unanswered queries within the domain of supply chain digital twin, encompassing the integration of IoT and sustainability, the utilization of digital twin technology in enterprise supply chains, and the analysis of the internal driving mechanism of the supply chain within manufacturing enterprises.

7.6.2 Urban Planning and Smart Cities

The integration of digital twin technology in conjunction with big data harbors the capability to revolutionize the realm of urban planning and the advancement of intelligent cities. This amalgamation empowers urban strategists to fabricate simulated replicas of metropolis regions, meticulously track instantaneous data from diverse origins, and judiciously deliberate to enhance urban infrastructure, ecological soundness, and quality of life. Here are examples of how digital twins and big data are applied to urban planning and smart cities:

1. **Traffic Management**
   - **Digital Twins**: A digital twin models a city's transportation system, including road networks, traffic signals, and public transportation infrastructure.
   - **Big Data Integration**: Data sources such as traffic cameras, GPS systems, social media feeds, and sensors on vehicles provide real-time data on traffic conditions, accidents, and public transportation schedules. This data is integrated into the digital twin.
   - **Use Case Example**: Predictive models within the digital twin can analyze historical and real-time traffic data to optimize traffic signal timings, reroute public transportation, and provide real-time traffic updates to commuters. This results in reduced congestion, improved traffic flow, and enhanced mobility in the city.
2. **Energy Efficiency**
   - **Digital Twins**: A digital twin replicates a smart city's energy grid, including power generation, distribution, and consumption.
   - **Big Data Integration**: Sensors on the grid, smart meters, and weather forecasts continuously provide real-time data on energy consumption, production, and weather conditions. This data is integrated into the digital twin.
   - **Use Case Example**: Predictive analysis within the digital twin can optimize energy distribution and storage, adjusting energy generation sources based on real-time data and weather forecasts. This reduces energy waste, increases the use of renewable energy sources, and enhances the stability of the energy grid.
3. **Urban Infrastructure Development**
   - **Digital Twins**: A digital twin models urban infrastructure, including buildings, utilities, and transportation systems.
   - **Big Data Integration**: Data from construction equipment, sensors, and satellite imagery provide real-time information on infrastructure development and usage. This data is integrated into the digital twin.
   - **Use Case Example**: Real-time monitoring within the digital twin can track the progress of construction projects, assess the integrity of buildings, and optimize the use of resources, ensuring timely project completion, safety, and cost efficiency.
4. **Public Safety and Emergency Response**
   - **Digital Twins**: A digital twin replicates a city's emergency response system, including law enforcement, fire departments, and medical services.
   - **Big Data Integration**: Data from emergency calls, sensors, and social media provide real-time information on incidents, response times, and resource allocation. This data is integrated into the digital twin.

- **Use Case Example**: Predictive models within the digital twin can analyze real-time data to improve emergency response coordination, optimize resource allocation during disasters, and ensure timely assistance to residents in need.

5. **Waste Management**
   - **Digital Twins**: A digital twin models waste collection and disposal processes in a smart city.
   - **Big Data Integration**: Sensors in waste bins and vehicles collect data on waste levels, collection routes, and disposal efficiency. This data is integrated into the digital twin.
   - **Use Case Example**: Real-time analysis conducted within the digital twin has the potential to enhance the optimization of waste collection routes, diminish fuel consumption, and ameliorate the overall efficiency of waste disposal services. Consequently, this will play a significant role in fostering cleaner and more sustainable urban environments.

Digital twin technology, when combined with big data, plays a pivotal role in the realm of urban planning and the advancement of smart cities. By harnessing the power of data understanding, analysis, and projection, digital twin models possess the capability to offer predictive insights into the future development of cities with regard to sustainability. These models integrate cutting-edge sensing technologies that rely on deep learning, algorithms that enhance spatial cognition, and the fusion of multiple sensors to create virtual spaces that are highly immersive and facilitate sustainable urban governance. Moreover, web-based visualization technology serves to enhance the visualization of vast amounts of spatial-temporal data, thereby making it a valuable tool for decision-making and management in the context of smart cities. The amalgamation of digital twins, the IoT, fifth-generation wireless systems, and AI technologies present immense potential for transforming urban governance and propelling cities toward becoming smart cities. To construct digital twins, methods such as data mining and neural network modeling are employed, which optimize processes and enable the prediction and analysis of urban challenges.

In the study by Barresi [38], the author explores the utilization of urban digital twins as a predictive model for fostering sustainable urban development. However, the author fails to explicitly address the integration of digital twin technology with big data within the realm of urban planning and smart cities.

In the work of Nica et al. [39], the authors delve into the potential of digital twin technology when combined with big data analytics, remote sensing systems, and geospatial tools. This integration facilitates intelligent and sustainable urban planning in the context of smart cities.

In their publication [40], the authors discuss the possibilities presented by web-based visualization technology in enhancing the presentation of extensive spatial-temporal data within urban digital twin projects. This advancement proves valuable for decision-making, planning, and management in smart cities.

Conversely, in another publication [41], the authors do not explicitly address the integration of digital twin technology with big data within the context of urban planning and smart cities.

Erken [42] examines various methodologies for constructing digital twins of smart cities. However, the author neglects to specifically mention the use of digital twin technology in conjunction with big data within the realm of urban planning and smart cities.

The notable weaknesses or drawbacks in the approaches used by these papers are as follows:

- The papers do not explicitly mention any weaknesses or drawbacks in their analyses.

Future research directions or unanswered questions suggested by these papers are as follows:

The papers propose various prospective avenues for research and unresolved inquiries in the realm of digital twins and smart cities. One potential avenue for future research involves examining the role of digital twins in expediting the digital transformation of smart cities. This includes investigating the emergence of digital twin cities and the utilization of digital twin technology in urban governance. Another possible direction for future research entails enhancing web-based visualization technology for urban digital twins, with a particular focus on effectively visualizing extensive spatial-temporal data. This encompasses exploring pivotal technologies, commonly employed tools, and diverse visualization methods for varying types of data. Furthermore, the papers advocate for additional research on integrating digital twin simulation tools, spatial cognition algorithms, and multi-sensor fusion technology into sustainable urban governance networks. This entails investigating how these technologies can optimize smart city environments based on the IoT and facilitate efficient urban governance. Moreover, the papers underscore the significance of digital innovation and data-driven planning technologies in realizing sustainable cities. Future research endeavors could concentrate on developing and implementing urban digital twins as predictive models for forecasting the sustainability-related development of cities. Together, these papers offer valuable insights into different research avenues and open questions in the domain of digital twins and smart cities, thereby demonstrating the potential for further exploration and progress in this field.

7.6.3 Healthcare and Personalized Medicine

Digital twin technology that is integrated with big data has the potential to revolutionize healthcare and personalized medicine. It enables the creation of virtual representations of patients and their medical conditions, allowing for real-time monitoring and predictive analysis. Here are a few examples of how digital twins and big data are applied in healthcare and personalized medicine:

1. **Patient Health Monitoring**
    - **Digital Twins**: A digital twin models a patient's physiology, medical history, and treatment plans.
    - **Big Data Integration**: Wearable sensors, electronic health records, and medical data continuously provide real-time information on the patient's

vital signs, medication adherence, and relevant medical data. This data is integrated into the patient's digital twin.
- **Use Case Example**: Predictive models within the digital twin can analyze real-time health data to forecast disease progression, detect early warning signs, and recommend adjustments to the treatment plan. This enables healthcare providers to deliver personalized care and improve patient outcomes.

2. **Surgical Planning**
   - **Digital Twins**: In surgical planning, a patient's digital twin models a specific body part or organ.
   - **Big Data Integration**: Medical imaging data, such as computed tomography (CT) or magnetic resonance imaging scans (MRIs), is integrated into the digital twin to create an accurate 3D representation of the organ or area that requires surgery.
   - **Use Case Example**: Surgeons can use the digital twin to simulate the surgical procedure, assess potential complications, and plan the optimal surgical approach, enhancing precision and reducing surgical risks.
3. **Drug Discovery and Development**
   - **Digital Twins**: In drug development, a digital twin models a specific disease or biological process.
   - **Big Data Integration**: Genomic data, clinical trial results, and molecular data are integrated into the digital twin to simulate disease mechanisms and test drug compounds.
   - **Use Case Example**: Pharmaceutical companies can use the digital twin to test potential drug candidates in a virtual environment, accelerating the drug discovery process and personalizing treatments for specific patient populations.
4. **Remote Patient Care**
   - **Digital Twins**: A digital twin models a patient's health status, home environment, and treatment devices.
   - **Big Data Integration**: IoT devices, telehealth data, and wearable technology continuously provide real-time data on the patient's condition, medication adherence, and treatment device performance. This data is integrated into the patient's digital twin.
   - **Use Case Example**: Predictive analysis within the digital twin can monitor the patient's health and the performance of medical devices, alerting healthcare providers to potential issues, enabling remote adjustments to treatment plans, and enhancing patient comfort and safety.
5. **Genomic and Personalized Medicine**
   - **Digital Twins**: A patient's digital twin models their genomic data, medical history, and responses to specific treatments.
   - **Big Data Integration**: Genomic sequencing data and clinical records are integrated into the digital twin to personalize treatment plans based on the patient's genetic makeup.

- **Use Case Example**: By analyzing genomic data and treatment outcomes, the digital twin can recommend personalized therapies, reducing adverse reactions and improving treatment effectiveness.

In these examples, digital twins combined with big data integration empower healthcare providers and researchers to personalize medical treatments, improve diagnostic accuracy, and optimize healthcare delivery. This synergy leads to more effective healthcare solutions, better patient outcomes, and the acceleration of medical research and innovation.

Digital twin technology, in conjunction with big data, has the potential to bring about a paradigm shift in the field of healthcare and personalized medicine. By harnessing the capabilities of digital twins, which serve as a conduit between the physical and virtual realms, healthcare professionals are able to optimize processes and make well-informed decisions without the need for immediate implementation of changes. The critical role played by big data in this transformation lies in its capacity to gather and analyze invaluable information, thereby enabling the creation of more refined digital twin models. This fusion of digital twin technology and big data has the capacity to yield enhancements in healthcare facilities, the generation of personalized HDTs, and the development of customized and sustainable therapies. By combining these technologies, real-time analytics, personalized medicine, early detection and intervention, and optimal surgical planning can become a tangible reality.

In their study, [43] presents a framework that proposes the utilization of digital twin technology in the healthcare domain through the organization of patient health data into a knowledge graph and the application of closed-form continuous-time liquid neural networks for real-time analytics. However, the study does not explicitly mention the incorporation of big data in healthcare or personalized medicine.

Contrarily, [44] does not explicitly address the integration of big data with digital twin technology in the context of healthcare and personalized medicine.

In [45], the authors delve into the application of digital twin technology in healthcare, including personalized medicine. The study highlights the utilization of big data in healthcare process design, resource allocation, and the prediction of medicinal activities.

The discourse presented in [46] revolves around the implementation of digital twin and big data in healthcare to enhance processes and establish a more individualized HDT.

Similarly, [47] explores the potential of digital twin and big data in healthcare for process improvement and the creation of a more personalized HDT.

The approach employed in these papers exhibits certain deficiencies and drawbacks. One limitation pertains to the high computational expenses and modeling intricacies that accompany the modeling of complex diseases using multi-modal patient data. This issue has impeded the widespread adoption of digital twin technology in the biomedical realm. Another shortcoming concerns the challenge of safeguarding patient privacy and ensuring data availability, particularly in sensitive healthcare domains such as lung cancer treatment, when generating synthetic data for digital twins. Moreover, there may exist limitations in the accuracy and dependability of the synthetic data generated for lung cancer patients, in comparison to real patient data. The integration of e-health systems with cloud technology to create intelligent objectives and solutions may also pose challenges in terms of privacy and security within cloud-based e-health systems. Furthermore, the implementation of digital twin technology in healthcare may necessitate substantial resources and infrastructure to support real-time analytics and personalized medicine.

The papers put forth several future research directions and unanswered questions within the field of digital twin technology in healthcare. One promising avenue for future research lies in exploring how to achieve personalized health monitoring throughout the lifespan of elderly patients, as well as how to integrate the material reality of healthcare with the digital reality to realize genuine, smart medicine. Another area that warrants investigation is the development of secure data-driven smart healthcare using digital twins. This would entail exploring, identifying, comprehending, and evaluating health information generated from electronic databases. The papers also suggest the need for further research on the integration of digital twin technology with virtual and augmented reality, as well as ML technologies, in order to enable personalized therapeutics and minimally invasive intervention procedures. Additionally, research is needed to investigate new hardware solutions in medical imaging and their advantages for organ-specific applications. This research should also involve comparing and evaluating different imaging methodologies and detector technologies. The papers also emphasize the importance of researching the structuring of patient health data as a knowledge graph and the utilization of closed-form continuous-time liquid neural networks for real-time analytics. This research aims to overcome the computational costs and modeling complexities associated with digital twin modeling in healthcare.

REFERENCES

[1] Cristina, Martínez-Ruedas., J.M., Flores-Arias., Isabel, M., Moreno-Garcia., Matias, Linan-Reyes., F.J., Bellido-Outeiriño. (2024). 1. A Cyber–Physical System Based on Digital Twin and 3D SCADA for Real-Time Monitoring of Olive Oil Mills. Technologies (Basel), doi: 10.3390/technologies12050060

[2] Maximilian, Kannapinn., M., Schafer., Oliver, Weeger. (2024). 1. TwinLab: a framework for data-efficient training of non-intrusive reduced-order models for digital twins. Engineering Computations, doi: 10.1108/ec-11-2023-0855

[3] E., Mohanraj., N, Eniyavan., S, Sidarth., Shankar, Sridharan. (2024). 1. Digital Twins for Automotive Predictive Maintenance. doi: 10.1109/icict60155.2024.10544392

[4] Zeyu, Sun., Runtong, Zhang., Xiaomin, Zhu. (2024). 1. The progress and trend of digital twin research over the last 20 years: A bibliometrics-based visualization analysis. Journal of Manufacturing Systems, doi: 10.1016/j.jmsy.2024.02.016

[5] Malithi, R., Abayadeera., Gamage, Upeksha, Ganegoda. (2024). 1. Digital Twin Technology: A Comprehensive Review Exploring the Potential, Evolution, Applications, and Future of Digital TwinTechnology. International Journal of Innovative Science and Research Technology, doi: 10.38124/ijisrt/ijisrt24jun425

[6] Vignesh, Murugan. (2024). 1. Unlocking the potential of digital twins: Transforming hospital practices for better care. Journal of Community Health Management, doi: 10.18231/j.jchm.2024.011

[7] Vignesh, Murugan. (2024). 1. Unlocking the potential of digital twins: Transforming hospital practices for better care. Journal of Community Health Management, doi: 10.18231/j.jchm.2024.011

[8] Michaela, Cellina, Maurizio, Cè, Marco, Alì, Giovanni, Irmici, Elena, Caloro, Deborah, Fazzini, Giancarlo, Oliva, Sergio, Papa. (2023). Digital Twins: The New Frontier for Personalized Medicine? Applied Sciences, doi: 10.3390/app13137940.

[9] Hui, Shan, Lock., Pui, Yee, Tan., Chi-Fai, Ng., J.W.L., Ooi. (2024). 1. Exploring the potential of digital twin technology as a training tool for new radiographers. Journal of Medical Imaging and Radiation Sciences, doi: 10.1016/j.jmir.2024.05.004

[10] Christina, Joseph, Jyothula., C., Kishor, Kumar, Reddy., Thandiwe, Sithole. (2024). 1. Digital Twins in Healthcare. Advances in medical technologies and clinical practice book series, doi: 10.4018/979-8-3693-5893-1.ch003

[11] Michelle, Callista, Hendra, Tjahyadi. (2022). Digital Twin and Big Data in Healthcare. doi: 10.1109/ICTIIA54654.2022.9935847

[12] Xingxing, Zhang, Jingchun, Shen, Puneet, Kumar, Saini, Marco, Lovati, Mengjie, Han, Pei, Huang, Zhihua, Huang. (2021). Digital Twin for Accelerating Sustainability in Positive Energy District: A Review of Simulation Tools and Applications. doi: 10.3389/FRSC.2021.663269

[13] Qinglin, Qi, Fei, Tao. (2018). Digital Twin and Big Data Towards Smart Manufacturing and Industry 4.0: 360 Degree Comparison. IEEE Access, doi: 10.1109/ACCESS.2018.2793265

[14] HanYong, Hao, Jinkui, Huang, Fucai, Luo, Shi, Liang, Zhang, Chu, Wang, Zhongfeng, Wang. (2022). Digital Power System Data Transmission Framework and Configuration Design. doi: 10.1109/sdpc55702.2022.9915863

[15] M. Yaseen, P. Durai, P. Gokul, S. Justin and A. J. Anand (2023). Artificial Intelligence Based Automated Appliances in Smart Home. 2023 7th International Conference on Image Information Processing (ICIIP), Solan, India, pp. 442–445.

[16] Leila, kamyabi, Tek, Tjing, Lie, Samaneh, Madanian. (2022). Applications of Digital Twins in Power Systems: A Perspective. TESEA, Transactions on Energy Systems and Engineering Applications, doi: 10.32397/tesea.vol3.n2.484

[17] Shunichi, Koshimura, Erick, Mas. (2023). Digital Twin Computing for Enhancing Resilience of Disaster Response System. doi: 10.5194/egusphere-egu23-11756

[18] Mei, Liu, Benkuan, Wang, Datong, Liu. (2020). A Digital Twin Modeling Method for Turbofan Engine Real-time Test Data Analysis and Performance Monitoring. doi: 10.1109/PHM-JINAN48558.2020.00086

[19] Agus, Hasan., Augie, Widyotriatmo., Eirik, Fagerhaug., Ottar, L., Osen. (2023). 1. Predictive digital twins for autonomous surface vessels. Ocean Engineering, doi: 10.1016/j.oceaneng.2023.116046

[20] Desai, A., Navaneeth, N. V., Samar, Adhikari, Subrata, Chakraborty. (2023). Enhanced Multi-Fidelity Modelling for Digital Twin and Uncertainty Quantification. arXiv.org, doi: 10.48550/arXiv.2306.14430

[21] D., J., Santosh, Kumar., Syed, B., Alam. (2024). 1. AI-driven non-intrusive uncertainty quantification of advanced nuclear fuels for digital twin-enabling technology. Progress in Nuclear Energy, doi: 10.1016/j.pnucene.2024.105177

[22] Sujing, Zhou. (2022). Numerical Analysis of Digital Twin System Modeling Methods Aided by Graph-Theoretic Combinatorial Optimization. Discrete Dynamics in Nature and Society, doi: 10.1155/2022/8598041

[23] Marco, Becattini., Davide, Borsatti., Armir, Bujari., Laura, Carnevali., Andrea, Garbugli., Hrant, Khachatrian., Theofanis, P., Raptis., Daniele, Tarchi. (2022). Digital Twin for Networking: A Data-driven Performance Modeling Perspective. (2022). https://doi.org/10.48550/arxiv.2206.00310

[24] Safiye, Turgay, Ömer, Bilgin, Necip, Akar. (2022). Digital Twin Based Flexible Manufacturing System Modelling with Fuzzy Approach. Advances in Computer, Signals and Systems, doi: 10.23977/acss.2022.060702

[25] Zheyu, Zhang, Jinyang, Jiang, Xiao, Cheng. (2022). Research on Digital Twin Model of Energy Equipment. doi: 10.1109/ICPECA53709.2022.9718834

[26] Meng, Zhang, Fei, Tao, Biqing, Huang, Ang, Liu, Lihui, Wang, Nabil, Anwer, Andrew, Y., Nee, C. (2021). Digital Twin Data: Methods and Key Technologies. doi: 10.12688/DIGITALTWIN.17467.1

[27] Ferranti, F., Flavio, Manenti, Vingerhoets, G., Mattia, Vallerio. (2021). Value Chain Planning Optimization: A Data Driven Digital Twin Approach. IFAC-PapersOnLine, doi: 10.1016/J.IFACOL.2021.08.303

[28] Shumin, Sun, Jiawei, Xing, Peng, Yu, Ya, Cheng, Ruiqi, Wang, Yuejiao, Wang. (2022). Energy Efficiency Optimization Technology of Integrated Energy System Based on Digital Twin Technology. Journal of Physics: Conference Series, doi: 10.1088/1742-6596/2409/1/012015

[29] Zheng, Liu, Erik, P., Blasch, Min, Liao, Chunsheng, Yang, Kazuhiko, Tsuda, Norbert, Meyendorf. (2023). Digital Twin for Predictive Maintenance. doi: 10.1117/12.2660270

[30] Valerio, Pulcini., Gianfranco, E., Modoni. (2024). 1. Machine Learning-Based Digital Twin of a Conveyor Belt for Predictive Maintenance. The International Journal of Advanced Manufacturing Technology, doi: 10.1007/s00170-024-14097-3

[31] Guc, F. A., Yangquan, Chen. (2022). Smart Predictive Maintenance Enabled by Digital Twins and Smart Big Data: A New Framework. doi: 10.1109/DTPI55838.2022.9998937

[32] S. Hemalatha, M. Mahalakshmi, V. Vignesh, M. Geethalakshmi, D. Balasubramanian and Jose A. A. (2023). Deep Learning Approaches for Intrusion Detection with Emerging Cybersecurity Challenges. 2023 International Conference on Sustainable Communication Networks and Application (ICSCNA), Theni, India, 2023, pp. 1522–1529.

[33] Christian, Avornu, Samuel, Dodzi, Akama, Eric, Kofi, Avornu, Valentine, Aveyom. (2022). Data Fusion for Predictive Maintenance of Industrial Assets Using Digital Twin. doi: 10.23919/EECSI56542.2022.9946575

[34] Feng, Wang, Cao, Zhe, Mingwei, Sun. (2023). A Study of Digital Twin-Based Digital Derivation Mechanisms for Manufacturing Companies. doi: 10.1109/CISCE58541.2023.10142669

[35] Dongsheng, Li, Jian, Li. (2023). Big Data of Enterprise Supply Chain Under Green Financial System Based on Digital Twin Technology. Kybernetes, doi: 10.1108/k-02-2023-0291

[36] Sameeruddin, Md., Khan, R., Chinnaiyan, S., Balachandar, Shamim, Ibrahim, N. S., Kalyan, Chakravarthy, Kalaiarasan, C. (2022). Centralized Digital Twin driven Supply Chain for Sustainable Smart Manufacturing Plants. doi: 10.1109/ICCMSO58359.2022.00077

[37] Nada, Ouahabi., Ahmed, Chebak., Oulaïd, Kamach., Oussama, Laayati, Mourad, Zegrari. (2024). 1. Leveraging digital twin into dynamic production scheduling: A review. Robotics and Computer-integrated Manufacturing, doi: 10.1016/j.rcim.2024.102778

[38] Alessandra, Barresi. (2023). Urban Digital Twin and urban planning for sustainable cities. Techne. Journal of Technology for Architecture and Environment, doi: 10.36253/techne-13568

[39] Elvira, Nica, Gheorghe, H., Popescu, Miloš, Poliak, Tomas, Kliestik, Oana, Matilda, Sabie. (2023). Digital Twin Simulation Tools, Spatial Cognition Algorithms, and Multi-Sensor Fusion Technology in Sustainable Urban Governance Networks. Mathematics, doi: 10.3390/math11091981

[40] Sanjay, Somanath., Vasilis, Naserentin., Orfeas, Eleftheriou., Daniel, Sjölie., Beata, Stahre, Wästberg., Anders, Logg. (2024). 1. Towards Urban Digital Twins: A Workflow for Procedural Visualization Using Geospatial Data. Remote sensing, doi: 10.3390/rs16111939

[41] Dechen, Peldon., Saeed, Banihashemi., Khuong, LeNguyen., Sybil, Derrible. (2024). 1. Navigating urban complexity: The transformative role of digital twins in smart city development. Sustainable Cities and Society, doi: 10.1016/j.scs.2024.105583

[42] Elif, Erken. (2023). Comparative Analysis of Methods and Algorithms for Building a Digital Twin of a Smart City. doi: 10.1007/978-3-031-34127-4_27

[43] Dechen, Peldon., Saeed, Banihashemi., Khuong, LeNguyen., Sybil, Derrible. (2024). 1. Navigating urban complexity: The transformative role of digital twins in smart city development. Sustainable Cities and Society, doi: 10.1016/j.scs.2024.105583

[44] Luigi, Landini, Vincenzo, Positano, Maria, Filomena, Santarelli, Simona, Celi. (2023). Digital Twin Technology: New Frontiers for Personalized Healthcare. Electronics, doi: 10.3390/electronics12081921

[45] Christina, Joseph, Jyothula., C., Kishor, Kumar, Reddy., Thandiwe, Sithole. (2024). 1. Digital Twins in Healthcare. Advances in medical technologies and clinical practice book series, doi: 10.4018/979-8-3693-5893-1.ch003

[46] Christina, Joseph, Jyothula., C., Kishor, Kumar, Reddy., Thandiwe, Sithole. (2024). 1. Digital Twins in Healthcare. Advances in medical technologies and clinical practice book series, doi: 10.4018/979-8-3693-5893-1.ch003

[47] Michelle, Callista, Hendra, Tjahyadi. (2022). Digital Twin and Big Data in Healthcare. doi: 10.1109/ICTIIA54654.2022.9935847

8 Unlocking Value
Big Data, Digital Twins, Exploring the Future Digital Transformation and Its Trends

G. Arun Sampaul Thomas, S. Muthukaruppasamy, J. Nandha Gopal and Saravanan Krishnan

8.1 INTRODUCTION

The world is undergoing a rapid digital transformation, driven by the increasing adoption of big data and digital twins. These technologies are having a profound impact on businesses, industries, and societies, enabling new levels of innovation, efficiency, and productivity. Big data is the collection, storage, analysis, and processing of large amounts of both structured and unstructured data. A digital twin is a digital copy of a real-life object or system that can be used for testing, analysis, and optimization purposes.

Together, digital twins and big data are unlocking new value in a wide range of solicitations, from manufacturing and healthcare to transportation and energy. This chapter explores the future of digital transformation and its trends, with a focus on the role of big data and digital twins. Anything, including people, houses, vehicles, buildings, and factories, can have a digital doppelganger. A famous example of this is the computerized model of an automobile's engine, brakes, and body used in engineering. Building a digital prototype and running data-driven two-dimensional (2D) and three-dimensional (3D) simulations are both made possible by this. One well-known application is object reproduction in video games, where a new generation is developing in these simulated and three-dimensional environments. Reproduction of items is a popular application in virtual fitting rooms for eyeglasses, lipstick, and furniture [1].

The idea of a digital twin has evolved to include the replication of any procedure, advertisement, product launch, customer habit, or abstract system. For the sake of running simulations, why not create a new consumer with all their past data and outcomes? An imaginary customer. The United States (US) is home to the earliest "Digital Twin of a Customer" studies and publications.

 DOI: 10.1201/9781003465904-10

The fundamental idea of creating a digital representation of the real world has been around for a long time, though. The digital twins are descended from agent-based models, also known as computer recreations and multi-agent structures. The purpose of artificial intelligence (AI) research from Alan Turing onwards has been to create machines that can do tasks traditionally associated with humans and other "intelligent" beings. Since the 1990s, people have been using the term an "intelligent agent" to describe anything or a system that can sense its surroundings through sensors, make judgments based on those perceptions, and then manipulate its surroundings using actuators to accomplish predetermined goals. The study of intelligent agents is a common definition of AI. An early intelligent agent, Shakey, developed by SRI International, navigates physical environments by first digitally modeling them and then determining the best route to take. Shakey the robot exemplified the kind of physical-digital synchronization that members of the digital twin group speak about. Like with modern digital twins modeling Internet of Things (IoT) sensors, early clarities of "intelligent agents" cited sensors as one of the most basic types of agents [2].

The digital twin makes it possible to conduct in-depth analyses by utilizing the power of big data. Assisting in the identification of prospective problems, the prevention of downtime, the testing of new business opportunities, the planning of future scenarios, and the customization of production based on client requirements are all benefits that may be gained via simulation.

The use of technology enables instant input regarding ongoing activities, which may then be utilized to implement adjustments in the future in a record amount of time. Digital twins are especially helpful for the maintenance of linked heavy machines and infrastructure that create and analyze massive volumes of data due to the capabilities that they possess. The applicability of digital twins, on the other hand, extends beyond this.

Simulations on computers have been around since the early days of computing, when Jon von Neumann, a renowned mathematician, and Stanislaw Ulam, a colleague of his, used computer simulations to determine how neutrons behave. The aerospace and military industries were the first to make extensive use of computer simulations. A growing number of scientific, engineering, and commercial domains began to use computer simulations as computing grew more pervasive. The idea of creating intelligent agents that could "imitate" real-world objects was central to agent-based simulations, which aimed to test such beings in a virtual setting by simulating their physical surroundings [3].

Both the modern idea of digital twins and the older, more traditional idea of intelligent agents have five essential functions.

Explanation: What occurred in the past can be better understood with the help of digital twins. With sensor-based digital twins, for instance, it is possible to examine the data for outliers and trends.

Future behavior can be anticipated with the help of digital twins. For instance, it is possible to forecast future values based on the past actions of a particular sensor.

Investigate: The digital twin really shines when it enables us to investigate virtual alternatives and assess them according to our specifications.

Revolution: The true potential of digital twins becomes apparent during the process of transforming the real world by applying the things learned in the virtual one.

While the investigation process has different possibilities with past data, it comes up with possible values to put the real world to the test. When dealing with data that is either missing or has significant uncertainty, digital twins provide a strong tool to work with. It appears that digital twins and agent-based simulations have seen a dramatic uptick in usage since the pandemic. This is likely because businesses have discovered the numerous advantages of investigating different possible courses of the Covid-19 virus and its economic effects, such as a V-shaped, U-shaped, or W-shaped recovery. Almost all respondents (96%) want to utilize AI simulations this year, according to PWC's 2022 AI Business Survey [3]. A great deal has changed throughout the digital transformation era for industries, thanks to the proliferation of new technologies that are being used in production and operations daily. However, the promise of these technologies is brought to light by a body of knowledge known as big data [5].

Utilizing big data in tandem with technologies such as the digital twins, the IoT, and AI-based machine learning (ML) has allowed for the efficient interpretation of massive data sets, leading to the production of useful knowledge. Big data approaches enable industries to leverage all available data to pinpoint potential bottlenecks, enhance processes, discover faults, and integrate knowledge across domains [6].

Modeling the physical environment is not a new concept; it has been around for quite some time. This is the essence of mathematical applications. As electronic devices became more commonplace, researchers began to use a variety of wave functions on electronic black boxes to determine their internal architecture and component makeup. Feynman [7] first suggested the concept of utilizing quantum computers to model the "physical interaction" of atoms in 1982. The advancements that have been made in computer technology, data management, ML, automation, and the IoT (sensors that link the real and virtual realities) have made it possible for us to imitate the real world in virtual reality presently. The AI-based simulations are carried out in the virtual world with little to no risk, and this trend is being fueled by the insatiable human demand for speed and safety [7].

The digital twin can now model a more complex and interconnected environment than ever before because of advancements in analytics, the processing power of big data engines, the vast and adaptable storage options provided by the aggregation area, and the integration with canonical data. A more accurate and advanced model might result from these advancements, which could also lead to cheaper software and hardware. Adaptability and scalability in message numbers, analytics, sensors, processing, etc., should be key considerations while designing the conceptual architecture. The design can then adapt quickly to the ever-changing market, which can experience exponential growth at times.

8.2 BIG DATA AND DIGITAL TWINS: THE FUTURE OF DIGITAL TRANSFORMATION

The process of reshaping an organization or business through the utilization of digital technologies is referred to as digital transformation. Big data and digital twins are two of the most essential technologies that are driving digital transformation in the modern era.

Big data is assisting businesses in gathering, storing, and analyzing an unprecedented amount of data now more than ever before. This information can be used to acquire fresh perspectives on the behavior of customers, the effectiveness of operations, and the trends in the market. The creation of virtual representations of physical items and systems is being accomplished using digital twins. Simulating, monitoring, and optimizing the performance of these items and systems can be accomplished with the help of digital twins.

Big data and digital twins are working together to provide businesses with the ability to improve their decision-making processes, increase their efficiency, and innovate. When businesses are working toward the development of skills for digital twins, it is helpful to take into consideration the complete spectrum of maturity of digital twins. It is possible to construct a data twin, which is the most basic type of digital twin. An asset, a consumer, or an organizational entity can be represented by a digital log known as a data twin. The data twin can be constructed with either synthetic data or data taken from the real world. Real-time and non-real-time data twins are both possible types of data twins that can be used. A good illustration of a digital twin would be an ordinal clone of a water level sensor that is used to provide flood alerts. Because of the way the sensor is constructed and the way the data is transmitted, this digital twin will be able to trace the water levels at various moments in time. Every second, minute, hour, and so on might be used to measure the levels, or they could be based on values that have been predetermined. Real-time or periodic transmission of this data may take place elsewhere, or it could be stored and transmitted elsewhere. It is not totally necessary for us to always have data from the real world to construct a data twin. In addition, synthesized data can be produced to reflect a certain distribution, such as the rising water levels that occur during a flood [8].

A digital twin's history log assists in the beginning stages of mimicking the physical behavior of an asset. The asset simulation twin is the name given to this. Companies operating in the realm of the IoT in connection with Industry 4.0 are more likely to adopt this form of digital twin. Specification, comprehension, and evaluation of the physics of most physical assets are all possible. This includes everything from straightforward sensors to intricate systems, such as aircraft engines. An understanding of aberrant patterns and the ability to anticipate possible machine faults can be gained through the utilization of historical data from digital twins. Based on past data from pressure and temperature sensors located at the tip of oil drilling equipment, for instance, it is possible to forecast the breakdown forms of these sensors and proactively schedule maintenance, thereby saving millions of dollars in terms of reduced downtime.

As they progress from digital twins of physical assets to decision-making choices of cognitive entities (such as customers or organizational entities), the world of traditional AI is reachable, often known as intelligent agents. These are referred to as choice simulation twins. To both explain and predict customer preferences and decisions, the modeling of customer choice functions has been the subject of substantial research in the field of economics. AI has mostly focused on the study of rational beings and how they make decisions in an environment that is both unpredictable and constantly changing and is composed of other agents. As an illustration, the developed consumer choice simulation twins comprehend, forecast, and alter the behavior of how people react to product campaigns. These choice simulation digital twins offer marketing professionals a vast playground in which to assess various product promotion techniques and the impact such strategies have on revenues and profitability [9].

By combining ML with the choice functions of cognitive agents or the physics of physical systems, cognitive learning can be created with more advanced twins. Not only are cognitive learning twins able to explain, forecast, and investigate alternate courses of action, but they can also adjust the way digital twins improve their performance over time. For example, it is possible to construct cognitive learning twins that can educate themselves on how customers react to competitor offers, and then modify those offers to maximize sales, margins, or any other criteria that are defined [10]–[11].

Cognitive learning twins are considered fully adaptive autonomous twins when they can adapt to changes in their surroundings—when they are able to learn not simply from the environment and from other digital twins, but also when they are able to change their learning strategy to adapt to new environmental uncertainties, and when they are able to do those things simultaneously. For example, it is feasible to build autonomous vehicles that can continuously learn and adapt to the dynamics of the road, which are always shifting. As shown in Figure 8.1, the maturity curve for digital twins can be seen [2].

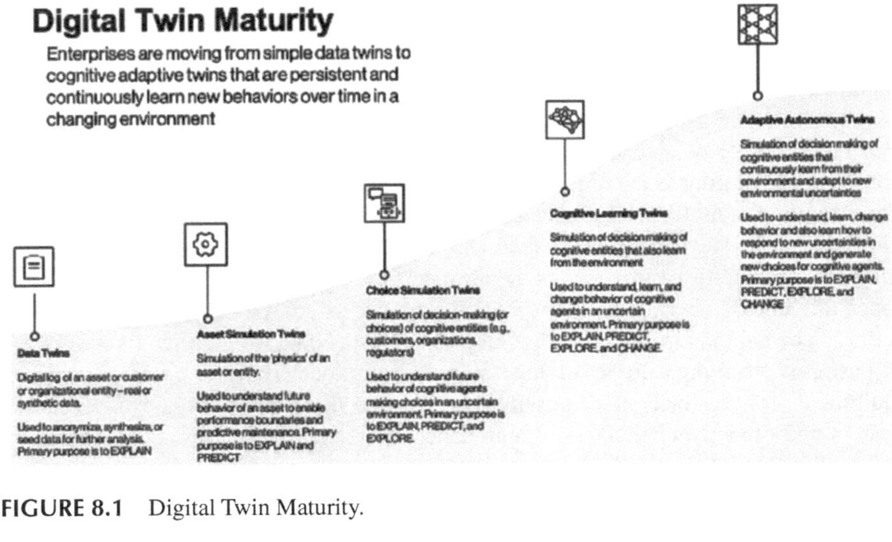

FIGURE 8.1 Digital Twin Maturity.

Big data and digital twins are working together to create a new reality within the framework of an industrial process, and there has been a lot of discussion about contextualizing data. It is common knowledge that data does not produce the knowledge that is necessary for decision-making by itself. To transform the vast amounts of data into information of significant worth, it is important to perform a contextualization activity. It is the digital twin's responsibility to ensure that there is a constant flow of information between virtually and physically held assets. Although the idea is well-known, the term "digital twin" does not only refer to a 3D model; rather, it encompasses a whole contextualization flow that is created through multiple layers of technology [5].

The digital twin is one of the most valuable allies of big data, and it is from its technology stacks that big data approaches can generate the contextualization of data during industrial operations. In addition, it is common knowledge that industries are notorious for incorporating dozens of various systems into their operations. In addition to creating the operational awareness that is required for a successful operation, the digital twin is also responsible for centralizing this information, which results in time savings and increases in productivity [12].

According to information provided by Statista [5], it is anticipated that the amount of data that will be produced by the year 2025 will amount to around 181 zettabytes. The acceleration of this process has been brought about by digital transformation, which has highlighted the significance of incorporating big data into industrial operations. In the same paper, it is stated that the profit that is anticipated to be made by big data in the year 2025 is estimated to be somewhere in the range of $65, with $16 billion referring to software programs that have the capacity to store and process massive volumes of data. Additionally, Gartner considers the digital twin to be one of the most recent technology trends in the field of digital transformation. Markets have estimated that the digital twin would have a market worth of $48.2 billion by the year 2026. Comprehended implementation of digital twins that are in sync with big data presents an opportunity that has the potential to be extremely advantageous for industries that are undergoing the process of digital transformation [13]–[14].

The adoption of digital twins in product development offers numerous benefits, including:

1. Improved Product Performance: Real-time monitoring and analysis enable proactive optimization and continuous improvement.
2. Cost Savings: Predictive maintenance, reduced downtime, and optimized operations result in significant cost savings over the product life cycle.
3. Faster time-to-market: Rapid prototyping, virtual testing, and streamlined iterations expedite the product development process.
4. Enhanced Innovation: Digital twins provide a platform for experimentation, simulation, and optimization, fostering innovation and creative problem-solving.
5. Data-driven insights: Accurate, real-time data empowers data-driven decision-making, facilitating strategic planning and improved product design.

6. Proactive Maintenance: Digital twins can forecast maintenance requirements, cut downtime, and take preventive measures to address problems by tracking the health and behavior of items or systems.
7. Streamline Development Process: Digital twins aid in quick prototyping, iterative design, and virtual testing, which lead to a quick product development cycle and a shorter time-to-market.

8.3 THE ROLE OF BIG DATA AND DIGITAL TWINS IN DIFFERENT INDUSTRIES

Digital twins are versatile and find applications in various industries [15]:

1. Manufacturing: Improves processes, monitors equipment performance, and drives quality improvements
2. Oil & Gas: Real-time view of processes and equipment, enabling predictive maintenance
3. Sustainable Power: Monitors renewable energy assets, predicts performance issues, and enhances network stability [16]
4. Healthcare: Creates a digital representation of a patient's body for informed decision-making
5. Retail: Analyzes consumer behavior to enhance the shopping experience
6. Sustainability: Optimizes physical equipment for maximum efficiency and reduces environmental impact

As the digital and physical worlds become more intertwined, digital siblings are assisting in bridging the divide between the two. The digital counterpart is a dynamic and virtual representation of a physical product that is equivalent to it. Despite their multifunctionality, these models maintain a robust connection with the product to facilitate design modifications and diagnostics. An increasing number of enterprises are adopting digital twins to enhance the functionality of their products in ways that were unattainable or unimaginable previously [4].

The utilization of big data and digital twins is significantly influencing an extensive range of industries. The manufacturing sector is implementing big data to increase production process efficiency and reduce costs. The generation of virtual representations of manufacturing facilities is one application of digital twins. By utilizing these representations, production workflows can be replicated, and their efficacy increased. To optimize patient diagnosis and treatment, the healthcare sector is implementing big data. Digital twins are being employed to facilitate the development of virtual patient representations that can be used to simulate and optimize treatment programs. The transportation industry is currently harnessing substantial volumes of data to optimize traffic flow and reduce congestion. The advancement of virtual representations of transit networks is being facilitated through the implementation of digital twin technology. Traffic patterns are simulated and optimized using these representations. The energy sector is currently employing substantial volumes of data to improve energy efficiency and reduce expenses. The development of digital counterparts is occurring for the construction of virtual representations of energy systems. Subsequently, these depictions used to be implemented to simulate and enhance energy consumption methodologies [17].

The word "avatar" means "digital twin". An individual digital counterpart, in addition to a use case, is distinct from the rest. A notable attribute of individuals who engage in video game consumption is that this is, in fact, the source of the entire phenomenon. Presently, all the necessary tools exist for the creation of an avatar, and each major technology company has developed its own avatar builder. A positive sign is that Google recently acquired an avatar tool developer for $100 million. A user-generated avatar is currently accessible across all platforms. This is comparable to a digital counterpart with social media and gaming applications. While on vacation, a weather speaker continues to deliver weather bulletins using a cloned voice that is safeguarded by intellectual property rights. The cloning of the speaker's voice and physique has occurred. A persona also serves as a protective barrier against the detrimental consequences of stereotypes that are prevalent in both offline and online environments, where every evaluation is accessible to the public—one that is amassing considerable momentum within Asian society, with a specific focus on China. Avatars can acquire virtual products as customers, thereby establishing a novel market. A 21-year-old professional at Space and Lemon Innovations consistently asserts that, "My avatar is an integral component of both my persona and online presence" [1].

When utilizing one of the numerous pre-built platforms to construct a digital counterpart, you need not confront an excessive number of challenges. To begin with, there are gaming and 3D platforms with millions of developers and enterprise solutions, such as Unreal Engine (Epic Games) and Omniverse (NVIDIA), which has recently introduced the "industrial metaverse" in partnership with Siemens. NVIDIA is a leader in the development of digital siblings for the industrial sector. The difficulty lies elsewhere; for example, in the procedures of gathering the necessary data, implementing monitors, and overcoming the mentality that "it is unnecessary; an Excel table suffices". Alongside the metaverse and the transition to 3D technology, the industry front runner in cloud computing offers innovative solutions, including Amazon Web Services (AWS) IoT TwinMaker and Microsoft Azure digital twins. The 3D game will feature the superior Apple ARKit 6, "other" 3D platforms, including Unity, and use case specialists. As an illustration, ANNEA, a solution originating from Hamburg that facilitates the generation of digital twins specifically designed for predictive apparatus maintenance in the renewable energy industry [1, 18].

Digital twins are the most adaptable tool for many industries. It gives a technical touch to assist businesses in providing flawless processes for systems and enhancing the performance of the product life cycle. It offers numerous opportunities, and the results are evident. Let's identify the industries at the forefront that are utilizing digital twin services [13].

8.3.1 Security Challenges with Digital Twins

Security usually receives less attention in the beginning when an innovative technology gains traction. As a result, firms are compelled to hastily address and resolve flaws that are exploited, resulting in the depletion of both time and income. Due to their cloud-based nature and lack of physical amenities, digital twins entail somewhat reduced security concerns compared to other system types. Nevertheless, the substantial quantities of data being gathered and employed are derived from multiple endpoints, each of which signifies a possible vulnerability. By 2023, 75% of digital twins are projected to be connected to at least five endpoints. In the future, connecting several digital twins to visualize intricate systems may be necessary [14].

With each new connection established and increased data transmission between devices and the cloud, the likelihood of a security breach rises. Hence, enterprises contemplating the implementation of digital twin technology must exercise caution and refrain from rapidly adopting it without thoroughly evaluating and enhancing their existing security processes. The areas of utmost significance encompass:

- Utilization of data encryption techniques
- Implementation of access privileges with clearly defined user roles
- Adherence to the principle of least privilege
- Mitigation of known device vulnerabilities
- Regular performance of security audits

While the implementation of digital twins is promising, it also faces challenges, such as data security, high initial costs, and the need for skilled professionals. However, as technology advances, these hurdles are gradually being overcome. The future of digital twins lies in their integration with technologies such as the IoT, AI, and augmented reality (AR), creating even more immersive and comprehensive models.

Digital twins have revolutionized various industries and have become an integral part of the digital transformation journey for many organizations. Here are some examples of how digital twins are currently being used:

Manufacturing and Industrial IoT: Digital twins help manufacturers to optimize the production process, increase efficiency, and reduce downtime. By creating a virtual replica of the entire manufacturing facility, manufacturers can simulate and analyze the production process, identify bottlenecks, and optimize the flow of materials, energy, and people.

Healthcare: Digital twins are being used in healthcare to simulate surgeries and test treatments on virtual patients. This helps medical professionals to make more informed decisions, reduce errors, and improve patient outcomes [19].

Smart Buildings and Infrastructure: Digital twins of buildings and infrastructure can be used to monitor and optimize energy consumption, detect maintenance issues, and improve safety and security [20].

Here are three statistics that showcase the growth and potential of digital twins,

According to a report by Grand View Research [5], the digital twins market size is expected to reach 22.8 billion USD by 2028, growing at a compound annual growth rate of 42.4% from 2021 to 2028.

A report by Marketsand Markets [7] predicts that the digital twin market will grow from 4.19 billion USD in 2020 to 18.8 billion USD by 2025, at a growth rate of 34.6% during the forecast period.

According to a survey by Accenture [21], 77% of companies are using or planning to use digital twins to improve their operations, and 72% of companies believe that digital twins will be crucial for their future success.

Here are a few of the subsequent ways digital twins are being implemented in our lives currently:

Home Automation: You can use digital twins to control and monitor your smart home devices, such as lighting, temperature, and security systems, from a single app [14].

Fitness Tracking: You can use digital twins to track your fitness goals, monitor your progress, and receive personalized recommendations based on your data.

Personal Finance: You can use digital twins to create a virtual replica of your financial situation, including your income, expenses, and assets. This can help you to make informed decisions about your finances and reach your financial goals.

Finally, here are a few prospect developments in digital twins look forward to:

AR and VR: Digital twins will be integrated with AR and VR to make intriguing experiences and simulations.

AI and ML: Digital twins will use AI and ML algorithms to automatically optimize and improve their performance over time [22].

Blockchains: Digital twins will be integrated with blockchain technology to increase security, transparency, and accountability in various industries and applications [14] [21]–[23].

8.3.2 Big Data Common Challenge with Real-Time Insights

Limited information technology (IT) funds are one of the most significant obstacles to gaining value from data, according to 50% of US executives and 39% of European executives. Big data implementation is relatively expensive, and a lot of forethought and money will have to go into this, which may not immediately pay off. In addition, the infrastructure needs to be upgraded to accommodate the ever-increasing data volumes. It is easy to lose sight of assets and the expense of managing them after a while. Flexera estimates that as much as 30% of cloud budgets go unused [8].

By keeping tabs on your infrastructure in real time, big data may alleviate most of the issues caused by increasing prices. You may monitor your data stack and the resources you use to store and manage it, find areas to reduce expenses, as well as the balance between scaling costs with the support of good DevOps and DataOps methods.

When you are constructing a data processing pipeline, you should first consider its cost. Data duplication across different stores can increase your expenses. Is it possible to reduce management expenses by organizing data into tiers based on its worth to the company? Do you often store information in a drawer and then completely lose track of it? You can devise a great plan to save money if you answer these questions.

Identify an inexpensive tool that suits your budget. You can usually get data stacks on the cloud and pay for them as you go. Put simply, the amount you pay is proportional to the amount of application programming interface (API) and data requests, as well as processing power, that you utilize. You can tailor your toolkit to your specific requirements and budget by taking advantage of the ever-expanding selection of new big data solutions.

8.3.2.1 Real-time Data Insights

The real-time data contains a wealth of information. However, information is useless unless it can be applied to real-world problems. Although some may consider the time it takes to extract and analyze data to be in real time, others may view it as instantaneous. Nevertheless, the main point is to gain a solid grasp of the advantages of endeavors, such as opening new channels for innovation and technology impact; for example:

- Achieving faster service delivery
- Cutting operational expenses
- Ground-breaking service offerings
- Fostering an environment driven by data

Generating timely reports and insights is one of the issues related to big data. Consequently, businesses are trying to find ways to invest in extract, transform, and load (ETL) tools and analytics that can process data in real time, so they can compete in the marketplace.

8.4 THE FUTURE OF DIGITAL TRANSFORMATION

The process of digital transformation is currently in its nascent phase, with ample opportunities for more advancements and innovations. With the ongoing advancement of big data and digital twins, you could anticipate the emergence of additional innovative and significant applications. The future of digital transformation is optimistic. By embracing big data and digital twins, organizations can unlock new value, improve efficiency, and innovate.

The emergence of a digital twin for consumers is a highly stimulating advancement. There are numerous potential applications for replicating a consumer's behavior in various scenarios, such as launching a new product, an advertising campaign, or a societal trend. The essence is in delineating, replicating, and foretelling a consumer's behavior, representing the epitome of marketing and a harbinger of its future.

Multiple firms seek to develop digital twin capabilities to thoroughly investigate the growing fields of IoT, Metaverse, and other novel technologies. However, in order to achieve this, organizations have to concentrate on several crucial areas [21].

Initially, organizations must establish and integrate an approach that focuses on the concept of systems throughout the entire organization. Companies must operate outside their conventional boundaries and silos. Firms should analyze the underlying causes and impacts of their plans and activities and consider numerous layers of connection. This will enable them to tackle business-level challenges by prioritizing choices and acts that fall within their jurisdiction, as well as those influenced by external factors. As such, they will be able to create suitable digital replicas.

Furthermore, it is beneficial for companies to integrate digital twins into their comprehensive IT infrastructure and enable them to interact with and integrate other analytics and ML models. The software and model development process should include the development of digital twins, and the systems should facilitate the construction, adaptation, expansion, and implementation of digital twins.

The digital twin is specifically engineered to get responses from sensors that gather data from a physical copy in the real world. The twin has the capability to accurately replicate the definite thing in real time, thereby offering valuable insights on its accomplishment and identifying any flaws. The twin can also be developed using a prototype of its physical equivalence. In this scenario, the twin can offer feedback while refining the product. It can even function as a prototype on its own before any physical version is constructed. The user's text is a reference to a source or citation [16].

The procedure is described extensively in a post by Eniram [24], a business specializing in creating digital replicas of large cargo ships that play a crucial role in global trade. This type of digital twin application is highly intricate. Nevertheless, a digital twin has the flexibility to be either complex or straightforward, and the level of data employed to construct and maintain it will dictate the accuracy of replicating a physical entity outlined in Figure 8.2.

Digital twins are valuable catalysts for driving innovation and enhancing performance. Imagine a scenario where highly skilled product technicians have access

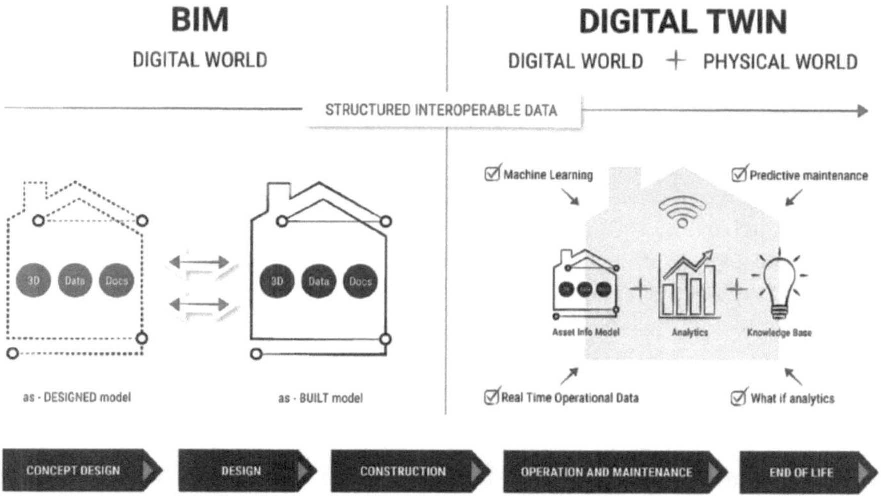

FIGURE 8.2 Digital Twin World.

to advanced tracking, computational, and forecasting abilities. In decades, digital twins can serve as an enormous number of objects, reaching into the billions. These substitutes, which represent the physical world, will create new possibilities for collaboration between experts in physical world products and data scientists, whose role is to analyze the insights provided by data regarding operations.

Technology for digital twins enables firms to enhance the customer experience by gaining greater knowledge of consumer requirements, improving existing products, processes, and services, and fostering the development of new business innovations. Digital twins provide a live representation of the status of physical assets, thereby significantly reducing the workload associated with maintenance. Companies apply digital twin technology for several purposes, such as enhancing current operations, training personnel, and conducting pre-launch testing of new goods or procedures. This approach enables them to address any difficulties cost-effectively and in a less complex manner before implementing them in the real world [24].

Regardless of where the physical facility is located, including in space, AI and ML are routinely used to evaluate the digital twin's operational model. Since the inception of space missions, NASA has employed pairing technology, which serves as a forerunner to digital twin technology. This technology has been utilized to remotely address the challenge of driving, preserving, and mending systems, without physical proximity. This is exactly how techies and cosmonauts on Earth selected the course of action for rescuing the Apollo 13 mission.

Currently, NASA employs digital twins to investigate and analyze advanced vehicles and planes.

8.4.1 Conducting Pre-Production Testing of New Systems

Firms can utilize digital twins to develop and evaluate systems, product concepts, and service methods before committing resources to construction or execution. If a model demonstrates efficacy, then its digital counterpart may potentially be connected to the actual entity for instantaneous supervision.

8.4.2 Enhancing Performance and Production

Forbes predicted in 2017 that the use of digital twins could result in a 30% increase in the pace of crucial processes. As per Gartner [3], industrial organizations have the potential to achieve a 10% enhancement in productivity. Digital twins are widely available and can be used in many ways. This helps businesses in almost every industry to figure out where they can organize and improve processes, which helps to reduce downtime through predictive maintenance.

8.4.3 Real-Time Asset Management

Employing digital twins to monitor everyday operations and optimize industrial processes minimizes avoidable machinery deterioration and notifies firm proprietors about potential cost-saving modifications, such as optimizing fuel consumption.

Enhanced efficiency in maintenance and repair enables organizations to sustain a competitive advantage by strengthening total productivity [12].

8.4.4 Enhancing Service by Comprehending Data

Aside from helping with problems, digital twins can also be used to help customers. Technicians can utilize virtual models to do tests for diagnosis remotely and guide consumers through the correct repair procedures, rather than solely depending on predefined protocols. The data collected from these sessions yields beneficial insights for future product planning and development, as depicted in Figure 8.3.

The organizational level in the built environment is where new and amazing possibilities are provided. By incorporating digital twins into hospitals or commercial real estate structures, such as implementing advanced technologies or systems, there is the potential to offer favorable results not only for the administrators or owners of the facilities, but also for the occupants therein. By adopting a people-centric strategy, you can first focus on individuals, and then analyze the issues and context, and ultimately include IT systems and linked devices to address significant challenges and generate enduring value.

The technology of digital twins has promising prospects. The quantity of data accessible for digital twin models will significantly increase as the IoT devices and sensors proliferate rapidly. The advancements in AI and ML will enable the creation

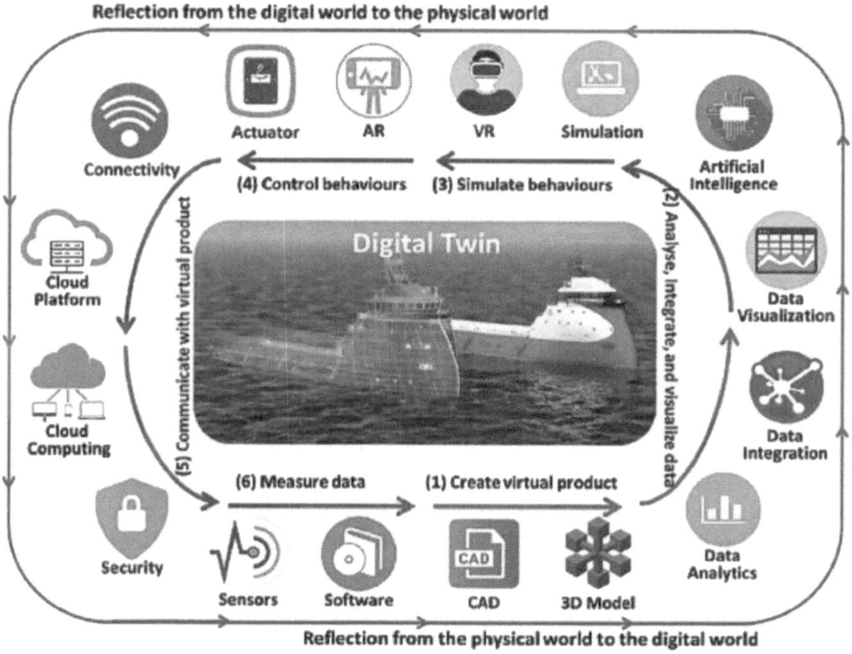

FIGURE 8.3 Digital World to a Physical World Use Case.

of intricate simulations, accurate forecasts, and autonomous decision-making using real-time data. The integration of digital twins will facilitate collaborations and ideas across different industries, beyond traditional limitations [14].

Digital twins represent the subsequent progression in their digital transformation for corporations and organizations already utilizing IoT. They can be used to enhance efficiencies, optimize operations, pre-emptively identify issues, and foster innovation for the future. If an organization seeks to achieve improved commercial outcomes and enhanced outcomes for all stakeholders, then it is worthwhile to consider studying digital twins [25].

The world is undergoing a transformative technological revolution, where the convergence of the IoT and digital twins is reshaping how we interact with our physical environment. These technologies are changing industries and influencing the future of business operations, healthcare, urban planning, and more. In this chapter, we have comprehensively delved into the world of digital twins, explored the concept of IoT, examined their synergy, and discussed their growing importance in shaping our future [24].

Digital twin technology is revolutionizing product development by providing virtual replicas that facilitate real-time monitoring, analysis, and simulation. Its applications span various industries, offering benefits such as performance optimization, predictive maintenance, streamlined development processes, and data-driven decision-making. As the future of digital twin technology unfolds, businesses stand to unlock new levels of innovation, efficiency, and competitiveness. Future business processes are awaiting and welcoming such innovations.

Digital twins can potentially revolutionize various industries and improve our lives in countless ways. The possibilities are endless, from optimizing production processes to tracking our fitness goals. So, let's embrace the future of digital replication and see how digital twins' innovative technology can help us to achieve our goals and improve the world.

8.5 CONCLUSION

Ultimately, it is imperative for firms to actively seek, retain, and nurture individuals with expertise in digital twin technology. While analytics and ML typically attract those with a scientific inclination, digital twin development is more intriguing to people with an engineering inclination. The combination of digital twins and big data revolutionizes how we interact with the physical world. These technologies improve efficiency, reduce costs, and enhance our ability to make informed decisions. As they continue to evolve and find applications in diverse fields, you can expect a future where your interactions with the physical world are more streamlined, sustainable, and data-driven. The future of digital twins and big data is bright, promising us a more intelligent, more connected world. Many organizations have only recently begun their journey with digital twins. However, this endeavor holds the potential to effectively merge the physical and digital realms, resulting in reduced operating expenses and improved decision-making capabilities. Digital twins represent a confluence of the virtual and physical worlds, offering unparalleled insights and opportunities across various industries. As we continue to embrace digital transformation, digital twins serve as a testament to human ingenuity and the endless possibilities of technology. They will undoubtedly play a pivotal role in shaping a smarter, more efficient, and sustainable world in the years ahead.

REFERENCES

[1] Medium, 2023. "Digital Twins, a breakthrough in 2023 and a new era," www.spaceandlemon.medium.com/digital-twins-a-breakthrough-in-2023-and-a-new-era-88555f1a7218

[2] Medium, n.d. "Evolution of Digital Twins," www.odsc.medium.com/evolution-of-digital-twins-5df92006878b

[3] PwC, 2022. "2024 AI Business Predictions," www.pwc.com/us/en/tech-effect/ai-analytics/ai-business-survey.html

[4] M. Prawin, "The power of Big Data and Digital Twin," Medium, 2020. www.medium.com/tilicholabs/role-of-digital-twin-in-the-age-of-industry-4-0-ab1d940ca9ef

[5] A. Andrade, "Role of Digital Twin in the age of Industry 4.0," Vidya Technology, 2021. www.vidyatec.com/blog/the-power-of-big-data-and-digital-twin/

[6] Medium, 2023. "Digital Twins and IoT: Shaping the Future," www.medium.com/@hamza.zellal/digital-twins-and-iot-shaping-the-future-f611f694dd11

[7] Medium, 2021. "The evolution of digital twin technology and its future role in enterprise digitalisation," www.medium.com/digital-bulletin/the-evolution-of-digital-twin-technology-and-its-future-role-in-enterprise-digitalisation-589b7cceb1d1

[8] Jayakrishnan, R., Balasubramanian, P., Anand, S., & Suri, S. (2021). Big data analytics and digital twins: A fusion for real-time decision-making in manufacturing. *Journal of Manufacturing Systems*, 62, 287–306.

[9] Tao, F., Zhang, L., Hu, C., & Zhang, Y. (2021). Digital twins for product design and manufacturing: A review. *Journal of Manufacturing Science and Engineering*, 143. pp. 134–143.

[10] Thillaikkarasi, R., Yaseen, M. M., Rameshbabu, R., Prabhakaran, R., Kesavan, R., & A. Jose Anand. (2023). Waysides inspection using wayside processing imaging and deep learning. *2023 3rd International Conference on Pervasive Computing and Social Networking (ICPCSN)*, pp. 359–365.

[11] Nair, Resmi R., Senthamizh, Selvi R., Harshaa, Vartini G., Jayashri, B., Suruthi, S., & Jose Anand, A. (2023). Maturity level classification of Moringa leaf using DNN algorithm. *2023 4th International Conference on Smart Electronics and Communication (ICOSEC)*, pp. 1667–1672.

[12] Kritzinger, W., Weiler, M., Umbrecht, D., Hockenjos, M., & Zaefferer, G. (2023). Digital twins in smart manufacturing. *CIRP Journal of Manufacturing Science and Technology*, 25, 100391.

[13] Zheng, Y., Liu, S., & Wu, Y. (2021). Digital twins and their potential applications in the construction industry. *Journal of Building Engineering*, 144, 103174.

[14] Arun Sampaul Thomas, S. Muthukaruppasamy. (2023). Green IoT use case approaches for blockchain technology taking industry 5.0 to the next level. *Elsevier Book Chapter- Green Blockchain Technology for Sustainable Smart Cities*, pp. 119–141, ISSN: 978-0-323-95407-5.

[15] Medium, 2024. "Demystifying digital twins: The future of replicated realities," www.medium.com/@hosteltrader/digital-twins-the-future-of-digital-replication-b6956ee66192

[16] Anuradha, T., Suganya, A., Muthalagu, R., Narender, M., & Jose Anand, A. (2023). Evaluation of a wireless sensor network for the detection of forest fires. *2023 2nd International Conference on Edge Computing and Applications (ICECAA)*, pp. 1021–1027.

[17] He, W., Liu, Y., & Xu, X. (2022). Digital twins for improving energy efficiency in buildings: A review. *Energy and Buildings*, 251, 111567.

[18] Arun Sampaul Thomas, G., Muthukaruppasamy, S., Saravanan, K., Muleta, N. (2023). Revolutionizing Smart Grids with Big Data Analytics: A Case Study on Integrating Renewable Energy and Predicting Faults. In: Kumar Sharma, D., Sharma, R., Jeon, G., Kumar, R. (eds) *Data Analytics for Smart Grids Applications—A Key to Smart City Development. Intelligent Systems Reference Library*, vol 247. Springer, Cham. https://doi.org/10.1007/978-3-031-46092-0_11.

[19] Jiang, P., Ding, K., & Mao, S. (2022). Digital twins for healthcare: A review. *IEEE Transactions on Biomedical Engineering*, 69(5), 1529–1541.

[20] Kang, H. G., Lee, J. Y., Choi, S. H., Kim, H., & Kim, H. C. (2022). Digital twins for smart cities: A survey and perspectives. *IEEE Transactions on Smart Cities*, 3(1), 1–16.

[21] Elhoseny, K., El-Haggan, A., & Ismail, M. (2023). Digital twins in the era of Industry 4.0: A comprehensive review. *Journal of Manufacturing Systems*, 68, 223–241.

[22] Arun Sampaul Thomas, G. (2020, November 26). Intelligent prediction approach for diabetic retinopathy using deep learning based convolutional neural networks Algorithm by means of retina photographs. *Computers, Materials & Continua Journal, Tech Science Press*, 66(02), ISSN: 1546–2218, SCI Index with IF 4.89, Scopus Indexed with IF 3.8.

[23] Arun Sampaul Thomas, G., & Harold Robinson, Y. (2020, February 26). IoT, big data, blockchain and machine learning besides its transmutation with modern technological applications. *Springer Book Chapter-Internet of Things and Big Data Applications-Part of the Intelligent Systems Reference Library book series* (ISRL, volume 180), pp. 47–63, ISSN: 978-3-030-39118-8.

[24] Zhang, H., Zhang, X., & Gu, G. (2023). Digital twins for urban infrastructure management: A review. *IEEE Transactions on Intelligent Transportation Systems*, 26(1), 1–21.

[25] G. Benliman, "### Exploring the Future with Digital Twins: Revolutionizing Industries One Virtual Model at a Time," Medium, 2023. www.medium.com/@benlimanghali/exploring-the-future-with-digital-twins-revolutionizing-industries-one-virtual-model-at-a-time-01bc6c82b80f

Part 3

Applications

9 Digital Twins in Industry 4.0 Automation

Muhammet Raşit Cesur, Elif Cesur and Ahmet Harun Çelik

9.1 INTRODUCTION

A digital twin is a virtual model simulating the behavior of a real system in various conditions regarding the system dynamics. This definition presents a general view of digital twins to determine the boundaries of the concept. In this context, we can categorize DTs in the industry into pre-production digital twin models, digital twin models utilized in the production process, and after-production digital twin models. These three types of digital twin models are considered from the perspective of five factors: application, life cycle, functions, architecture, and components [1]–[6]. On the other hand, in the view of business, this concept contributes to various goals of the industry, as presented in Capgemini Research Institute's report [7]. The first goal is to improve sustainability, as it constitutes one component of Industry 5.0, as presented by the European Commission [8], aligning with the target of green production. Due to their ability to simulate real-world conditions and deliver real-time information, organizations can optimize resource usage, reducing carbon emissions and improving supply and transportation networks. Also, digital twins contribute to the sustainable maintenance field, which is another contribution to sustainability.

The second goal is to improve interoperability, which enables digital twin models to be implemented with various industrial hardware and manufacturing systems. In this case, digital twin models must be adapted to servers and historian databases for integration with industrial automation. The integration must be bi-directional because a digital twin model can provide data to the industrial automation system via virtual sensors defined in the model. The most prominent technologies in the Industry 4.0 concept where digital twins are utilized for interoperability are the industrial Internet of Things (IIOT) and blockchains [9]–[10]. Because sensor data is required to build and run digital twin models, IIOT has a significant role, especially in the execution process of DT. With digital twin and IIOT technology, industrial automation has evolved into industrial robotics. Collaborative robots called cobots are the first step in this transformation, and they are prominent samples of cyber-physical systems [11]. One example of cobots is autonomous mobile robots (AMR), which demonstrate robot-to-robot collaboration [12]. Another example of cobots is

DOI: 10.1201/9781003465904-12

the new generation of robot arms, which exemplify human-robot collaboration [13]. These cases show that industrial transformation led to the use of intelligent robots as production equipment to increase flexibility and adaptability in the context of digital twins [14]. Consequently, the interoperability goal of digital twins contributes to improving the integration level of manufacturing automation over digital twin models of production systems.

The third goal is to improve the product development process by including digital twins in the pre-production phase and after-production phase (especially in the maintenance process). In the production design process, the digital twin contributes computer-aided design (CAD) software functionality, such as finite element analysis [15]. Also, various tests and physical simulations can be performed on the digital model of the product via digital twin models. Product-related issues such as maintenance, product life cycle, and after-production of the product can be managed via digital twin models [16]. Thus, digital twin technology increases the efficiency of the product development phase by decreasing the cost and time requirements of the design process. Digital twins also provide an optimal product design via simulations and differential calculations to the product design phase, which is not feasible to find during trial and error in real life [17]. Also, product design contributes to sustainability with process optimization in the design phase or designing sustainable materials to decrease carbon emissions [18]–[19].

To summarize, digital transformation comprises several essential elements, including DT. Through its role in managing cyber-physical systems, helping to achieve sustainability goals, and facilitating the optimization of the product and production process throughout the product design stage, digital twins play a significant role in the industry's digital transformation. With digital twin models, it is possible to simulate production processes and products under a variety of real-world scenarios in order to determine the best configurations, as they are based on the dynamic behavior of both. Industrial automation heavily relies on digital twin models because of this quality, which makes them a useful tool for the industry.

9.2 LITERATURE REVIEW

There are now many academics and practitioners that value design thinking more than ever, and many design thinking publications include case studies from the manufacturing sector [20], such as the automotive industry, aerospace industry [21]–[23], industry of energy [25], and various manufacturing companies [26], [56], [6]. Numerous technologies and methodologies are employed in conjunction with industrial automation in the development and use of digital twins in these industrial sectors. These technologies and methodologies are simulations [28], [23], AI [21], [30]–[31], big data analytics [32]–[33], the IIoT [34]–[35], physical systems modeling [36], and data visualization [37]–[38]. However, there is still a dearth of research on how digital twins may be applied to enhance the utilization of manufacturing capacity. Thus, we categorized digital twin studies based on functionality and its contribution to continuous improvement. The first category is product design [17], [39]–[40]. Digital twins especially contribute to product life-cycle management and production planning-oriented product design. The second category is production planning, in which digital twins model the behavior of the production system. The digital twin is only one of the technologies for production system modeling [41]–[42]. Production planning consists of various functionalities, such as sustainable planning, maintenance planning, scheduling, autonomous system planning, quality control, testing, and verification.

Sustainability has been a major concern over the past few decades, and many studies focusing on energy efficiency have been carried out by academics to improve sustainability in manufacturing. Most energy-efficient scheduling studies focus on determining the ideal machine speed or dispatching for minimizing total energy consumption [43]–[45]. Digital twins are used to model energy consumption under various machine operating conditions. Energy efficiency-oriented planning studies focus not only on reducing the energy consumption of machines, but also on minimizing fuel consumption, especially in transportation planning for production systems to achieve sustainability and energy efficiency [46]–[48]. Fuel consumption models of various engines are constructed using ML algorithms, such as data-driven digital twin models [49]–[51]. Predicative maintenance is another field in which digital twin is utilized [52]–[53]. Support vector machines (SVMs), XGBoost, artificial neural networks (ANN), and deep learning (DL) are the most common methods that are utilized for constructing digital twin models. These models are mostly used for fault diagnosis and remaining useful life prediction. In scheduling, digital twin models are utilized to optimize system dynamics in a detailed manner, including reliability and overall carbon emission [54]–[56]. Digital twins are also utilized to improve human-robot collaboration and coordination of cyber-physical systems in the perspective of improving industrial automation [57]–[58]. This approach improves the utilization level of the production system. Another way that the digital twin contributes to industrial automation is quality control, which also improves the level of utilization [59]. Finally, the last category is prognostics and health management, for which digital twin models are developed [60]–[61].

We can conclude from the literature research that digital twins are a common instrument that helps production systems across a product's entire life cycle. At all levels of the life cycle, various ML algorithms are utilized to construct digital twin models. In this case, we noticed that most digital twin models rely on physical calculations of sensor data to gain the output, and ML models operate on inputs to calculate the output. Similar to digital twin studies, there are several ML applications in industrial processes, including process control maintenance, and product design [62]–[65]. The ML models built in the industrial projects/studies, which fulfill their aims by simulating system dynamics, are referred to as data-driven digital twin models [49]. Physical model-based digital twins, on the other hand, are based on models that depend on differential equations and physical relations or physical equations themselves [66]. Lastly, hybrid digital twin models are those that rest on both ML and physical interactions. We may argue that ML algorithms have been the basis for the development of most digital twin model types.

9.3 THE STRUCTURE OF A DIGITAL TWIN MODEL

Mass customization situations such as pandemics and financial performance indicators (e.g., contribution margin, profitability rate, and cash turnover rate) make it a must to eliminate humans from factories. Humans cannot adapt to a very high variety of production because laborers must learn the production details of every product, and product designers must define the quality and production standards of every product to achieve a complete manufacturing process. From this perspective, industrial automation has transformed from computer-controlled machines to intelligent robotics, which is a functional alternative to humans in the industry. Thus, cyber-physical systems are an indispensable component of a dark (autonomous) factory, which helps to increase the level of automation in the industry. The digital twin is a popular tool of the Industry 4.0 era because cyber-physical systems require a digital duplicate for real-time control and planning.

Digital twins can model a detailed industrial process, including system dynamics. With digital twins, production systems can be simulated more realistically. Thus, digital twins provide us with the capability of efficiently managing and organizing components in the dark factory. The digital twin model of a cyber-physical system simulates every action of the system in various real-world environments. That is why the digital twin will calculate every state in the production environment before it occurs. The digital twin contributes to finding the best action for the cyber-physical system to achieve the overall system goal. This feature of the digital twin is utilized in production design and after-sales support in the digital environment. In this way, digital duplicates of products and production environments are constructed and utilized during a product's life cycle to optimize industrial processes.

To construct a DT, the data about the dynamics of a physical object is required for modeling. Most of the required data is collected with sensors. Some of them are calculated regarding one or multiple physical sensor data, which is called a virtual sensor. The collected data is used with physical equations, or ML models, to figure out the physical system's characteristics, such as energy consumption, operation duration, failure, etc. When the dynamics of the system are calculated by physical equations or physical relations-based differential equations in a digital twin model, it is called a physical-model-based digital twin, as provided in Figure 9.1a. If an ML model is expected to figure out the relation between sensor data and system dynamics in a digital twin model, then it is called a data-driven digital twin, as shown in Figure 9.1b. If physical equations and ML models are both used in a digital twin model, then it is a hybrid DT, which is given in Figure 9.1c. Hybrid models can be constructed in three ways. The first is that the outcomes of physical calculations are used as inputs of an ML model to increase the performance of the digital twin model. In the second way, which is similar to the first, the outcomes of the ML model are used as parameters of physical or differential equations in the digital twin model. In the first two approaches, physical equations or ML models, which provide inputs to the digital twin model, are utilized as virtual sensors. The third approach is to apply both physical models and data-driven models to calculate/predict different outcomes.

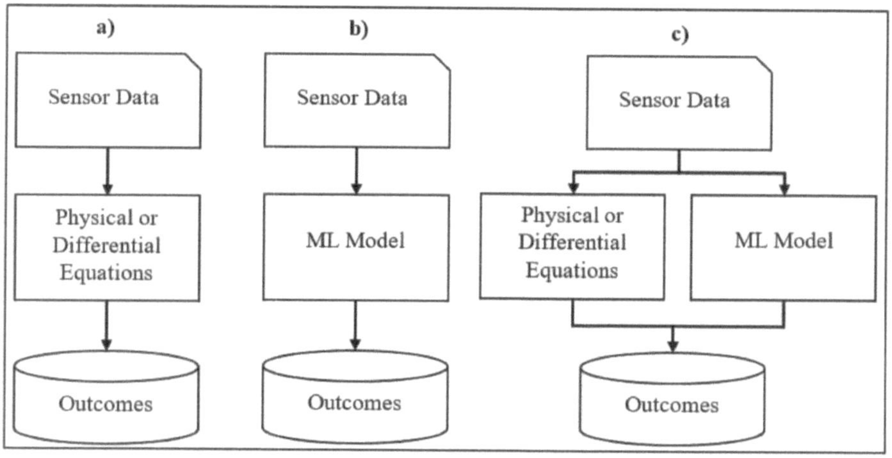

FIGURE 9.1 (a) Physical Model-Based DT; (b) Data-driven DT; and (c) Hybrid DT.

9.3.1 Industrial Hardware for Digital Twins

Industrial automation hardware is used both to collect data from sensors for digital twin models and to operate digital twin models. Smart sensors, programmable logic controllers (PLCs), industrial computers, and industrial embedded systems are utilized in the data collection process. All of these categories of hardware contribute to the digital transformation in the industry. For hardware, this transformation occurs in the change from flexible manufacturing systems to robotic manufacturing systems. The new generation of robotic automation systems is collaborative, and this attribute of the system is the reason behind the cyber-physical system concept. Because digital twins are one important component of cyber-physical systems, robotic automation hardware contributes to the build and execution of digital twin models.

9.3.1.1 Smart Sensors

Smart sensors differ from normal sensors in that their controlling unit provides a simple or uncomplicated decision structure. Smart sensors can detect environmental effects and noise to eliminate them from measurements. Additionally, some smart sensors are able to calibrate themselves regarding the attributes of measured materials. Smart sensors with these properties make more accurate measurements in the industrial environment. Smart sensors are one of the crucial components of industrial robotic systems because of their robustness and measurement performance. Accurate measurements, purified from the effect of noise and environmental factors, help robotic systems to make accurate real-time decisions. Thus, smart sensors constitute a smart infrastructure for digital twin models [67]–[68].

Positioning, dimensions measurements (2D/3D), gap and edge detection, displacement detection, product inspection, vibration and current measurement, visual

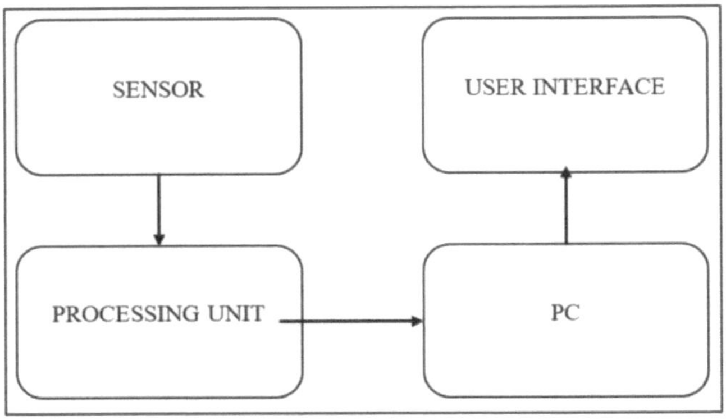

FIGURE 9.2 Structure of a Smart Sensor.

inspection, and other applications are all made possible by smart sensors. The sensor makes decisions about its validity, its corrected value, and whether to send an autonomous signal to the automation system's control unit during a measurement. The user interface of a smart sensor determines the measurement methods or conditional judgments, as shown in Figure 9.2. When a smart sensor is linked to a PC, the user interface operates. Ultimately, the detected signals are processed by processing units and sensor measurements.

9.3.1.2 Computing Interfaces

Three alternative computing interfaces widely used in industrial automation systems are PLCs, industrial computers, and embedded single-board computers (SBCs). PLCs are computing units used in factory production lines to regulate machinery and operations. Because of its operational durability, robustness, and maintenance advantages, PLCs have grown in significance and are now a major component of the industrial sector [69]–[70]. As you can see in Figure 9.2, industrial automation is achieved when a PLC uses preprogrammed control software to process sensor input and provide the necessary outputs. These outputs can be used as instructions to start or stop a machine, or as a way to sound an alarm when a process deviates from the acceptable range. Conveyor belt speed control, light control to indicate a particular condition on the production line, temperature setting of the heating or cooling systems, and warning system activation, when specific conditions are met (e.g., when something exceeds the limit), are just a few of the tasks that PLCs can perform. It is utilized for sophisticated system programming, as it can carry out a variety of control functions. They greatly lower the manufacturing margin of error and improve system stability by automating production operations. An interface is provided by input-output units between the final control components, sensors and the PLC. Data from or to other PLCs is sent via the communication interface. PLCs use network, serial interfaces, and digital I/O units for communication and integration. A PLC application structure is provided by Keyence [71] in Figure 9.3.

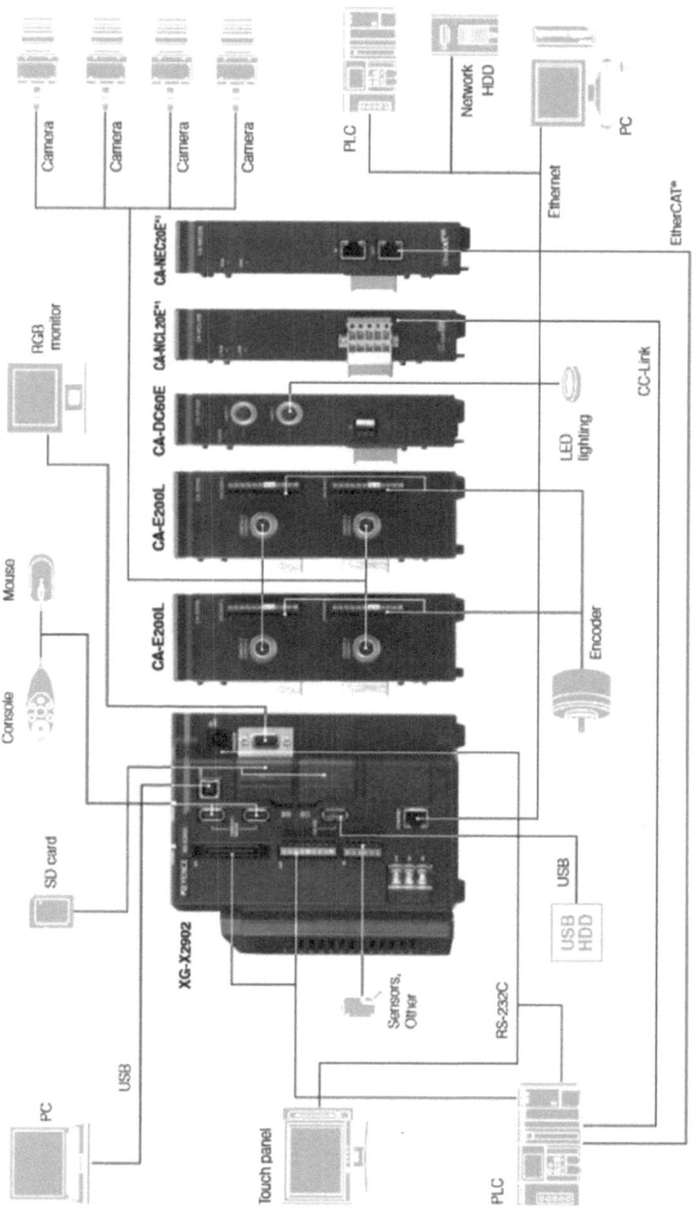

FIGURE 9.3 A PLC Application Structure by Keyence [71].

Because PLC systems have limited flexibility, more customizable control units are required to satisfy the needs of Industry 4.0 [69]–[70]. Thus, the industrial PC (IPC) is utilized in automation control. IPC has a general-purpose input/output (GPIO) interface, serial communication interfaces such as RS-232, and ethernet-like PLC systems. To ensure a leakproof structure and compliance with standards such as IP67, a cap must be placed on every socket and a heatsink on the chassis for passive cooling. With all its features, an IPC provides a flexible control of automation, complex integration, and a higher-level information security than a PLC system [72]–[73]. Another alternative is SBCs with Industry 4.0 and smart production concepts. Edge intelligence is especially enabled with SBCs, which use industrial internet for both communication and robotics. PICMG, AAEON PICO-ITX, AAEON Gene, AAEON EPIC, and Nvidia Jetson Orin boards are well-known industrial SBCs.

9.3.2 Required Software Libraries

Many software libraries are utilized for digital twin generation. The libraries are categorized as simulation libraries, multi-physics engines, data management libraries, computational mathematics libraries, ML libraries, and image processing libraries. The list of libraries is provided in Table 9.1.

TABLE 9.1
Software Libraries

Category	Library
Simulation	OMPython
	PySimulator
	SimPy
	PyTwin
Multi-physics Engine	PyChrono
	Pymunk
	perfect_physics
Data Management	PyData
	pandas
Computational Mathematics	NumPy
	SciPy
	SymPy
ML	sci-kit-learn
	TensorFlow
	Keras
	TinyML
Image Processing	OpenCV
	BoofCv

9.4 CASE STUDIES

To explain digital twin applications in the industry, we presented two case studies. The first case study is about modeling the test duration of automobiles manufactured in Daimler's Mercedes-Benz factory to improve sustainability. The contribution of the model is in predicting test durations in a search function to find the best test combination. The second case study is about identifying sections of processed furniture parts, such as holes, edges, and surfaces, and generating a digital twin of the actual product to compare it with technical drawings to detect dimensional problems.

9.4.1 Mercedes-Benz's Green Manufacturing

Mercedes-Benz vehicles made by Daimler are among the best luxury vehicles available in the market. Customers may pick their dream Mercedes-Benz with an enormous array of features and extras. The experts at Daimler have designed a thorough and comprehensive testing method for all customized cars in order to guarantee the dependability and safety of each and every unique car configuration prior to them being driven. However, without a strong algorithmic approach, maximizing the speed of their testing system for the countless features and combinations is difficult and time-consuming. To overcome this situation, we developed a digital twin model to predict the test duration for using them in an iterative optimization algorithm as a goal function to minimize. Mercedes-Benz provided a dataset [74] with 376 features to predict test duration. A sample of ten rows of data with 16 features is shown in Table 9.2. We eliminated the first eight categorical features (X0-X8) to construct a deep neural network (DNN) model.

We built the DNN model after partitioning the dataset. There are five layers in the model. Thirty nodes with the rectified linear unit (ReLU) activation function make up the first layer. With the ReLU activation function, 80 nodes make up the second layer. There are 40 nodes with a ReLU activation function in the third layer. A dropout layer with a 50% dropout rate makes up the fourth layer. The output layer, which has

TABLE 9.2
The Test Duration Dataset

Y	X0	X1	X2	X3	X4	X5	X6	X8	X10	X11	X12	X13	X14	X15
130.81	K	v	at	a	d	u	j	o	0	0	0	1	0	0
88.53	K	t	av	e	d	y	l	o	0	0	0	0	0	0
76.26	Az	w	n	c	d	x	j	x	0	0	0	0	0	0
80.62	az	t	n	f	d	x	l	e	0	0	0	0	0	0
78.02	az	v	n	f	d	h	d	n	0	0	0	0	0	0
92.93	t	b	e	c	d	g	h	s	0	0	0	0	1	0
128.76	al	r	e	f	d	f	h	s	0	0	0	0	1	0
91.91	o	l	as	f	d	f	j	a	0	0	0	0	1	0
108.67	w	s	as	e	d	f	i	h	0	0	0	0	1	0
126.99	j	b	aq	c	d	f	a	e	0	0	0	0	1	0

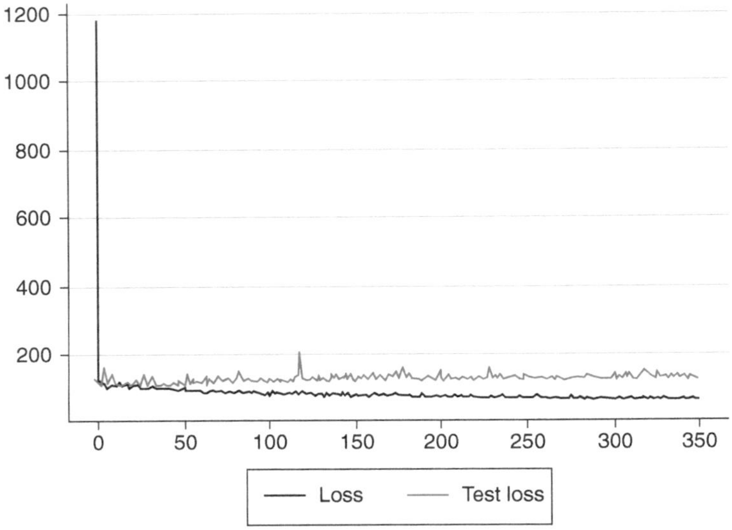

FIGURE 9.4 Generalization Gap of DL Model.

one node and a ReLU activation function, is the last layer. Adam optimizer is used in training with a 0.005 learning rate, and training that took 350 epocs. The size of the test dataset is set to 20% of the entire dataset for the training process. Mean squared error (MSE) is used as the loss function. The accuracy of the model is 94.25%, the mean standard error of the model is 124.8285, and the mean absolute error is 6.2370. Finally, we investigated the generalization gap by drawing the loss function plot during training. After a few training sessions, we realized that the difference between the loss value of the test dataset and the loss value of the training dataset increases after 350 epochs. For this reason, we terminate training after 350 epochs, as seen in Figure 9.4.

9.4.2 Inspection of Dimensional Problems in Furniture Manufacturing

In this study, we aimed to generate a technical drawing of the processed part of the furniture by detecting edges, surfaces, and holes in the part. In this way, we generate the digital twin of the produced product to compare it with design data for inspection. We combined the convolutional layers of convolutional neural networks with an SVM and a random forest (RF) algorithm. Our intelligent model consists of convolution, pooling, and flattening layers; however, we did not use neural networks as the final layer for classification. We used SVM and RF algorithms at the last step to decrease training time. The hybrid structure is given in Figure 9.5.

For pointwise classification, we marked training images. RGB-based features are extracted with 3x3, 4x4, 8x8, 16x16 kernels. Since we applied an RGB-based kernel, 3D convolution was applied to the images. Three classes were used in the training model: hole, surface, and outer space. The outer space is used to detect the outer edge

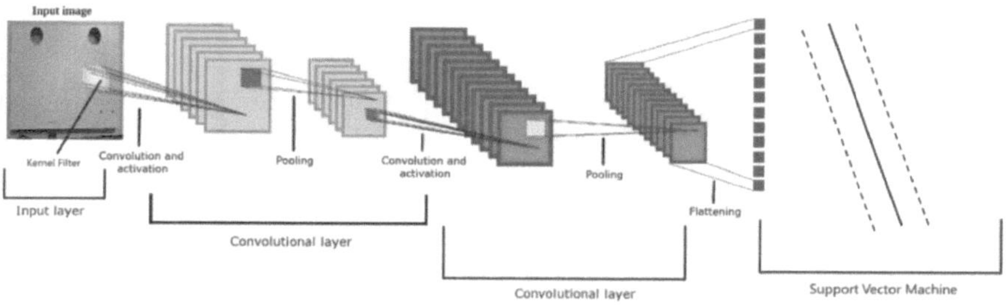

FIGURE 9.5 Combined DNN Layers with SVM.

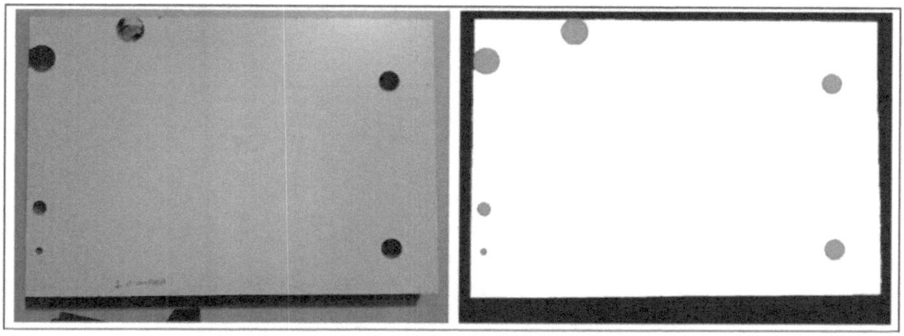

FIGURE 9.6 Labeling Images for Training.

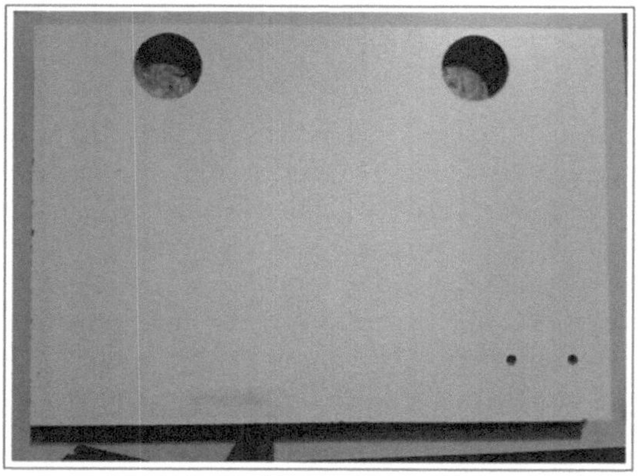

FIGURE 9.7 A Part Image from Test Dataset.

points of the parts. Images are labeled by colorizing features so that black represents space, green indicates holes, and white represents surfaces, as seen in Figure 9.6.

After constructing the model, the classification of a part image from the test dataset is presented. Figure 9.7 is classified with both SVM and RF.

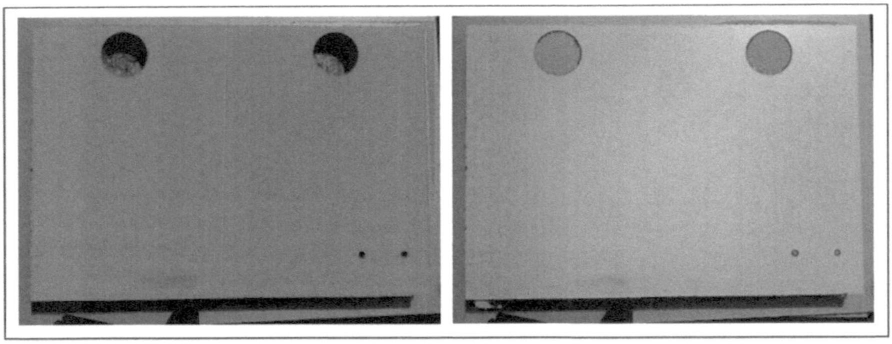

FIGURE 9.8 Classification Outcomes of the SVM Models.

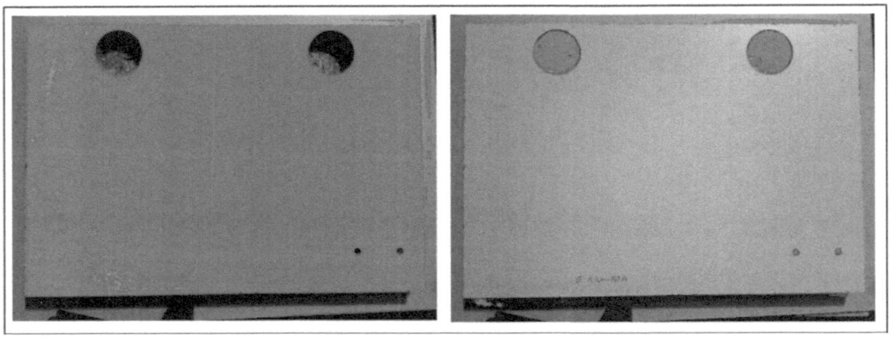

FIGURE 9.9 Classification Outcomes of the RF Models.

The classification results of the SVM models are shown in Figure 9.8. The radial basis function is used as a kernel in SVM models. The regularization parameter is set to one, the degree is set to three, and the gamma function is set to scale. The proposed model marks some of the outer space as a surface, which is unwanted for outer edge detection. However, holes are detected with satisfactory performance.

The classification results of RF models are seen in Figure 9.9. The RF model is constructed with 100 estimators. The Gini algorithm is used to generate decision trees, and boosting is enabled to enhance the classification performance. While the RF model is better for the detection of the outer edge, the SVM is better for hole detection.

Because the pointwise classification is performed, the accuracy is measured by the number of points classified correctly. Overall, the performance of both systems is closer to the classification process, as seen in Table 9.3.

After the classification process, we applied Gaussian blur and Canny algorithms to detect edges. First, we applied dilation to the detected outer space points to intersect outer space with the edge points of the part. When points of outer space intersect with a large number of edge points, we mark the inner boundary of the class as outer edge

TABLE 9.3
Performance of the ML Models

ML Algorithm	Accuracy	Precision	Recall	F1-Score
SVMs	0.9092	0.9086	0.9987	0.9515
RF	0.9056	0.9057	0.9984	0.9498

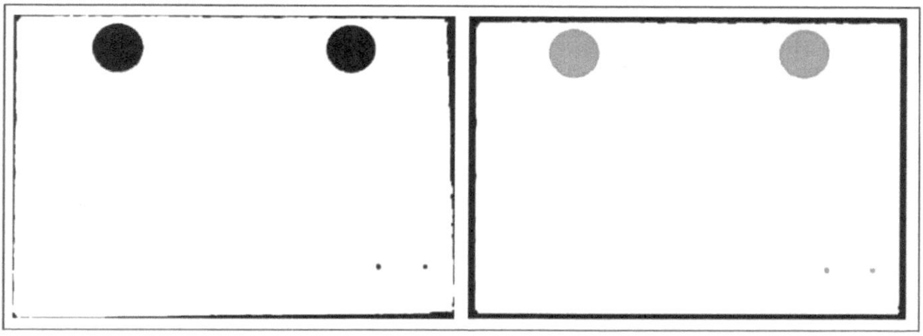

FIGURE 9.10 Reconstruction of Fields.

points. We detected edge lines for the second in the same way. Finally, we applied a perspective transformation to determine fields, as seen in Figure 9.10.

9.5 CONCLUSION

Automation systems used in production may now respond with greater adaptability and flexibility thanks to the digital transformation. The digital transition has numerous benefits, but dark factories continue to face development challenges. The dark factory's goal cannot be achieved without the use of digital twin technology. Numerous pieces of hardware are available to assist digital twin initiatives in the industry, offering benefits in terms of cost, maintenance, and technology. This circumstance facilitates the industrialization of DTs. For this reason, digital twins are used in the majority of industrial processes, including production planning, sustainable manufacturing, product life-cycle management, process planning, and product design.

From the standpoint of a cyber-physical system, we described the hardware utilized to execute digital twin models, as well as the structure of digital twin applications in the industry. Our methodology divided the digital twin models into three categories: digital twin models used during production, digital twin models used during pre-production, and digital twin models used post-production. Three different digital twin models that are employed at every stage of the product life cycle were presented. Physical model-based digital twin is the first type, data-driven digital twin is the second, and hybrid digital twin is the third. All of the digital twin model structures that depend on using sensor data to simulate system dynamics were discussed.

We concluded by presenting two case examples. In the first case study, Mercedes-Benz's modeling test time is discussed. The optimum test case may be determined using the suggested data-driven digital twin model, which predicts test length based on the test setup. As a result, the concept helps to reduce costs, lengthen processes, and increase sustainability. The suggested model has a 94.25% prediction accuracy. In the second paradigm, dimensional control is made possible by producing technical drawings of equivalent data from component photos. In order to get DNN performance in the short training period of SVM and RF, we integrated 3D convolutional layers into these algorithms. The F1-score of SVM and RF classifier models, respectively, are 95.15% and 94.98%. Following the part's field classification, we used the Canny edge detector, digital image processing operators, and perspective transformation to reconstruct the results for comparison with technical drawings.

REFERENCES

[1] M. Grieves, "Digital Twin: Manufacturing excellence through virtual factory replication," *White Paper 1*, pp. 1–7, March, 2014.

[2] R. Zhang, F. Gu, H. Mansaf, T. Wang, and A. D. Ball, "Gear wear monitoring by modulation signal bispectrum based on motor current signal analysis," *Mech Syst Signal Process*, vol. 94, pp. 202–213, Sep. 2017, doi: 10.1016/j.ymssp.2017.02.037.

[3] M. W. Grieves, "Product lifecycle management: The new paradigm for enterprises," *Int J Prod Develop*, vol. 2, no. 1–2, 2005, doi: 10.1504/ijpd.2005.006669.

[4] K. T. Park *et al.*, "Design and implementation of a digital twin application for a connected micro smart factory," *Int J Comput Integr Manuf*, vol. 32, no. 6, 2019, doi: 10.1080/0951192X.2019.1599439.

[5] Q. Qi, F. Tao, Y. Zuo, and D. Zhao, "Digital Twin service towards smart manufacturing," *Procedia CIRP*, 2018. doi: 10.1016/j.procir.2018.03.103.

[6] I. Onaji, D. Tiwari, P. Soulatiantork, B. Song, and A. Tiwari, "Digital twin in manufacturing: Conceptual framework and case studies," *Int J Comput Int Manufact.* Taylor and Francis Ltd., 2022. doi: 10.1080/0951192X.2022.2027014.

[7] R. Gya *et al.*, "Digital Twins: Adding intelligence to the real world," 2022. [Online]. Available: www.prod.ucwe.capgemini.com/wp-content/uploads/2022/05/Capgemini-Research-Institute_DigitalTwins_Web.pdf

[8] European Commission, "Industry 5.0," Research and innovation. [Online]. Available: www.research-and-innovation.ec.europa.eu/research-area/industrial-research-and-innovation/industry-50_en

[9] G. Kou, H. Dinçer, S. Yüksel, and M. Deveci, "Synergistic integration of digital twins and sustainable industrial internet of things for new generation energy investments," *J Adv Res*, no. xxxx, 2023, doi: 10.1016/j.jare.2023.11.023.

[10] M. A. Mohammed *et al.*, "Industrial Internet of water things architecture for data standarization based on blockchain and digital twin technology☆," *J Adv Res*, no. xxxx, 2023, doi: 10.1016/j.jare.2023.10.005.

[11] S. Comari *et al.*, "Mobile cobots for autonomous raw-material feeding of automatic packaging machines," *J Manuf Syst*, vol. 64, no. April, pp. 211–224, 2022, doi: 10.1016/j.jmsy.2022.06.007.

[12] M. Eyuboglu and G. Atali, "A novel collaborative path planning algorithm for 3-wheel omnidirectional Autonomous Mobile Robot," *Rob Auton Syst*, vol. 169, p. 104527, 2023, doi: 10.1016/j.robot.2023.104527.

[13] C. H. Chu and Y. L. Liu, "Augmented reality user interface design and experimental evaluation for human-robot collaborative assembly," *J Manuf Syst*, vol. 68, no. November 2022, pp. 313–324, 2023, doi: 10.1016/j.jmsy.2023.04.007.

[14] Z. Zhang, Y. Ji, D. Tang, J. Chen, and C. Liu, "Enabling collaborative assembly between humans and robots using a digital twin system," *Robot Comput Integr Manuf*, vol. 86, no. November 2023, p. 102691, 2024, doi: 10.1016/j.rcim.2023.102691.

[15] C. Moussa, K. Ai-Haddad, B. Kedjar, and A. Merkhouf, "Insights into Digital Twin Based on Finite Element Simulation of a Large Hydro Generator," in *IECON 2018 – 44th Annual Conference of the IEEE Industrial Electronics Society*, 2018, pp. 553–558, doi: 10.1109/IECON.2018.8591653.

[16] C. Cimino, E. Negri, and L. Fumagalli, "Review of digital twin applications in manufacturing," *Comput Ind*, vol. 113, Dec. 2019, doi: 10.1016/j.compind.2019.103130.

[17] C. K. Lo, C. H. Chen, and R. Y. Zhong, "A review of digital twin in product design and development," *Adv Eng Inf*, vol. 48, no. March 2021, 2021, doi: 10.1016/j.aei.2021.101297.

[18] D. Ntamo *et al.*, "Industry 4.0 in action: Digitalization of a continuous process manufacturing for formulated products," *Digital Chem Eng*, vol. 3, no. April, p. 100025, 2022, doi: 10.1016/j.dche.2022.100025.

[19] P. Pascoal-Faria, D. P. da Silva, A. Mateus, and G. R. Mitchell, "Digitalization of material science – Improving product design in the context of Industry 4.0," *Mater Today Proc*, no. xxxx, 2023, doi: 10.1016/j.matpr.2023.06.237.

[20] P. Aivaliotis, K. Georgoulias, and G. Chryssolouris, "The use of Digital Twin for predictive maintenance in manufacturing," *Int J Comput Integr Manuf*, vol. 32, no. 11, 2019, doi: 10.1080/0951192X.2019.1686173.

[21] A. Theissler, J. Pérez-Velázquez, M. Kettelgerdes, and G. Elger, "Predictive maintenance enabled by machine learning: Use cases and challenges in the automotive industry," *Reliab Eng Syst Saf*, vol. 215, Nov. 2021, doi: 10.1016/j.ress.2021.107864.

[22] Y. H. Son, K. T. Park, D. Lee, S. W. Jeon, and S. do Noh, "Digital twin–based cyber-physical system for automotive body production lines," *Int J Adv Manufact Technol*, vol. 115, no. 1–2, 2021, doi: 10.1007/s00170-021-07183-3.

[23] R. K. Phanden, P. Sharma, and A. Dubey, "6 A review on simulation in digital twin for aerospace, manufacturing and robotics," in *Materials Today: Proceedings*, Elsevier Ltd, 2020, pp. 174–178. doi: 10.1016/j.matpr.2020.06.446.

[24] M. Mazzoleni, M. Scandella, F. Previdi, and G. Pispola, "Data on the first endurance activity of a Brushless DC motor for aerospace applications," *Data Brief*, vol. 29, Apr. 2020, doi: 10.1016/j.dib.2020.105153.

[25] A. E. Onile, R. Machlev, E. Petlenkov, Y. Levron, and J. Belikov, "Uses of the digital twins concept for energy services, intelligent recommendation systems, and demand side management: A review," *Energy Rep*, vol. 7. 2021. doi: 10.1016/j.egyr.2021.01.090.

[26] L. Lattanzi, R. Raffaeli, M. Peruzzini, and M. Pellicciari, "12 Digital twin for smart manufacturing: A review of concepts towards a practical industrial implementation," *Int J Comput Integr Manuf*, vol. 34, no. 6, pp. 567–597, 2021, doi: 10.1080/0951192X.2021.1911003.

[27] L. Li, B. Lei, and C. Mao, "Digital twin in smart manufacturing," *J Ind Inf Integr*, vol. 26 , Mar. 2022, doi: 10.1016/j.jii.2021.100289.

[28] L. Pan, X. Guo, Y. Luan, and H. Wang, "Design and realization of cutting simulation function of digital twin system of CNC machine tool," *Procedia Comput Sci*, 2021. doi: 10.1016/j.procs.2021.02.057.

[29] C. H. dos Santos, J. A. B. Montevechi, J. A. de Queiroz, R. de Carvalho Miranda, and F. Leal, "11 Decision support in productive processes through DES and ABS in the Digital Twin era: A systematic literature review," *Int J Prod Res*. Taylor and Francis Ltd., 2021. doi: 10.1080/00207543.2021.1898691.

[30] C. Kammerer, M. Gaust, M. Küstner, P. Starke, R. Radtke, and A. Jesser, "Motor classification with machine learning methods for predictive maintenance," *IFAC-PapersOnLine*, vol. 54, no. 1, pp. 1059–1064, 2021, doi: 10.1016/j.ifacol.2021.08.126.

[31] H. A. Gohel, H. Upadhyay, L. Lagos, K. Cooper, and A. Sanzetenea, "Predictive maintenance architecture development for nuclear infrastructure using machine learning," *Nucl Eng Technol*, vol. 52, no. 7, pp. 1436–1442, Jul. 2020, doi: 10.1016/j.net.2019.12.029.

[32] H. Singh, "Big data, industry 4.0 and cyber-physical systems integration: A smart industry context," *Mater Today Proc*, 2021. doi: 10.1016/j.matpr.2020.07.170.

[33] M. M. Rathore, S. A. Shah, D. Shukla, E. Bentafat, and S. Bakiras, "The role of AI, machine learning, and big data in digital Twinning: A systematic literature review, challenges, and opportunities," *IEEE Access*, vol. 9, 2021. doi: 10.1109/ACCESS.2021.3060863.

[34] J. C. P. Cheng, W. Chen, K. Chen, and Q. Wang, "Data-driven predictive maintenance planning framework for MEP components based on BIM and IoT using machine learning algorithms," *Autom Constr*, vol. 112, Apr. 2020, doi: 10.1016/j.autcon.2020.103087.

[35] S. Ayvaz and K. Alpay, "Predictive maintenance system for production lines in manufacturing: A machine learning approach using IoT data in real-time," *Expert Syst Appl*, vol. 173, Jul. 2021, doi: 10.1016/j.eswa.2021.114598.

[36] J. Wilhelm, C. Petzoldt, T. Beinke, and M. Freitag, "16 review of Digital Twin-based interaction in smart manufacturing: Enabling cyber-physical systems for human-machine interaction," *Int J Comput Integr Manuf*, vol. 34, no. 10, pp. 1031–1048, 2021, doi: 10.1080/0951192X.2021.1963482.

[37] Q. Zhang, Z. Yang, J. Duan, Z. Liu, and J. Qin, "Three-dimensional visualization interactive system for digital twin workshop," *J Southeast Univ (Engl Ed)*, vol. 37, no. 2, 2021, doi: 10.3969/j.issn.1003-7985.2021.02.003.

[38] Z. Han, Y. Li, M. Yang, Q. Yuan, L. Ba, and E. Xu, "Digital twin-driven 3D visualization monitoring and traceability system for general parts in continuous casting machine," *J Adv Mech Design Syst Manufact*, vol. 14, no. 7, 2020, doi: 10.1299/jamdsm.2020jamdsm0100.

[39] L. Arnemann, S. Winter, N. Quernheim, P. Grieser, R. Anderl, and B. Schleich, "Information model to return data of Digital Twins into product design," *Procedia CIRP*, vol. 116, pp. 173–178, 2023, doi: 10.1016/j.procir.2023.02.030.

[40] K. Y. H. Lim, P. Zheng, C. H. Chen, and L. Huang, "A digital twin-enhanced system for engineering product family design and optimization," *J Manuf Syst*, vol. 57, no. August, pp. 82–93, 2020, doi: 10.1016/j.jmsy.2020.08.011.

[41] A. M. Madni, C. C. Madni, and S. D. Lucero, "Leveraging digital twin technology in model-based systems engineering," *Systems*, vol. 7, no. 1, 2019, doi: 10.3390/systems7010007.

[42] Q. Liu, B. Liu, G. Wang, and C. Zhang, "A comparative study on digital twin models," in *AIP Conference Proceedings*, 2019. doi: 10.1063/1.5090745.

[43] X. Wen, Y. Sun, H. L. Ma, and S. H. Chung, "Green smart manufacturing: energy-efficient robotic job shop scheduling models," *Int J Prod Res*, 2022, doi: 10.1080/00207543.2022.2112989.

[44] Z. Wei, W. Liao, and L. Zhang, "Hybrid energy-efficient scheduling measures for flexible job-shop problem with variable machining speeds," *Expert Syst Appl*, vol. 197, Jul. 2022, doi: 10.1016/j.eswa.2022.116785.

[45] D. B. M. M. Fontes, S. M. Homayouni, and J. C. Fernandes, "Energy-efficient job shop scheduling problem with transport resources considering speed adjustable resources," *Int J Prod Res*, 2023, doi: 10.1080/00207543.2023.2175172.

[46] G. Jinquan, H. Hongwen, L. Jianwei, and L. Qingwu, "Driving information process system-based real-time energy management for the fuel cell bus to minimize fuel cell engine aging and energy consumption," *Energy*, vol. 248, Jun. 2022, doi: 10.1016/j.energy.2022.123474.

[47] D. A. Howard, Z. Ma, C. Veje, A. Clausen, J. M. Aaslyng, and B. N. Jørgensen, "Greenhouse industry 4.0 – digital twin technology for commercial greenhouses," *Energy Inf*, vol. 4, Jan. 2021, doi: 10.1186/s42162-021-00161-9.

[48] H. Dündar, M. Soysal, M. Ömürgönülşen, and A. Kanellopoulos, "A green dynamic TSP with detailed road gradient dependent fuel consumption estimation," *Comput Ind Eng*, vol. 168, p. 108024, Jun. 2022, doi: 10.1016/J.CIE.2022.108024.

[49] J. Friederich, D. P. Francis, S. Lazarova-Molnar, and N. Mohamed, "A framework for data-driven digital twins for smart manufacturing," *Comput Ind*, vol. 136, Apr. 2022, doi: 10.1016/j.compind.2021.103586.

[50] W. Kritzinger, M. Karner, G. Traar, J. Henjes, and W. Sihn, "Digital Twin in manufacturing: A categorical literature review and classification," in *IFAC-PapersOnLine*, Elsevier BV, Jan. 2018, pp. 1016–1022, doi: 10.1016/j.ifacol.2018.08.474.

[51] R. J. Somers, J. A. Douthwaite, D. J. Wagg, N. Walkinshaw, and R. M. Hierons, "Digital-twin-based testing for cyber–physical systems: A systematic literature review," *Information and Software Technology*, vol. 156. Elsevier B.V., Apr. 01, 2023, doi: 10.1016/j.infsof.2022.107145.

[52] C. Chen, H. Fu, Y. Zheng, F. Tao, and Y. Liu, "The advance of digital twin for predictive maintenance: The role and function of machine learning," *J Manuf Syst*, vol. 71, no. June, pp. 581–594, 2023, doi: 10.1016/j.jmsy.2023.10.010.

[53] J. Liu, X. Liu, J. Vatn, and S. Yin, "A generic framework for qualifications of digital twins in maintenance," *J Automation Intell*, vol. 2, no. 4, pp. 196–203, 2023, doi: 10.1016/j.jai.2023.07.002.

[54] M. Macchi, L. Ragazzini, and E. Negri, "A conceptual framework for Digital Twins in production scheduling and control," *IFAC PapersOnLine*, vol. 56, no. 2, pp. 5661–5666, 2023, doi: 10.1016/j.ifacol.2023.10.491.

[55] K. Alexopoulos, N. Nikolakis, E. Bakopoulos, S. Vasilis, and M. Panagiotis, "Machine learning agents augmented by Digital Twinning for smart production scheduling," *IFAC PapersOnLine*, vol. 56, no. 2, pp. 2963–2968, 2023, doi: 10.1016/j.ifacol.2023.10.1420.

[56] Z. Li and Y. Chen, "Dynamic scheduling of multi-memory process flexible job shop problem based on digital twin," *Comput Ind Eng*, vol. 183, no. October 2022, p. 109498, 2023, doi: 10.1016/j.cie.2023.109498.

[57] L. Ante, "Digital twin technology for smart manufacturing and industry 4.0: A bibliometric analysis of the intellectual structure of the research discourse," *Manuf Lett*, vol. 27, no. October 2020, pp. 96–102, 2021, doi: 10.1016/j.mfglet.2021.01.003.

[58] A. A. Malik and A. Brem, "Digital twins for collaborative robots: A case study in human-robot interaction," *Robot Comput Integr Manuf*, vol. 68, no. September 2019, p. 102092, 2021, doi: 10.1016/j.rcim.2020.102092.

[59] J. Liu *et al.*, "A digital twin-driven approach towards traceability and dynamic control for processing quality," *Adv Eng Inf*, vol. 50, no. August 2020, p. 101395, 2021, doi: 10.1016/j.aei.2021.101395.

[60] W. He *et al.*, "Structural performance prediction based on the digital twin model: A battery bracket example," *Reliab Eng Syst Saf*, vol. 229, no. April 2022, p. 108874, 2023, doi: 10.1016/j.ress.2022.108874.

[61] Z. Wu and J. Li, "A framework of dynamic data driven digital twin for complex engineering products: The example of aircraft engine health management," *Procedia Manuf*, vol. 55, no. C, pp. 139–146, 2021, doi: 10.1016/j.promfg.2021.10.020.

[62] M. Perani, S. Baraldo, M. Decker, A. Vandone, A. Valente, and B. Paoli, "Track geometry prediction for Laser Metal Deposition based on on-line artificial vision and deep neural networks," *Robot Comput Integr Manuf*, vol. 79, no. September 2022, p. 102445, 2023, doi: 10.1016/j.rcim.2022.102445.

[63] J. S. Devagiri, S. Paheding, Q. Niyaz, X. Yang, and S. Smith, "Augmented reality and artificial intelligence in industry: Trends, tools, and future challenges," *Expert Syst Appl*, vol. 207, no. January, p. 118002, 2022, doi: 10.1016/j.eswa.2022.118002.

[64] K. Velmurugan, S. Saravanasankar, and S. Bathrinath, "Smart maintenance management approach: Critical review of present practices and future trends in SMEs 4.0," *Mater Today Proc*, vol. 62, pp. 2988–2995, 2022, doi: 10.1016/j.matpr.2022.02.622.

[65] Y. Wang, S. Addepalli, and Y. Zhao, "Recurrent neural networks and its variants in remaining useful life prediction," *IFAC-PapersOnLine*, vol. 53, no. 3, pp. 137–142, 2020, doi: 10.1016/j.ifacol.2020.11.022.

[66] T. G. Ritto and F. A. Rochinha, "Digital twin, physics-based model, and machine learning applied to damage detection in structures," *Mech Syst Signal Process*, vol. 155, p. 107614, 2021, doi: 10.1016/j.ymssp.2021.107614.

[67] D. Gürdür Broo, M. Bravo-Haro, and J. Schooling, "Design and implementation of a smart infrastructure digital twin," *Autom Constr*, vol. 136, no. February, p. 104171, 2022, doi: 10.1016/j.autcon.2022.104171.

[68] A. K. Ghosh, A. S. Ullah, R. Teti, and A. Kubo, "Developing sensor signal-based digital twins for intelligent machine tools," *J Ind Inf Integr*, vol. 24, p. 100242, 2021, doi: 10.1016/j.jii.2021.100242.

[69] Y. J. Lin, Y. Y. Hsieh, and C. Y. Huang, "Ontology-based Manufacturing Control Systems (MCS)," *Procedia Manuf*, vol. 39, no. 2019, pp. 1906–1912, 2019, doi: 10.1016/j.promfg.2020.01.242.

[70] Y. J. Lin, Y. H. Lin, and C. Y. Huang, "Using cyber PLC to link physical operations with cyber control decisions," *Procedia Manuf*, vol. 39, no. 2019, pp. 294–299, 2019, doi: 10.1016/j.promfg.2020.01.450.

[71] Keyence, "Keyence Linescan Camera Control Unit." [Online]. Available: www.keyence.com/mykeyence/?ptn=001

[72] J. M. Horcas, M. Pinto, and L. Fuentes, "A modular metamodel and refactoring rules to achieve software product line interoperability," *J Syst Software*, vol. 197, p. 111579, 2023, doi: 10.1016/j.jss.2022.111579.

[73] S. Gupta, "An edge-computing based Industrial Gateway for Industry 4.0 using ARM TrustZone technology," *J Ind Inf Integr*, vol. 33, no. June 2022, p. 100441, 2023, doi: 10.1016/j.jii.2023.100441.

[74] Daimler, "Mercedes-Benz Greener Manufacturing," [Online]. 2021. Available: www.kaggle.com/c/mercedes-benz-greener-manufacturing.

10 Leveraging Digital Twins for Optimal Automation and Smart Decision-Making in Industry 4.0

Revolutionizing Automation and Efficiency

Anita Mohanty, Ambarish G. Mohapatra, Subrat Kumar Mohanty, Nitaigour Premchand Mahalik and A. Jose Anand

10.1 INTRODUCTION

In the era of Industry 4.0, the convergence of cutting-edge technologies has paved the way for unprecedented advancements in automation and efficiency within industrial processes. At the forefront of this transformative landscape is the concept of digital twins [1]—a revolutionary paradigm that enables the creation of virtual replicas of physical assets and systems. Leveraging digital twins has emerged as a pivotal strategy for optimizing automation and facilitating smart decision-making in various industries [2]. These digital duplicates give a data-driven, real-time reflection of their physical counterparts, providing a thorough grasp of status, performance, and possible problems. Organizations are now able to anticipate and proactively handle difficulties in addition to precisely monitoring and controlling their operations, thanks to the seamless integration of sensors, IoT devices, and advanced analytics. Industry 4.0's paradigm shift toward digital twins not only improves operational efficiency but also creates opportunities for creative problem-solving, proactive maintenance, and quick decisions, changing the industrial automation environment. Digital twins can revolutionize company operations and ensure a more resilient, intelligent, and adaptable future. Industries are beginning to realize this. An overview of Industry 4.0 and digital twins technology is portrayed in Figure 10.1.

DOI: 10.1201/9781003465904-13

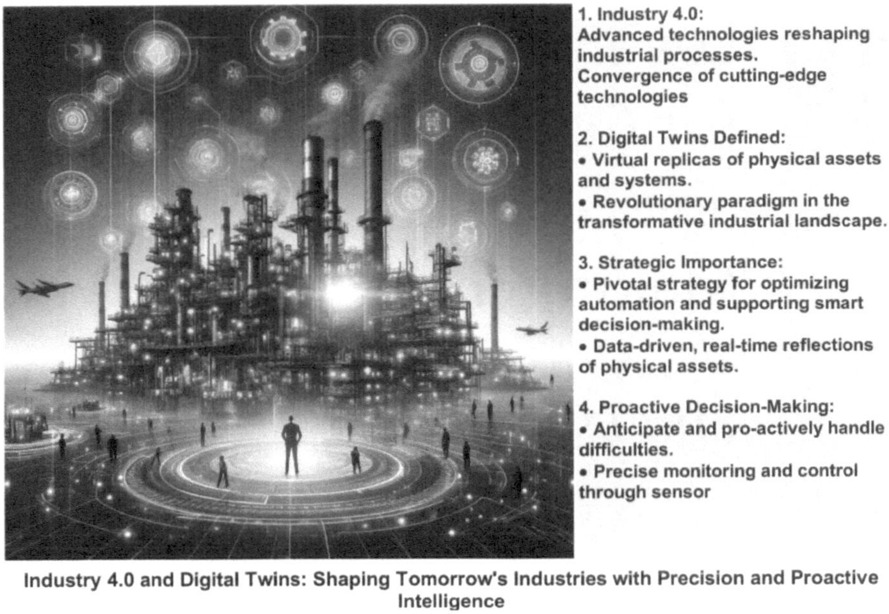

FIGURE 10.1 Overview of Industry 4.0 and Digital Twins Technology.

10.1.1 Outline of Industry 4.0 and the Role of Digitalization in Modern Manufacturing

As a paradigm change in production, Industry 4.0 is defined by the incorporation of digital technology to produce intelligent, networked, and self-governing systems. Using innovations such as the Internet of Things (IoT), artificial intelligence (AI), big data analytics, and cyber-physical systems, this fourth industrial revolution expands on the automation base. The objective is to establish a manufacturing environment that is more responsive, adaptable, and efficient. Digitalization is at the core of Industry 4.0 and is vital to contemporary manufacturing [3]. Converting conventional procedures, goods, and services into digital formats allows for smooth decision-making, data interchange, and communication. We call this process "digitalization." Digital twins, smart factories, and connected supply chains are examples of how manufacturing is becoming more digitalized. Digital twins—virtual replicas of physical assets—allow for real-time observation, analysis, and optimization of manufacturing processes. IoT devices and data analytics are used by smart factories to increase automation, boost productivity, and enable predictive maintenance [4]. Collaboration and agility are fostered via connected supply chains, which guarantee a seamless information flow from suppliers to manufacturers to customers. A new era of competitiveness and responsiveness in the global industrial landscape is being ushered in by the integration of digitalization into modern manufacturing, which not only improves productivity and quality but also makes innovation, customization, and sustainability easier.

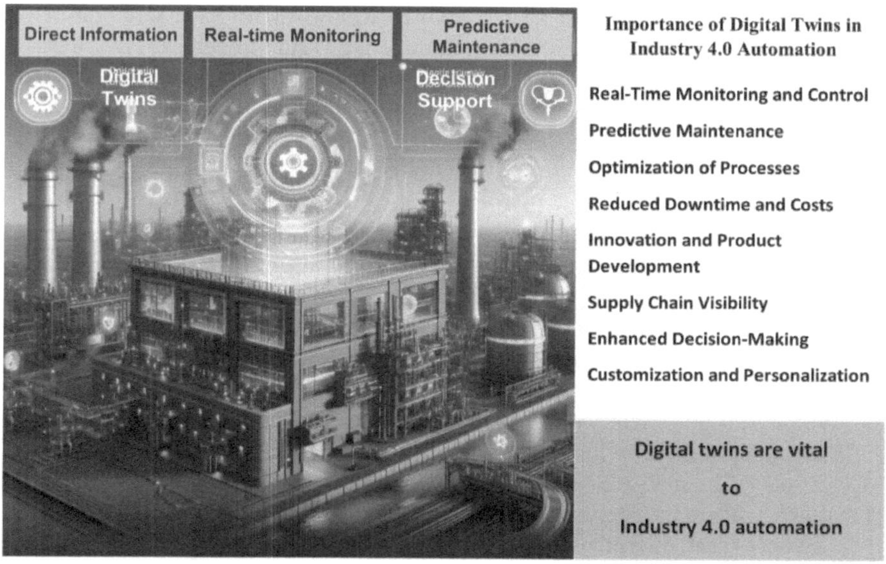

FIGURE 10.2 Importance of Digital Twins Technology in Industry 4.0.

Figure 10.2 highlights the importance of digital twin technology in Industry 4.0-based implementations. To summarize, digital twins are a fundamental element of Industry 4.0, offering the technology framework required for intelligent automation, data-driven decision-making, and process optimization. Their significance is seen up the value chain, where they help the industrial industry to become more inventive, efficient, and flexible.

10.1.2 Importance of Digital Twins in Industry 4.0 Automation

Digital twins are vital to Industry 4.0 automation because they are virtual copies of existing industrial processes that are revolutionizing manufacturing [5]. Here are a few salient features emphasizing their importance:

a. Real-Time Monitoring and Control: Digital twins offer up-to-date information on the condition and effectiveness of tangible assets and procedures. This makes it possible for constant observation and management, empowering operators to act quickly and decisively. Digital twins' dynamic properties guarantee that the virtual and physical representations are always in sync.
b. Predictive Maintenance: Digital twins enable predictive maintenance by continuously gathering and evaluating data from the physical asset. Proactive maintenance interventions are made possible by ML algorithms that can forecast when equipment is going to fail. This prolongs the life of the machinery and reduces downtime [6].

c. Optimization of Processes: By modeling various situations and configurations, digital twins make it possible to optimize industrial processes. This makes it possible to detect inefficiencies, bottlenecks, and areas for potential improvement. Organizations can improve their overall operational efficiency by implementing data-driven modifications.
d. Reduced Downtime and Costs: Unplanned downtime is significantly reduced when equipment failures may be predicted and prevented with the help of digital twins. This directly affects production continuity and saves businesses money on related expenses, such as unplanned maintenance and lost productivity [7].
e. Innovation and Product Development: By offering a platform for virtual prototyping and testing, digital twins support innovation in product development. This quickens the process of developing new products by enabling quick iteration and optimization before the construction of actual prototypes. It also encourages the development of more sophisticated and dependable goods.
f. Supply Chain Visibility: Industry 4.0 digital twins encompass the complete supply chain in addition to individual assets. Better synchronization, responsiveness, and coordination throughout the supply network are made possible by this visibility. Businesses can adjust to shifting customer needs and streamline their logistics to increase productivity.
g. Enhanced Decision-Making: The wealth of data generated by digital twins empowers decision-makers with comprehensive insights. This data-driven decision-making process enables organizations to respond quickly to changing conditions, build strategic choices, and keep competitive in a fast-evolving industrial landscape [8].
h. Customization and Personalization: Digital twins are supporting the move toward more personalized and customized products. Manufacturers can readily customize designs, settings, and processes to match individual client requests by simulating goods or production lines virtually.

Digital twins have evolved from static computer models to dynamic, real-time digital duplicates that are intricately entwined with their physical surroundings [9]. Technological developments in computers and connection, as well as an increasing awareness of the revolutionary possibilities of digital twins in terms of increasing productivity, cutting expenses, and promoting innovation across industries.

The historical evolution of digital twins unfolds through key developmental stages, each contributing to their current significance in Industry 4.0. Table 10.1 outlines the timeline from early modeling and simulation to the integration of IoT, capturing this transformative journey. However, it is essential to acknowledge the limitations faced in the contemporary landscape. Challenges such as adapting to dynamic changes, integrating real-time data, and ensuring security and privacy persist. Understanding these historical contexts and limitations provides a comprehensive perspective on the remarkable trajectory and ongoing advancements in the realm of digital twins, shaping the intelligent, networked systems of the future.

TABLE 10.1
Evolution of Digital Twins: Historical Stages, Timeline, and Addressing Limitations in Industrial Applications

Development Stage	Timeline
Early Modeling and Simulation (1950s-1980s)	The emergence of digital twins began with early computer-aided design (CAD) and simulation tools. They provided a static representation that lacked real-time connectivity and had a limited ability to adapt to dynamic real-world changes.
Advancements in CAD and computer-aided engineering (CAE) (1980s-2000s)	With progress in CAD and CAE, more accurate models were produced. However, there was a lack of a constant feedback loop, and these models remained isolated from the real world, resulting in limited integration with real-time data and dynamic environments.
Integration of IoT and Sensors (2000s-2010s)	The acceleration of technology occurred with the widespread use of sensors and IoT, leading to dynamic interaction between the virtual and physical worlds and real-time data integration. However, challenges arose in terms of data security, privacy, and standardization.
Rise of Industry 4.0 (2010s-Present)	This represented a paradigm shift toward intelligent, networked production, which has been boosted by the capabilities of cloud computing, big data analytics, and AI. Despite these advancements, limitations in dealing with complex, unstructured data persist.
Expansion Beyond Manufacturing (2010s-Present)	The concept has spread to various industries (e.g., infrastructure, smart cities, healthcare), allowing for the modeling and optimization of complex systems beyond manufacturing. Nonetheless, there remains difficulty in achieving universal standards for digital twin implementation.
Current State and Future Prospects	Digital twins are now an essential component of Industry 4.0, playing a crucial role in data-driven decision-making, process optimization, and predictive maintenance. Ongoing challenges include technological refinement, interoperability, and addressing security and privacy concerns. Their predicted significance lies in shaping intelligent, networked systems across industries.

10.2 DIGITAL TWINS IN MANUFACTURING

In the manufacturing industry, digital twins have become a disruptive force that is altering conventional procedures and optimizing the whole product life cycle. Digital twins are dynamic virtual copies of real assets that are used in manufacturing, such as production lines, machinery, and finished goods. These digital equivalents provide a thorough and accurate portrayal, as they are updated in real time with data from sensors and IoT devices embedded in the physical entities. Digital twins are used by manufacturers for many different tasks, such as process optimization, quality control, predictive maintenance, and product design and prototyping [10]. Digital twins improve operational efficiency, lower downtime through predictive maintenance, and enable fast decision-making by modeling and analyzing various scenarios [11]. By providing production processes with a platform for quick prototyping and testing, the incorporation of digital twins increases productivity while simultaneously encouraging innovation. Digital twins are essential to the smooth coordination and optimization of industrial processes in the era of Industry 4.0, from design and production to maintenance and eventual decommissioning [12].

Throughout the production process, digital twins have a wide range of significant applications that improve productivity, creativity, and decision-making. The following are a few important manufacturing application areas [13]:

a. Product Design and Development: To generate virtual prototypes early in the process, digital twins are employed. This enables producers to test features, simulate and evaluate various design iterations, and spot possible problems before creating actual prototypes.
b. Smart Manufacturing and Production: By offering real-time insights into the operation of production lines and machines, digital twins facilitate smart manufacturing during the production phase. This makes quality assurance, predictive maintenance, and process optimization in manufacturing more efficient.
c. Supply Chain Management: By providing visibility and coordination between suppliers, manufacturers, and distributors, digital twins cover the entire supply chain. This increases the overall efficiency of the supply chain by facilitating the timely delivery of components, effective management of inventory, and optimization of logistics.
d. Asset Management: By monitoring a machine or piece of equipment's condition, performance, and usage trends, digital twins assist in managing individual assets. Utilizing data from digital twins, predictive maintenance reduces downtime and increases machinery longevity.
e. Quality Control and Assurance: Manufacturers use digital twins to apply cutting-edge quality assurance techniques. Real-time data analysis from the production process makes it possible to identify deviations and flaws early on, guaranteeing that the final product meets strict quality requirements.
f. Energy Management: By tracking and examining energy usage trends in manufacturing plants, digital twins help to optimize energy consumption. When a digital twin points out opportunities for efficiency gains, it promotes cost-effective and sustainable practices.
g. Worker Training and Safety: Digital twins can be used to construct virtual versions of manufacturing facilities that are useful for providing safe, regulated training for employees. This reduces the dangers connected with on-the-job training, ensures safety compliance, and helps to improve skills.
h. Life-cycle Management: From design and manufacture to operation and maintenance, digital twins encompass a product's whole life cycle. Throughout the life cycle, this all-encompassing strategy guarantees that the insights gleaned from the digital twin will contribute to ongoing innovation and development.

Digital twins—which provide a comprehensive and data-driven approach to product creation, production, and overall operational management—have essentially become a crucial component of contemporary manufacturing [14]. In the era of Industry 4.0, their diverse applications support the manufacturing industries' resilience, efficiency, and competitiveness.

10.3 CREATING A DIGITAL TWIN

A methodical approach to capturing, modeling, and linking physical objects in the real world to their virtual equivalents is required to create a digital twin [15]–[16]. Installing sensors and data-generating equipment on the actual object or system of interest is the first stage. These sensors gather data in real time on several different parameters, including movement, pressure, temperature, and other pertinent factors. The next important step is using this data to create a detailed virtual model. Sophisticated technologies are used to create a precise and dynamic representation of the physical object, including 3D modeling, simulation, and data analytics. Subsequently, the digital twin is updated in real time via communication architecture that frequently makes use of the IoT. The bidirectional flow guarantees that the behavior and present condition of the physical counterpart is reflected in the virtual model. At last, the digital twin is included in a larger system, enabling stakeholders to communicate with the virtual version and gain insightful knowledge from it. In addition to technological know-how, the development of a digital twin necessitates a thorough comprehension of the precise goals and features that are needed, whether the applications are for process optimization, predictive maintenance, or other industry-specific uses.

10.3.1 Data Collection and Integration from Physical Assets and Sensors

Creating a digital twin starts with gathering and integrating data from sensors and physical assets. There are multiple crucial phases in the process:

a. Sensor Deployment: Installing sensors and data-generating equipment on the physical assets is the initial phase. Depending on the type of asset and the criteria that require monitoring, these sensors are positioned strategically to collect pertinent data points. Depending on the kind of information needed, sensors can include accelerometers, temperature sensors, pressure sensors, cameras, and more.
b. Data Acquisition: Real-time data is continuously gathered by sensors from the physical assets. Numerous variables, including temperature, humidity, vibration, energy consumption, and other pertinent metrics, may be included in this data. The frequency and level of detail in the data that is gathered depends on the specific requirements and objectives of the digital twin application.
c. Communication Infrastructure: Data transmission from sensors to the digital twin platform requires a strong communication infrastructure. IoT technologies, wireless networks, and other connectivity options that facilitate dependable and easy data transfer are frequently used in this infrastructure.

d. Data Processing and Aggregation: After being gathered, sensor raw data is cleaned, filtered, and aggregated. Before being integrated into the digital twin, this procedure helps to guarantee the accuracy and consistency of the data. The aggregated data offers a more thorough understanding of the behavior and performance of the physical asset.
e. Integration with Digital Twin Platform: Following data processing, the digital twin platform is integrated. Mapping the data points to appropriate virtual model elements is the integration process. Application programming interfaces (APIs) and middleware are examples of advanced data integration techniques and technologies that are used to provide a smooth transition between the digital and physical worlds.
f. Real-Time Updates: The integrated data is updated continually in real time to ensure the accuracy of the digital twin. The digital twin updates the virtual model whenever the physical asset changes or suffers variations in conditions, guaranteeing that the virtual model always remains a dynamic and precise depiction of the actual asset.
g. Analytics and Insight Generation: Analytics and insight generation rely on integrated data from the digital twin. AI, ML algorithms, and advanced analytics technologies are used to examine trends, forecast behavior, and extract valuable information from the data that has been gathered. Making educated decisions and maximizing the physical asset's performance depends on this stage.
h. Visualization and User Interface: A user-friendly interface presents the processed and analyzed data, enabling stakeholders to see the insights obtained from the digital twin. Using visualization tools, users may engage with the virtual model, keep an eye on real-time data, and learn more about the performance and condition of the physical asset.

10.3.2 Data Modeling and Simulation

Digital twin development and operation require both data modeling and simulation [17]. To facilitate research, testing, and optimization, these procedures entail the virtualization of actual systems, assets, or operations. An overview of data modeling and simulation about digital twins is provided below:

a. **Data Modeling**
   - **Definition of a Data Model:** The first stage is creating a data model that depicts the relationships and structure of the pertinent data for the system or physical asset. This entails determining the important variables, parameters, and relationships between them.
   - **Data Mapping:** Corresponding elements in the data model are mapped to the sensor data that has been gathered. This mapping makes sure that the virtual model faithfully replicates the physical world's circumstances and the behavior of the physical asset.

b. **Simulation**
   - **Dynamic Representation:** Based on the data model, simulation entails producing a dynamic representation of the real asset or process. This virtual model mimics the behavior of its physical counterpart in real life by taking into account the relationships and interactions between various components.
   - **Real-Time Updates:** As fresh information is gathered from the physical asset, the simulation is updated continually in real time. In doing so, it ensures that the virtual model stays in sync with the real world, providing an accurate representation of the situation as it is at that moment.

c. **Advanced Analytics and ML**
   - **Algorithm Integration:** Advanced analytics and ML algorithms are frequently incorporated into data modeling and simulation. These algorithms use real-time and historical data analysis to spot trends, forecast behavior, and enhance efficiency.
   - **Predictive Capabilities:** By combining analytics and simulation, a digital twin may anticipate possible problems or deviations in the physical asset. Proactive decision-making is made possible by this predictive skill, such as planning maintenance before a failure happens.

d. **Visualization and User Interaction**
   - **User-Friendly Interface:** A user-friendly interface is frequently used to convey the outcomes of data modeling and simulation. Stakeholders can engage with the virtual model, see data patterns, and learn more about how well the actual asset is performing thanks to visualization tools.
   - **Decision Support:** Using the insights from the digital twin, the visual depiction helps decision-makers to comprehend complex relationships and make well-informed judgments.

e. **Iterative Refinement**
   - **Continuous Improvement:** Iterative refinement is a continuous improvement process that is facilitated by data modeling and simulation. The digital twin can be improved to include more factors, adjust to changes in the physical world, and increase accuracy when new data is gathered and new insights are discovered.

Organizations can forecast outcomes, optimize processes, and have a better understanding of their assets when data modeling and simulation are incorporated into the digital twin framework. Applications of this technology may be found in a variety of sectors, such as infrastructure, manufacturing, healthcare, and energy. These applications promote cost savings, enhanced productivity, and better decision-making.

10.4 DIGITAL TWIN TECHNOLOGIES

An extensive ecosystem of digital twin technologies enables the construction and administration of dynamic virtual copies of real systems or assets. At its core, the IoT facilitates real-time data collection through sensor networks, establishing a constant flow of information between the physical and virtual realms. Advanced data analytics and ML processes extract valuable insights from this data, predicting behaviors and patterns [18]. Cloud computing ensures scalability and efficient data storage, while edge computing enables real-time processing and reduced latency. AR and virtual reality (VR) technologies enhance visualization and immersive experiences. Blockchains ensure data security and integrity, especially in contexts requiring trust and transparency. The digital thread integrates data across the product life cycle, supporting collaboration and continuity [19]. The digital and physical components are closely connected via cyber-physical systems, and specialized digital twin platforms offer integrated settings for easy management and analysis. When considered as a whole, these technologies enable businesses in a variety of sectors to innovate, improve decision-making, and streamline processes as a result of the revolutionary potential of digital twins.

10.4.1 IoT and Sensor Networks

Modern connectivity is mostly dependent on the IoT and sensor networks, which allow for the seamless integration of physical devices and the interchange of data across a wide range of applications. The IoT and sensor networks are essential in the digital twin context because they enable real-time data to be collected from physical assets and to continuously transfer information between the virtual and physical worlds [20].

The IoT is a network of networked objects that may trade and gather data because they are frequently equipped with sensors, actuators, and communication modules. When it comes to digital twins, these gadgets act as the eyes and ears of the physical world, continuously delivering data that represents the condition of assets or operations in real time.

In the context of digital twins, sensor networks are essential IoT facilitators. These networks are made up of sensors that are carefully positioned on tangible assets to record particular data points, such as vibration, temperature, pressure, or other pertinent characteristics. The sensors serve as data sources, sending information to the digital twin platform continually.

Complete data collection and monitoring are made possible by the interplay between sensor networks and the IoT. This combination enables organizations in manufacturing, healthcare, infrastructure, and other sectors to obtain precise and fast data regarding the functionality, state, and behavior of physical assets. IoT-enabled bidirectional communication supports applications, such as predictive maintenance, process optimization, and scenario testing. It does this by keeping the digital twin synchronized with its physical counterpart.

Ultimately, integrating sensor networks and the IoT with digital twins improves visibility, control, and decision-making across many industries, encouraging a more linked and astute method of overseeing physical assets and operations.

10.4.2 Cloud Computing and Big Data Analytics

Digital twins and other data-driven applications are greatly impacted by the capabilities and efficacy of cloud computing and big data analytics—two interconnected technology pillars. Digital twins generate massive amounts of data, which cloud computing can handle by providing the infrastructure, storage, and computational power needed. Big data analytics then processes and distills valuable insights from this data. When combined, they provide a dynamic synergy that enables organizations to fully utilize digital twins for improved decision-making, optimization, and creativity [21].

Cloud computing provides scalable and on-demand resources to store and handle large data sets in the context of digital twins. It offers the processing power required for analytics, simulations, and real-time updates related to digital twin applications. Cloud solutions promote the continuous flow of information between physical assets and their virtual counterparts by facilitating the smooth integration of several data sources.

Big data analytics draws insightful conclusions from the enormous data sets created by digital twins, which is a beneficial addition to cloud computing. This data is processed by ML algorithms and sophisticated analytics tools, which look for trends, abnormalities, and patterns. These observations are crucial for forecasting behavior, making well-informed decisions, and maximizing the efficiency of physical assets.

Several significant benefits result from the integration of big data analytics and cloud computing in digital twins:

a. **Scalability:** Cloud computing gives organizations the flexibility to grow their computing and storage capacity in response to the changing requirements of their digital twin applications. Digital twins can manage expanding data sets and rising complexity due to their scalability.
b. **Real-Time Processing:** Data processing made possible by cloud computing allows for real-time data updates to the digital twin. This capacity is essential for staying in sync with the physical asset's ever-changing conditions.
c. **Cost Efficiency:** Cloud services provide an economical approach, as they enable enterprises to pay only for the resources they utilize, negating the need for large upfront infrastructure investments. This increases the accessibility and adaptability of digital twin deployments to different workloads.
d. **Advanced Analytics:** With the use of cloud resources and big data analytics, organizations may extract useful information from the massive amounts of data that digital twins gather. This covers anomaly detection, optimization suggestions, and predictive maintenance.
e. **Collaboration and Accessibility:** By offering a centralized platform that is available to pertinent stakeholders, cloud-based digital twins improve collaboration. This encourages smooth cooperation between groups, divisions, and even geographical locations.

To support the dynamic requirements of digital twins, cloud computing and big data analytics work together to build a flexible and robust ecosystem that promotes efficiency and innovation in a variety of industries, including smart infrastructure, manufacturing, and healthcare. With these technologies together, digital twins are guaranteed to be a useful tool in the data-driven decision-making and Industry 4.0 era.

10.4.3 AR and VR in Digital Twins

Immersion technologies, such as AR and VR, improve the interaction and visualization of digital twins, providing users with a more realistic and captivating experience. AR and VR enhance comprehension, instruction, and decision-making when they are used with digital twins [22]. The following describes how these technologies are used in the domain of digital twins:

a. **Immersive Visualization**
   - Virtual Replication: AR and VR technologies enable the production of virtual, immersive duplicates of real-world objects or operations that are shown in digital twins.
   - Spatial Understanding: By viewing and interacting with the digital twin in three dimensions, users can better comprehend the real counterpart's spatial relationships.

b. **Training and Simulation**
   - Interactive Training: Using AR and VR, users can replicate real-world situations inside the digital twin's virtual environment.
   - Hands-On Experience: This increases the efficacy of training, particularly in fields involving intricate apparatus or procedures.

c. **Maintenance and Repair**
   - Guided Maintenance: By displaying digital data in the real environment, AR helps maintenance professionals to repair processes by highlighting parts and giving step-by-step guidance.
   - Remote Assistance: AR makes remote help easier by enabling professionals to mentor field technicians in real time through AR-enhanced images.

d. **Visualization of Data and Analytics**
   - Real-Time Data Overlay: By superimposing real-time data and analytics on tangible assets, AR provides operators with instant access to information about performance, conditions, and important metrics.
   - Data Interaction: By interacting with data points in the AR environment, users can grasp complex information more naturally.

e. **Collaborative Decision-Making**
   - Virtual Collaboration: By enabling numerous people to engage with the same digital twin concurrently, even if they are physically separated, AR and VR promote collaborative decision-making.
   - Spatial Context: Discussing and analyzing data in a spatial context enables collaborators to solve problems and communicate more effectively.

f. **User-Friendly Interfaces**
   - Intuitive Controls: AR and VR offer user-friendly interfaces that enable users to move around and communicate with the digital twin through controllers, voice commands, and gestures.
   - Enhanced User Experience: By offering a more organic and immersive method of interacting with the virtual representation, these technologies improve the user experience.

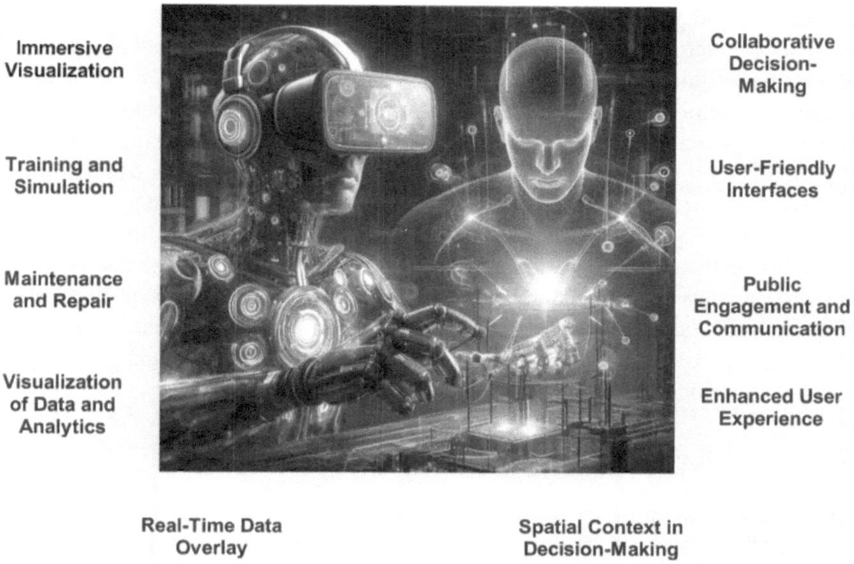

FIGURE 10.3 Enriching Digital Twins: AR and VR Integration for Enhanced Visualization, Training, Maintenance, and Collaborative Decision-Making Across Industries.

Integrating AR and VR into digital twins enriches the user experience, making complex information more accessible and actionable. Whether it is used for training, maintenance, or collaborative decision-making, these immersive technologies contribute to the broader goal of leveraging digital twins for enhanced efficiency and innovation across various industries. The integration of AR and VR into digital twins marks a significant leap forward in transforming user experiences and functionalities across various industries. As depicted in Figure 10.3, these immersive technologies play a pivotal role in enhancing the visualization of digital twins, providing users with realistic and captivating experiences. From spatial understanding and interactive training to guided maintenance and real-time data overlay, AR and VR contribute to a more dynamic and user-friendly interaction with digital replicas. This transformative technology extends to collaborative decision-making, offering virtual collaboration possibilities and facilitating effective discussions within a spatial context. Furthermore, integrating AR and VR enables public engagement and communication by visualizing large-scale projects, making them accessible for educational and stakeholder awareness purposes. Overall, Figure 10.3 encapsulates the diverse applications of AR and VR in digital twins, showcasing their potential to revolutionize how we perceive, interact with, and benefit from virtual representations of real-world systems.

10.4.4 Challenges and Considerations

Implementing digital twins is accompanied by several challenges and considerations that organizations must navigate for successful deployment. Issues such as data security and privacy loom large, demanding robust measures to protect sensitive information. Interoperability hurdles between diverse systems necessitate establishing industry-wide standards, while ensuring scalability becomes vital to handle the increasing volumes of data generated. The complexity of implementation alongside the need for specialized expertise can pose obstacles, and organizations must carefully weigh the upfront costs against the long-term benefits [23]. Maintaining the accuracy and reliability of data collected from sensors, addressing regulatory compliance, and fostering a cultural shift toward data-driven decision-making are additional challenges. Ethical considerations, limited industry standards, and potential environmental impacts further underscore the need for a holistic approach that balances technological advancements with ethical, legal, and organizational considerations [24]–[25]. The successful integration of digital twins requires strategic planning, continuous evaluation, and adaptive responses to evolving challenges.

10.5 INDUSTRY 4.0 AND DIGITAL TWIN SYNERGY

The synergy between Industry 4.0 and digital twins represents a transformative force reshaping the modern manufacturing landscape and beyond. Industry 4.0, which is characterized by integrating digital technologies into industrial processes, converges seamlessly with digital twins. Together, they form a powerful alliance that enhances operational efficiency, agility, and innovation. Industry 4.0 technologies, such as IoT, big data analytics, and AI, provide the foundation for the data-driven ecosystem essential to digital twins [26]. Digital twins, in turn, leverage real-time data and simulations to optimize processes, predict maintenance needs, and facilitate smart decision-making. This synergy empowers organizations to move beyond traditional manufacturing paradigms, offering dynamic insights into asset performance, enabling predictive capabilities, and fostering a more adaptive and responsive industrial ecosystem [27]. The collaboration between Industry 4.0 and digital twins exemplifies a holistic approach to digital transformation, ushering in an era where physical and virtual realms coalesce to drive unprecedented advancements in efficiency, sustainability, and overall operational excellence.

10.5.1 Leveraging AI and ML for Predictive Maintenance and Optimization

Using ML and AI for predictive maintenance [28] and optimization has changed the game for many industries. To forecast when equipment is likely to break, predictive maintenance uses AI and ML algorithms to evaluate past and current data from sensors, digital twins, and other sources. These technologies enable organizations to schedule maintenance proactively, preventing unscheduled downtime and increasing the lifespan of equipment by spotting patterns and abnormalities.

Additionally, by continuously analyzing data to find areas for development, AI and ML support optimization. These technologies can find efficiencies, streamline processes, and improve overall performance in various areas, including energy usage, supply chain logistics, and industrial processes. By adapting and learning from fresh data, ML algorithms enable organizations to adjust their procedures in response to evolving circumstances.

In addition to improving operational efficiency, the integration of AI and ML with predictive maintenance and optimization reduces costs and better allocates resources. By switching from reactive maintenance models to proactive techniques, organizations can minimize interruptions, reduce downtime, and eventually improve the overall productivity and reliability of their systems. Industry 4.0 will continue to spur innovation and competitiveness across a range of industries using AI and ML in predictive maintenance and optimization techniques.

10.5.2 Human–Machine Collaboration and Decision Support

Decision support and human-machine collaboration are examples of a dynamic partnership where the strengths of both parties are utilized to create more efficient and well-informed decision-making. In this collaborative paradigm, machines with the help of sophisticated analytics and AI analyze enormous volumes of data, spot trends, and produce insights. Human decision-makers are then given access to these insights, which give them a data-driven basis on which to make strategic decisions [29].

Collaboration between humans and machines entails combining human knowledge, ingenuity, and contextual awareness with computer power and analytical skills. Through the combination of human intuition and experience with computer speed and pattern recognition, this partnership improves decision-making.

Decision support systems make use of this cooperation by offering intuitive user interfaces that display practical insights obtained from machine learning. These tools help decision-makers to comprehend complicated situations, consider several options, and forecast results using data-driven models.

The synergy between humans and machines is particularly potent in complex and dynamic environments, such as Industry 4.0. In manufacturing, for example, human-machine collaboration can optimize production processes, predict maintenance needs, and improve overall operational efficiency. In healthcare, it can aid in diagnosis, treatment planning, and personalized patient care.

Effective human-machine collaboration requires clear communication, trust, and a user-friendly interface that enables humans to understand and validate machine-generated insights. As technology advances, the evolving landscape of human-machine collaboration and decision support continues to shape a future where the combined intelligence of humans and machines creates innovative solutions and drives progress across diverse domains.

10.6 FUTURE TRENDS AND DEVELOPMENTS

Anticipating the future trajectory of technology is inherently intricate, yet various directions point to the evolution of digital twins and their related technologies. Foreseeable trends include the integration of more advanced AI algorithms into digital twins, enhancing decision-making, anomaly detection, and optimization, which may lead to more autonomous and intelligent systems. The prevalence of

TABLE 10.2
Charting the Future of Digital Twins: Trends, Developments, and Impactful Emerging Technologies

Future Trends and Developments	Emerging Technologies Impacting Digital Twins
Integration of advanced AI algorithms for enhanced decision-making	AI Advancements: Explainable and interpretable AI in digital twins
Rise of edge computing in digital twin applications	Quantum Computing: Revolutionizing simulations in digital twins
Growing integration of AR and VR with digital twins	Edge Computing Enhancements: Reducing latency and improving real-time processing
Increasing use of blockchain for data security in digital twin exchanges	Extended Reality (XR): Enhancing visualization and collaboration in digital twins
The potential impact of quantum computing on digital twin capabilities	Blockchain for Data Security: Securing transactions in the digital twin environment
Widespread adoption of 5G networks for improved connectivity	5G Connectivity: Enhancing speed and reliability of data transfer in digital twin ecosystems
Expanded application of digital twins in healthcare	Generative Design: Optimizing and automating design processes in digital twins
Digital twins playing a pivotal role in smart cities	Digital Thread Integration: Enhancing continuity and traceability in digital twin ecosystems
A cross-industry collaboration involving digital twins	Robotics and Automation: Advanced simulations and optimizations in manufacturing
Used for optimizing resource consumption and sustainability	NLP: Improving human-machine interactions in digital twins
Potential development of standardized rules for data privacy and ethics	Advanced Sensor Technologies: Enhancing accuracy and richness of data in digital twins
Dynamic future characterized by innovation and adaptation	Exponential Growth in IoT Devices: Contributing to comprehensive monitoring and data collection

edge computing in digital twin applications is expected to rise, promising reduced latency, improved real-time processing, and heightened overall system responsiveness [30]. Additionally, the growing integration of AR and VR with digital twins holds potential for enhanced visualization, training, and maintenance experiences. Blockchain technology may see increased use for securing and validating data exchanges between physical assets and their digital counterparts, addressing concerns surrounding data integrity and security. The advent of QC could significantly impact digital twin capabilities, offering new avenues for the simulation and analysis of complex data sets. The widespread adoption of 5G networks is anticipated to enhance connectivity and data transfer speeds critical for real-time updates and communication within digital twin ecosystems. The application of digital twins in healthcare is poised to expand, supporting personalized medicine, more accurate treatment simulations, and improved patient outcomes. In the realm of smart cities, digital twins may play a pivotal role, integrating with urban infrastructure to enhance city planning, optimize resource utilization, and elevate the overall quality of life. A cross-industry collaboration involving digital twins is expected to increase, with insights gained in one sector informing advancements in others, such as supply chain logistics or predictive maintenance practices. The use of digital twins to optimize resource consumption, energy efficiency, and environmental impact across a variety of businesses is probably motivated by sustainability concerns. Standardized rules and guidelines covering data privacy, security, and ethical considerations may potentially be developed in the future for the use of digital twins. It is imperative to acknowledge that these trends are theoretical, and that their precise course depends on breakthroughs in technology, industrial demands, and legislative developments. Digital twins are expected to have a dynamic future characterized by constant innovation and adaptation to new possibilities and challenges.

Table 10.2 outlines the anticipated future trends and the corresponding emerging technologies that are poised to significantly impact the evolution and capabilities of digital twins across various industries.

10.6.1 Emerging Technologies That Will Impact Digital Twins

Several emerging technologies may impact the evolution and capabilities of digital twins significantly. These technologies can potentially enhance the efficiency, accuracy, and scope of digital twin applications across various industries. A few noteworthy emerging technologies include:

a. AI Advancements: As AI becomes more integrated into digital twins, efforts to make AI systems explainable and interpretable will gain prominence. This is crucial for understanding the decision-making processes of AI algorithms within digital twin applications.
b. QC: QC's unparalleled processing capabilities could revolutionize the complexity and scale of simulations within digital twins, allowing for more accurate and intricate modeling of physical systems.
c. Extended Reality (XR): The integration of AR, VR, and mixed reality into digital twins can enhance visualization, collaboration, and training experiences, providing more immersive interactions with virtual replicas.
d. Blockchain for Data Security: The application of blockchain technology to secure and verify transactions in the digital twin environment can enhance data security and integrity, which is critical for maintaining trust in the accuracy of digital twin representations.
e. 5G Connectivity: The widespread adoption of 5G networks will significantly improve the speed and reliability of data transfer between physical assets and their digital twins, facilitating real-time updates and communication.
f. Generative Design: AI-driven generative design tools can be integrated into digital twins to optimize and automate the design process, creating more efficient and innovative solutions for physical assets.

g. Robotics and Automation: The synergy between digital twins and robotics can lead to more advanced simulations and optimizations in manufacturing, allowing for efficient planning and control of robotic processes.
h. Natural Language Processing (NLP): Integrating NLP capabilities into digital twins can enhance human-machine interactions, enabling users to interact with and query digital twin systems using natural language.

10.7 CONCLUSION

Finally, digital twins offer a revolutionary connection between the digital and physical worlds, marking a paradigm shift in the field of Industry 4.0. The idea of digitally modeling physical assets, processes, or systems has developed from a technological breakthrough to a fundamental component of contemporary industrial processes. With the help of digital twins, businesses can monitor, simulate, and analyze in real time, enabling data-driven decision-making, operational optimization, and creative thinking. When cutting-edge technologies such as IoT, AI, and ML are combined, digital twins become more powerful. This makes predictive maintenance, enhanced agility, and higher efficiency possible. Digital twins are proving their adaptability and versatility by finding use in a wide range of industries, such as smart cities, healthcare, and manufacturing.

As digital twins continue to advance, their role in advancing sustainability through resource optimization, waste reduction, and environmental impact reduction will only grow. Their importance is further enhanced by their ability to work in tandem with cutting-edge technologies such as blockchain and 5G, which guarantee data confidentiality, integrity, and real-time communication. Obstacles such as data protection, scalability, and legal issues highlight the necessity of an all-encompassing and methodical implementation strategy.

Digital twins have a bright future ahead of them, one that will be characterized by constant innovation, cross-industry partnerships, and a dedication to solving societal issues. Digital twins are becoming increasingly integrated across many sectors as businesses realize how they can improve sustainability, resilience, and efficiency. Digital twins are not simply a technological advance, but also the catalyst for a new era of data-driven, intelligent, and networked decision-making that will fundamentally alter industry landscapes for years to come.

10.7.1 The Significance of Digital Twins in Industry 4.0 Automation

Digital twins are essential to Industry 4.0 automation because they provide a revolutionary way to innovate and operate across industries. Digital twins are essential because of their ability to create virtual replicas of real-world assets, systems, or processes while also providing timely insights and helpful data. In the industrial landscape of Industry 4.0, where automation and connectivity are critical, digital twins bridge the digital and physical worlds. They enhance decision-making, enable predictive maintenance, and optimize operations by using cutting-edge technologies such as AI and the IoT. This improves operating efficiency and reduces costs and downtime. The emergence of digital twins has contributed to developing smart factories and automated settings that foster innovation, adaptability, and agility. Their role in sustainability, resource optimization, and seamless interaction with other Industry 4.0 technologies highlights their significance as a cornerstone of the industrial landscape's future automation.

10.7.2 Encourage Further Research and Adoption

To fully realize the potential of digital twin technology and promote innovation across a range of industries, more study and implementation are needed. To encourage further research and application, consider the following important points:

a. Innovation Catalyst: Showcase digital twins' revolutionary effects on product creation, sustainability, and operational efficiency to position them as an innovation catalyst. Highlight their contribution to advancing technology and creating an environment that values ongoing development.
b. Cross-Industry Collaboration: Promote joint ventures between technology suppliers, research institutes, and industry. Collaborating across industries helps to establish optimal methodologies, benchmarks, and mutual understandings, thereby fostering a standard comprehension of digital twin applications.
c. Knowledge Sharing: Promote the exchange of knowledge through publications, conferences, and workshops. To promote a community-driven strategy for the growth of digital twins, encourage researchers, practitioners, and thought leaders to contribute their experiences, case studies, and research findings.
d. Education and Training Programs: Create training programs and educational activities to increase knowledge of digital twin technologies. Give experts, engineers, and students the tools they need to use and benefit from digital twins in numerous fields.
e. Demonstration Projects: Encourage and present practical demonstration projects that illustrate the quantifiable advantages of digital twins. These initiatives act as real-world illustrations of practical implementation, encouraging other businesses to investigate and utilize this technology.
f. Government and Industry Support: Promote industry and government support for digital twin-related research and development projects. This assistance can take the shape of financial resources, legal guidelines, and financial rewards that motivate businesses to spend money on digital twin technology.
g. Highlight Success Stories: Provide use cases and success stories highlighting the advantages of adopting digital twins. Emphasizing observable outcomes, including financial savings, increased productivity, and environmentally friendly procedures, might encourage others to act.
h. Interdisciplinary Research: Promote multidisciplinary studies that investigate how digital twins interact with other cutting-edge technology. Examine how

the capabilities and uses of digital twins can be further enhanced by integrating blockchains, 5G, AI, and other developments.

i. Open-Source Initiatives: Support open-source initiatives that promote collaboration and the development of standardized digital twin frameworks. Open-source platforms encourage innovation, facilitate interoperability, and they make digital twin technologies more accessible to a broader audience.
j. Ethical Considerations: Prioritize discussions on ethical considerations, data privacy, and security in digital twin research and adoption. Establish guidelines and best practices to ensure responsible and ethical use of data, fostering trust among stakeholders.

Encouraging research and adoption of digital twin technology requires a collective effort from academia, industry leaders, policymakers, and technology enthusiasts. By fostering a collaborative and supportive ecosystem, we can propel digital twins from a transformative concept to a ubiquitous and integral component of the technological landscape, driving progress and innovation across various sectors.

REFERENCES

[1] A. Sharma, E. Kosasih, J. Zhang, A. Brintrup, A. Calinescu. 2022. Digital Twins: State of the art theory and practice, challenges, and open research questions. *Journal of Industrial Information Integration*. 30: 1–15.
[2] A. Fuller, Z. Fan, C. Barlow. 2020. Digital Twin: Enabling technologies, challenges and open research. *IEEE Access*. 8: 108952–108971. doi: 10.1109/ACCESS.2020.2998358
[3] E. R. Silva, A. C. Shinohara, C. P. Nielsen, E. P. Lima, J. Angelis. 2020. Operating Digital manufacturing in Industry 4.0: The role of advanced manufacturing technologies. *Procedia CIRP*. 93: 174–179.
[4] W. Hu, T. Zhang, X. Deng, Z. Liu, J. Tan. 2021. Digital twin: A state-of-the-art review of its enabling technologies, applications and challenges. *Journal of Intelligent Manufacturing and Special Equipment*. 2(1): 1–34. https://doi.org/10.1108/JIMSE-12-2020-010
[5] M. Javaid, A. Haleem, R. Suman. 2023. Digital Twin applications toward Industry 4.0: A review. *Cognitive Robotics*. 3: 71–92.
[6] J. Leng, D. Wang, W. Shen, et al. 2021. Digital twins-based smart manufacturing system design in Industry 4.0: A review. *Journal of Manufacturing Systems*. 60(9): 119–137.
[7] T. Uslander, et al. 2022. Symbiotic evolution of digital twin systems and dataspaces. *Automation*. 3(3): 378–399.
[8] L. Schweiger, L. Barth. 2023. Properties and characteristics of Digital Twins: Review of industrial definitions. *SN Computer Science*. 4 (436). https://doi.org/10.1007/s42979-023-01937-4
[9] H. van der Valk, et al. 2022. Archetypes of Digital Twins. *Business & Information Systems Engineering*. 64: 375–391.
[10] I. Onaji, D. Tiwari, P. Soulatiantork, B. Song, A. Tiwari. 2022. Digital twin in manufacturing: Conceptual framework and case studies. *International Journal of Computer Integrated Manufacturing*. 35(8): 831–858.

[11] M. Soori, B. Arezoo, R. Dastres. 2023. Digital twin for smart manufacturing: A review. *Sustainable Manufacturing and Service Economics*. 2: 100017. https://doi.org/10.1016/j.smse.2023.100017

[12] M. Atalay, U. Murat, B. Oksuz, et al. 2022. Digital twins in manufacturing: Systematic literature review for physical–digital layer categorization and future research directions. *International Journal of Computer Integrated Manufacturing*. 35(11): 1–17.

[13] M. Perno, L. Hvam, A. Haug, 2022. Implementation of digital twins in the process industry: A systematic literature review of enablers and barriers. *Computers in Industry*. 134: 103558. https://doi.org/10.1016/j.compind.2021.103558

[14] M. Balla, O. Haffner, E. Kucera, J. Ciganek. 2023. Educational case studies: Creating a Digital Twin of the production line in TIA portal, unity, and Game4Automation framework. *Sensors*. 23(10): 4977. https://doi.org/10.3390/s23104977

[15] C. Ruzsa. 2021. Digital twin technology – external data resources in creating the model and classification of different digital twin types in manufacturing. *Procedia Manufacturing*. 54: 209–215.

[16] J.F. Yao, Y. Yang, X.C. Wang, X. Zhang. 2023. Systematic review of digital twin technology and applications. *Visual Computing for Industry, Biomedicine, and Art*. 6(10). https://doi.org/10.1186/s42492-023-00137-4

[17] G. Schroeder, C. Steinmetz, C. E. Pereira, D. B. Espindola. 2016. Digital Twin data modeling with automation ML and a communication methodology for data exchange. *IFAC-Papers OnLine*. 49(30): 12–17.

[18] H. Liu, M. Xia, D. Williams, J. Sun, H. Yan. 2022. Digital Twin-driven machine condition monitoring: A literature review. *Journal of Sensors*. 2022: 6129995. https://doi.org/10.1155/2022/6129995

[19] D.M. Botin-Sanabria, A.-S. Mihaita, R.E. Peimbert-García, M.A. Ramírez-Moreno, R.A. Ramírez-Mendoza, J.d.J. Lozoya-Santos 2022. Digital Twin technology challenges and applications: A comprehensive review. *Remote Sensing*. 14(6): 1335. https://doi.org/10.3390/rs14061335

[20] L. Gopal, H. Singh, P. Mounica, N. Mohankumar, N.P. Challa, P. Jayaraman, 2023. Digital twin and IoT technology for secure manufacturing systems. *Measurement: Sensors*. 25: 1–6. https://doi.org/10.1016/j.measen.2022.100661

[21] X. Li, H. Liu, W. Wang, Y. Zheng, H. Lv, Z. Lv. 2021. Big data analysis of the Internet of Things in the digital twins of smart city based on deep learning. *Future Generation Computer Systems*. 128: 167–177. https://doi.org/10.1016/j.future.2021.10.006

[22] O. Spyrou, W. Hurst, C. Verdouw. 2023. Virtual reality-based Digital Twins: A case study on pharmaceutical Cannabis. *Big Data and Cognitive Computing*. 7(2).

[23] M. Attaran, B. G. Celik. 2023. Digital Twin: Benefits, use cases, challenges, and opportunities. *Decision Analytics Journal*. 6: 1–10. https://doi.org/10.1016/j.dajour.2023.100165

[24] G. Sirigu, B. Carminati, E. Ferrari. 2022. Privacy and Security Issues for Human Digital Twins. IEEE International Conference on Trust, Privacy and Security in Intelligent Systems and Applications (TPS-ISA), Atlanta, GA, USA.

[25] H. Yi. 2023. Improving cloud storage and privacy security for digital twin based medical records. *Journal of Cloud Computing*. 12(151). https://doi.org/10.1186/s13677-023-00523-6

[26] J. Leng, et al. 2021. Digital twins-based smart manufacturing system design in Industry 4.0: A review. *Journal of Manufacturing Systems*. 60(9): 119–137.

[27] D. Ntamo, E. Lopez-Montero, J. Mack, C. Omar, M.I. Highett, D. Moss, N. Mitchell, P. Soulatintork, P.Z. Moghadam, M. Zandi. 2022. Industry 4.0 in action: Digitalization

of a continuous process manufacturing for formulated products. *Digital Chemical Engineering*. 3: 100025. https://doi.org/10.1016/j.dche.2022.100025

[28] R. van Dinter, B. Tekinerdogan, C. Catal. 2022. Predictive maintenance using digital twins: A systematic literature review. *Information and Software Technology*. 151: 107008. https://doi.org/10.1016/j.infsof.2022.107008

[29] R. Zhang, J. Lv, J. Bao, Y. Zheng. 2023. A digital twin-driven flexible scheduling method in a human–machine collaborative workshop based on hierarchical reinforcement learning. *Flexible Services and Manufacturing Journal*. 35(1): 1116–1138.

[30] N. Karnik, U. Bora, K. Bhadri, P. Kadambi, P. Dhatrak. 2022. A comprehensive study on current and future trends towards the characteristics and enablers of industry 4.0. *Journal of Industrial Information Integration*. 27. https://doi.org/10.1016/j.jii.2021.100294

11 Digital Twins for Smart Manufacturing

Elif Cesur, Muhammet Raşit Cesur and Elif Alptekin

11.1 INTRODUCTION

In recent times, digital twins stand out as a leading technology, particularly within the domain of smart manufacturing [1]. The growing field of digital twin applications finds its foundation in artificial intelligence (AI) studies, which are rapidly gaining traction [2]. The development of a digital twin within the sphere of AI is pivotal for enhancing industrial operations, standardizing methodologies, ensuring repeatability, and fostering sustainability. Presently, applications categorized under industrial AI are steadfastly working toward these pivotal objectives [3].

On the other hand, developing models that can be used with any industry dataset is a barrier for AI applications, making it challenging to set up repeatable procedures. Furthermore, deep programming, statistical, and optimization skills are needed for AI applications.

AI applications in the industry are generally considered to be based solely on ML algorithms, leading to inefficiencies in applications [4]–[5]. Additionally, to enhance algorithm performance, they often rely on expert opinions and trial-and-error approaches. In contrast, by aiming to improve decision-making processes, it is possible to open the door to more automated and reliable methods.

In this study, focusing primarily on fault classification within digital twin applications, we first identified the decision tree algorithm that yielded the best results across seven distinct performance metrics using AutoML libraries. Subsequently, to further enhance the algorithm's performance, we implemented four advanced ensemble methods: Bootstrap Aggregating (bagging), boosting, blending, and stacking models, and then compared their outcomes. The contribution of this study is to ensure more efficient and high-accuracy operations, thereby creating an effective digital twin process in the realm of repeatable industrial AI.

Upon examining the structure of the paper, an extensive literature review is conducted initially. Subsequently, in the Methodology section, advanced ensemble models are introduced, and the logic behind these models is explained. The performance metrics to be used are also detailed in the Methodology section. Under the fourth heading, a case study related to fault classification is presented. The dataset is elaborately described, and using PyCaret AutoML, results are obtained with four different ensemble models. Finally, the findings are compared, and the paper concludes with a conclusion section.

DOI: 10.1201/9781003465904-14

11.2 LITERATURE REVIEW

The literature review is organized under four main headings: 1) A general framework related to the digital twin and studies involving digital twin and ML algorithms are introduced. 2) Research on ensemble models is summarized without specific categorization. 3) Studies specifically focusing on ensemble models within the smart manufacturing domain are identified and summarized. 4) Research within the realm of fault prediction is also summarized. This approach helps to determine the position of our study within the existing literature and underscores the gaps that need addressing.

The inception of digital twin technology occurred in 2010 when NASA initially applied it to replicate the activities of its real-world counterpart, such as flying physical entities. Since then, the scope of digital twin literature has considerably broadened, encompassing diverse fields: manufacturing, aviation, and health. The work needed to create and maintain digital twin infrastructures may be greatly decreased by combining ML algorithms with digital twin technology [6]. This is especially true for the automobile sector, since DTs may be applied to provide a traffic environment model that is more precise and trustworthy [7]. There has been a great deal of emphasis on the application of DTs in model-based systems engineering (MBSE) and the advantages of combining DTs with the Internet of Things (IoT) and system simulation [8]. However, there are several difficulties in building DTs, such as the requirement for real-time prediction and optimization, which may be overcome by utilizing big data, AI, and sophisticated computational pipelines [9]. Kaur et al. emphasize how the IoT and AI have been combined to create digital twin technologies that can be used to turn data into insights, which can be put to use [10]. Thus, the research listed above examines the possibility of integrating ML and digital twin in a variety of fields.

Ensemble models, which combine multiple individual models to improve generalization performance, have been extensively reviewed in recent literature. There has been a noticeable increase in research focusing on ensemble methods. Various approaches, including bagging and boosting algorithms, have been employed for this purpose. Ganaie et al. provide a comprehensive overview of deep ensemble models, categorizing them into various types and discussing their applications [11]. Mohammed et al. focus on ensemble forecasting, highlighting its ability to improve accuracy and robustness in various fields, including ensemble deep learning [12]. Tanveer et al. offer a detailed analysis of ensemble deep learning architectures, fusion strategies, and algorithms, with a focus on their impact on bias and variance in the field of signal processing [13]. These reviews collectively underscore the potential of ensemble models in enhancing predictive accuracy and reliability across different domains.

These studies summarize the critical role of ensemble models in AutoML and the potential of various methods to enhance its effectiveness. Optimizing ensemble models within an ML algorithm model is crucial for establishing an automated, repeatable, and executable process within the realm of industrial AI. Lee et al. identifies key components of industrial AI, which encompass analytics technology, big data technology, cloud or cyber technology, domain expertise, and evidence. According to

Peres et al., the majority of industrial AI projects focus on areas such as process optimization, quality control, predictive maintenance, human-robot collaboration, and ergonomics [15]. Within the context of industrial AI, there is a notable exploration of automated digital twin projects in the literature. Lugaresi and Matta introduced an automated digital twin model designed to swiftly create digital twins for various processes with minimal external intervention. Testing the model in two cases and a production line demonstrated its successful functionality [16].

In this study, we aim to conduct fault prediction. Upon reviewing the existing literature in this domain, many studies have explored the use of digital twin technology in fault classification. Zhang et al. have proposed methods for fault diagnosis, focusing on bearing faults, while and Guo et al. focus on production line faults [17]–[18]. Both studies achieve high-accuracy rates, with the former reaching 100% and the latter achieving 97.8%. Gao et al. extend this work to instrument fault diagnosis, using a three-layer system and achieving rapid feedback [19]. Xu et al. introduce a two-phase method for fault diagnosis, using deep transfer learning to migrate a trained diagnosis model from the virtual to the physical space [20].

When examining the studies conducted in the literature, fault prediction is typically carried out using classical AI or ML algorithms. In contrast to other studies that aim to fill this gap in the literature, our paper integrates ensemble techniques into the fault prediction process within the context of DT, demonstrating the potential for further improvement in performance metrics. The fundamental contribution of this study lies in making the digital twin concept more applicable with a reduced coding infrastructure, utilizing ensemble models with AutoML libraries. Thus, digital twin technology will become accessible not only to researchers, but also to various sectors within the industrial domain.

11.3 THEORETICAL BACKGROUND

Within this specific part, the foundational concepts and principles behind digital twin research and studies are thoroughly elaborated. Recently, several digital twin studies have been conducted. The primary goal of this study is to figure out how the systems work. To create digital twin models, hybrid methodologies, statistical methods, traditional AI, deep learning, and dynamic data modeling techniques are applied [21]. In the development and implementation of DT, these critical enabling technologies and their integration are required. These technologies can be divided into six categories for this study:

1. **Simulation methods:** Although the simulation model is the key technology for building and operating a DT, a key difference between the digital twin approach and traditional simulation approaches is the real-time virtual-physical integration [22]–[24].
2. **Physical systems modeling:** This category includes constructing digital representations or models of real-world physical systems, such as equipment, gadgets, structures, or other physical assets [25].

3. **Industrial Internet of Things:** (IIoT) is a transmitter consisting of many different types of sensors to create a dynamic and virtual physical system that provides data streams to the digital twin to make an actual physical and virtual model [26]–[27].
4. **Big Data analytics:** Data analytics is becoming increasingly important with the growing availability of data streams and the expansion of historical data [4], [28].
5. **AI:** The integration of AI algorithms with digital twin models is expected to expand the application field for DT. Machine learning (ML) is one of the most exciting developments in AI [29]–[32].
6. **Visualization techniques**: Recent research has focused on improving the digital twin system by using virtual reality (VR), augmented reality (AR), or mixed reality (MR) technologies to increase design efficiency [33]–[34].

In this study, we examined AI technology and specifically addressed ensemble models from ML techniques to enhance digital twin technology. Several studies in scientific literature present digital twin models of machine sub-systems predicting failures and performance indicators of tasks before beginning processing [35]–[37]. Digital twin models are developed via AI or ML models in these studies. Wang et al. [35] used gray relations (GR) to analyze malfunction risk on the workbench and emphasized deep learning (DL) would contribute more to the proposed model. [36] used the support vector machine (SVM) algorithm to classify machine equipment by its level of availability. [37] studied quality control with digital twin by placing the accelerometer and gyroscope on the tool head of the computer numerical control (CNC). Sensor data is matched with solid model data by a sliding window algorithm. Matched data is analyzed by AI.

We formulate a structured approach for developing a digital twin model, as shown in Figure 11.1. Initially, we collected data collection, followed by particular data preprocessing to facilitate AI model training. Subsequently, the AI model undergoes training. The conclusion of our methodology involves validating the AI model. While the outlined procedures align with established frameworks prevalent in digital twin literature, this study endeavors to introduce a digital twin model tailored explicitly for practical applications, emphasizing fault classification.

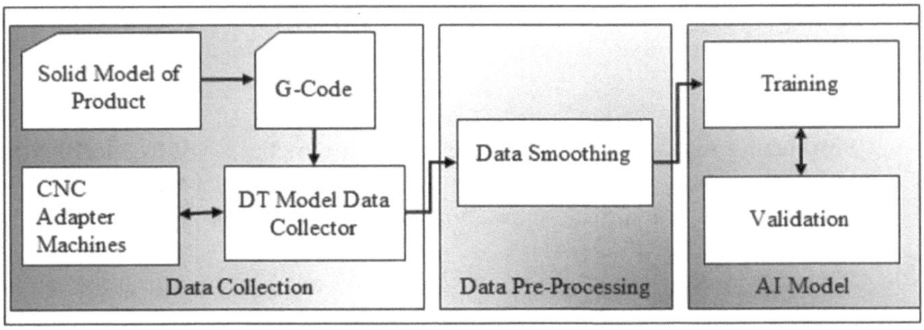

FIGURE 11.1 Digital Twin Model Procedures [38].

Digital twin technology, which concentrates on modeling the dynamics of manufacturing system components, aims to create comprehensive digital twin models. A crucial aspect of these models involves fault identification within the predictive maintenance framework. Tzanis et al. [39] developed an algorithm that utilizes ML and real-time estimations to detect and anticipate potential issues. Early detection allows for prompt interventions, preventing significant failures.

In a related manner, Luo et al. [25] explored how digital twin models assist in predicting and diagnosing faults in CNC milling machines. They introduced techniques to enhance CNC performance, minimize unexpected failures, and boost system stability.

Furthermore, Xu et al. [20] presented a method called the "Digital Twin-assisted Fault Diagnosis with Deep Transfer Learning". This two-step approach first utilizes a high-fidelity virtual model to identify potential problems, followed by training a deep neural network for diagnostic purposes. This technology surpasses prior methods in fault prediction accuracy, as evidenced in its application to aero-engine bearings [40]. Wang et al. [41] further enriched this domain by demonstrating enhanced defect prediction accuracy by training convolutional neural networks.

Lastly, Qiao et al. [42] investigated the potential of digital twin technology in improving predictive maintenance for industrial machines, emphasizing the condition of cutting tools. Their experiments, based on vibration data from milling machine tools, confirmed the efficacy of digital twin models in accurately forecasting tool wear.

Although the number of scholars and practitioners that appreciate digital twin models for fault diagnosis has grown exponentially over time, and numerous papers on digital twin offer case studies from the manufacturing environment mentioned above, there is still a lack of studies that explore how AI methods can enhance digital twin models in the realm of fault classification.

11.4 METHODOLOGY: ADVANCED ENSEMBLE METHODS

The purpose of the ensemble methods is to apply multiple learning machines instead of a single method to solve classification and regression problems. Ensemble methods are aimed to increase performance metrics and create a more reliable ML model. The logic of ensemble methods is shown in Figure 11.2. Each learner

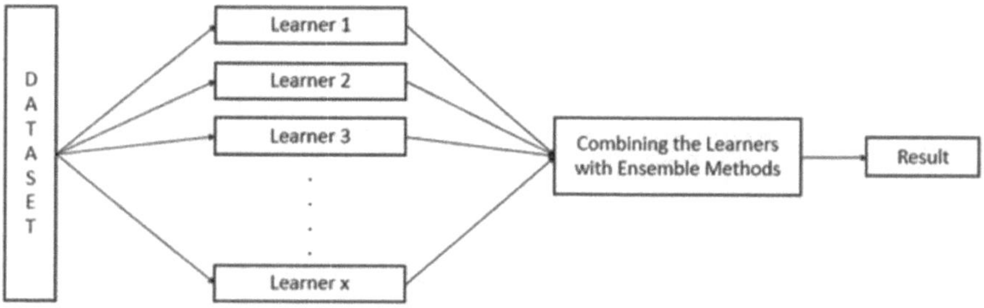

FIGURE 11.2 Logic of Ensemble Models.

represents a separate ML algorithm, such as a decision tree or k-nearest neighbors (KNN). If there is one type of learning algorithm, then it is called a homogeneous ensemble. Most of the ensemble methods are homogeneous ensembles. Additionally, some methods use more than one type of learners, which is called heterogeneous ensembles.

11.4.1 Stacking

In this technique, there are two levels of learners. The first-level learners comprise individual ML algorithms. Stacking is a heterogeneous ensemble method, implying that the first-level learners can be diverse types of ML algorithms. These algorithms generate various results from the original training dataset. The outcomes from the first-level learners are then aggregated or stacked. The second-level learner, which is also known as the meta learner, leverages these stacked results to construct a more efficient model. The stacking method is sometimes referred to as super learning. To mitigate the risk of overfitting, techniques such as cross-validation and leave-one-out can be employed to create datasets for first-level learners.

11.4.2 Blending

Blending and stacking ensemble methods share many similarities. Like the stacking method, the blending method also uses two levels of learners. The underlying logic of these methods is essentially the same. In the initial step, the first-level learners, or the base learners, make predictions that serve as input for the second-level learner, known as the meta learner. The primary distinction between the stacking and blending methods lies in how they allocate training data for the metamodel. Specifically, the blending method relies on a separate holdout set for the metamodel, whereas the stacking method utilizes the entire dataset, incorporating both complete and folded data.

11.4.3 Bagging

Bagging is an ensemble method designed to create a more accurate model and reduce overfitting. In this method, multiple base models are trained individualistically. Each base model is trained on different subsets generated arbitrarily, with replacement, from the training dataset. To make the final prediction, various methods are applied depending on the characteristics of the problem. For regression cases, the prediction is the average of the results produced by the base models. For classification problems, the majority voting method is employed. The structure of bagging method is shown in Figure 11.3. Using different datasets for each base model helps to mitigate the overfitting problem.

One of the advantages of the bagging method is its flexibility. This approach can be applied to various ML algorithms. It can also be executed efficiently on large datasets. Due to each base model operating on different data subsets, the variance decreases, resulting in a more robust and accurate model. The bagging method finds

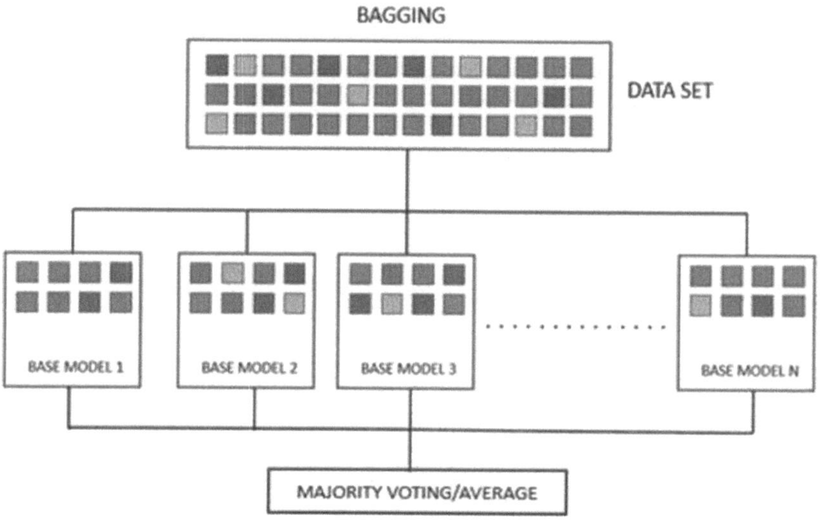

FIGURE 11.3 Structure of the Bagging Method.

applications in numerous areas, such as fraud detection, spam filtering, credit scoring, image classification, and natural language processing.

11.4.4 Boosting

Boosting is an ensemble method that is designed to create a robust model by utilizing weak models iteratively. It operates as a supervised ML approach. The boosting method aims to improve upon the results of the previous base models. This iterative approach distinguishes boosting from other ensemble methods, as it focuses on learning from its prediction errors.

Initially, each row of training data is assigned an equal weight. The first base model is trained using these equal weights, and its results are evaluated. Examples that are predicted incorrectly are identified. In subsequent iterations, the weights of these mispredicted examples are increased, while the weights of correctly predicted examples are decreased. This iterative adjustment aims to enhance the model's accuracy over time. The distinction between bagging and boosting methods is illustrated in Figure 11.4. Several different boosting algorithms have been discussed in the literature. AdaBoost, gradient boosting, stochastic gradient boosting, LPBoost, and totalBoost are a few well-known examples of boosting algorithms.

Performance metrics are employed to evaluate and validate the model's success, providing analytical results to assess and compare its outcomes. These metrics utilize the training dataset results alongside the actual outcomes to generate evaluative scores for the model. This paper uses metrics such as accuracy, area under the ROC curve (AUC), recall, precision, F1 score, Cohen's kappa, Matthew's correlation coefficient

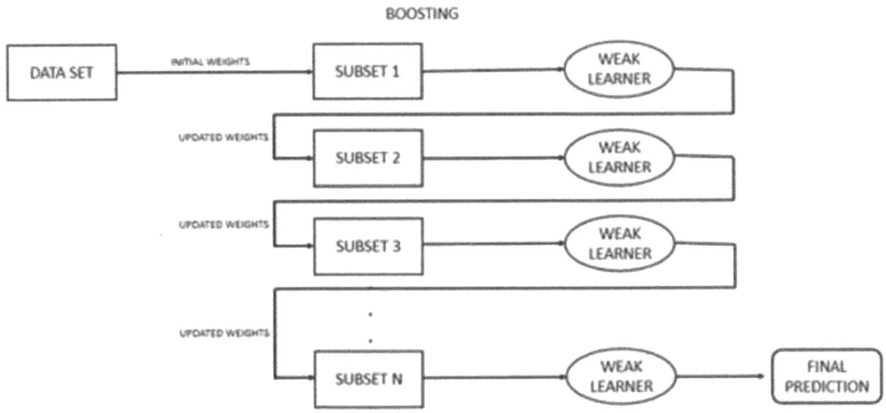

FIGURE 11.4 Architecture of Boosting Methods.

TABLE 11.1
Confusion Matrix

	Predicted Positive (PP)	Predicted Negative (PN)
Actual Positive (AP)	TP	FN
Actual Negative (AN)	FP	TN

(MCC) scores, and the confusion matrix to determine the model's efficiency. This section elaborates on these performance metrics. Table 11.1 displays a confusion matrix, which is a common tool used for evaluating classification models, facilitating an analysis of the model's performance by contrasting its predictions with the actual results. The cells in the confusion matrix are filled as follows.

True Positives (TP): Instances correctly predicted as positive. These are cases where the model predicted positive, and the actual outcome was indeed positive.
False Negatives (FN): Instances incorrectly predicted as negative. In these cases, the model predicted a negative, but the actual outcome was positive.
False Positives (FP): Instances incorrectly predicted as positive. In these cases, the model predicted a positive, but the actual outcome was negative.
True Negatives (TN): Instances correctly predicted as negative. These are cases where the model predicted negative, and the actual outcome was negative.

These metrics are crucial for calculating various performance metrics, such as accuracy, precision, recall, and F1 score, providing a comprehensive understanding of how well a classification model is performing.

Accuracy Score: Accuracy score is the ratio of correct predictions to all predictions. It takes values between 0 and 1.

$$Accuracy\,Score = \frac{TP + TN}{TP + FP + TN + FN}$$

AUC: An ROC graph must be drawn to compute the AUC-ROC score. The area under this curve will give the AUC-ROC score. The following equations are used to draw the ROC graph.

$$TPR = \frac{TP}{TP + FN} \qquad FPR = \frac{FP}{FP + TN}$$

TPR (true positive rate) represents the proportion of correctly predicted positive class instances. The combined count of TP and FN provides the total number of positive class instances. False positive rate (FPR) indicates the proportion of negative class instances incorrectly predicted as positive. The aggregate count of FP and TN yields the total number of negative class instances.

The ROC curve plots these values across various thresholds. The area beneath this curve determines the AUC-ROC score. An AUC-ROC score of 0.7 signifies that the model has a 70% chance of correctly predicting both positive and negative classes. An AUC-ROC score of 0.5 suggests that the model's predictive capability is no better than random guessing. As the AUC-ROC score approaches zero, the model increasingly misclassifies negative instances as positive and vice versa.

Recall (Sensitivity or TPR): The recall metric represents the ratio of correctly predicted positive instances to all actual positive instances. This is equivalent to the TPR defined earlier. As the recall value approaches 1, the model's ability to accurately predict the entire positive class improves. The recall is also referred to as "sensibility".

$$R = \frac{TP}{TP + FN}$$

Precision (Positive Predictive Value): The precision metric is the ratio of correctly predicted positive instances to all instances predicted as positive by the model.

$$P = \frac{TP}{TP + FP}$$

Precision is associated with a Type-I error, which represents the probability of rejecting a true hypothesis. On the other hand, recall is linked to a Type-II error, indicating the likelihood of accepting a false hypothesis.

F1 Score: The F1 score is the harmonic mean of recall and precision. A high F1 score indicates both high recall and high precision. Conversely, a low F1 score suggests that performance is suboptimal at a particular threshold, making it unclear whether the issue lies with recall or precision alone.

$$F1 = \frac{2}{\frac{1}{Precision} + \frac{1}{Recall}}$$

Kappa: The Kappa score assesses the agreement between the model's predictions and the actual data. In the context of binary classification, Cohen's kappa statistic is utilized.

$$k = \frac{2 * (TP * TN - FN * FP)}{(TP + FP) * (FP + TN) + (TP + FN) * (FN + TN)}$$

MCC: The MCC quantifies the correlation—akin to the Pearson product-moment correlation coefficient—between the model's predictions and the actual dataset outcomes. Its values range between -1 and 1. A high MCC score indicates that the model effectively predicts both positive and negative classes.

$$MCC = \frac{(TP * TN) - (FP * FN)}{\sqrt{(TP + FP) * (TP + FN) * (TN + FP) * (TN + FN)}}$$

In this paper, ML algorithms are evaluated based on these performance metrics and the time required for their execution.

Total Training Time in Seconds (TT sec): TT sec represents the total time required for the model training process, measured in seconds. This metric offers insights into the computational efficiency of the training phase. Smaller TT values are preferable, as they signify quicker training times.

11.5 CASE STUDY

In this case study, we aimed to investigate whether tasks or jobs created from different units, such as robot arms, machines, CNC, and adapters, were successful or unsuccessful based on their unique characteristics when allocated to various serving machines. To provide a clearer explanation, Figure 11.5 has been included below. Firstly, how tasks and jobs are created from various machines are generated on the right side. Moving to the left side of the figure, it shows how these tasks are assigned to specific machines based on factors such as constraints, priorities, and hardware capabilities such as CPU, memory, and disk space. The results from this setup tell us whether these tasks were completed successfully or if they encountered issues,

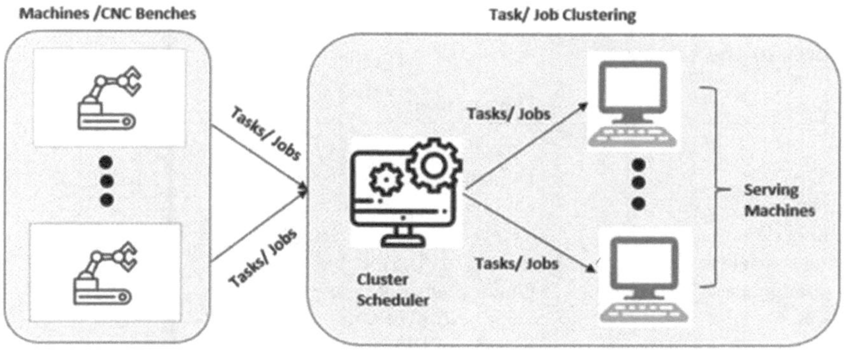

FIGURE 11.5 Structure of Task Scheduling.

labeled as outcomes such as "evict", "fail", "finish", or "kill". When tasks fail, we identify them as faults. Our objective is to use data-driven digital twin technology to analyze task details and improve our fault diagnosis accuracy using advanced ensemble models.

11.5.1 Data Analysis

The goal of the case study is to perform classification on a dataset comprising 11 columns and 77,005 rows. The primary task is classifying entries based on the "Event Type" column. The primary focus of the analysis is on the "Event Type" column, which serves as the target variable for classification. The remaining ten columns are features that influence the classification outcome, as shown in Table 11.2. In this comprehensive study, we employ advanced ensemble techniques to enhance fault detection in a classification problem.

Table 5.1 contains information about the failure classification dataset related to tasks and jobs in a scheduling system. Here are detailed explanations for each attribute:

1. **Time:** Indicates the arrival time of a task or job. This attribute helps in tracking when a particular task becomes available for scheduling.
2. **Job_ID:** Provides unique identifiers for jobs. Each job is assigned a distinct Job_ID, facilitating the identification and tracking of individual jobs in the scheduling system.
3. **Task_index:** Offers unique identifiers for tasks. Task_index distinguishes different tasks associated with a job and aids in their identification and organization.
4. **Machine_ID:** Represents the machine to which the job or task has been scheduled. This attribute indicates the machine on which the job or task will be executed.

TABLE 11.2
Features of the Dataset

Attributes	Explanations
Time	Arrival time of task
Job_ID	Unique identifiers for jobs
Task_index	Unique identifiers for tasks
Machine_ID	Represents the machine that the job or task was scheduled
Different_machines_restriction	Restriction is available or not
Scheduling_class	Latency-sensitivity level of task
Priority	Priority level of task
CPU	Maximum CPU that the job is permitted to use
Memory	Maximum memory that the job is permitted to use
Disk_space	Maximum disk space that the job is permitted to use
Event_Type	Task types

5. **Different_machines_restriction:** Indicates whether there is a restriction on using different machines for a job or task. It provides information on whether the scheduling system enables or restricts the allocation of a job or task to different machines.
6. **Scheduling_class:** Represents the latency-sensitivity level of a task. This attribute categorizes tasks based on their sensitivity to latency, providing insights into the priority of tasks in terms of response time. It takes values between 0 and 3. "3" means the job's latency-sensitive level is high.
7. **Priority:** Reflects the priority level assigned to a task. This attribute helps to determine the order in which tasks are scheduled, ensuring that higher-priority tasks are addressed first. The dataset has values between 0 and 11.
8. **CPU:** Indicates the maximum CPU resources permitted for a job. This attribute defines the upper limit on the computational resources that the job is allowed to use during execution.
9. **Memory:** Specifies the maximum memory allocation allowed for a job. This attribute sets the upper limit on the amount of memory that the job can consume during its execution.
10. **Disk_space:** Denotes the maximum disk space permitted for a job. This attribute establishes the upper limit on the storage space that the job is allowed to use during its execution.
11. **Event_Type:** Represents the types of tasks. This attribute categorizes tasks based on their specific characteristics or nature, providing insights into the diversity of tasks handled by the scheduling system. There are nine distinct event types. The objective is to predict the event type of a job based on the values in the preceding columns. The explanations for the event types are as follows, and transitions between event types are depicted in Figure 11.6.
    - Schedule (1): This denotes that a job has been successfully scheduled on a machine.

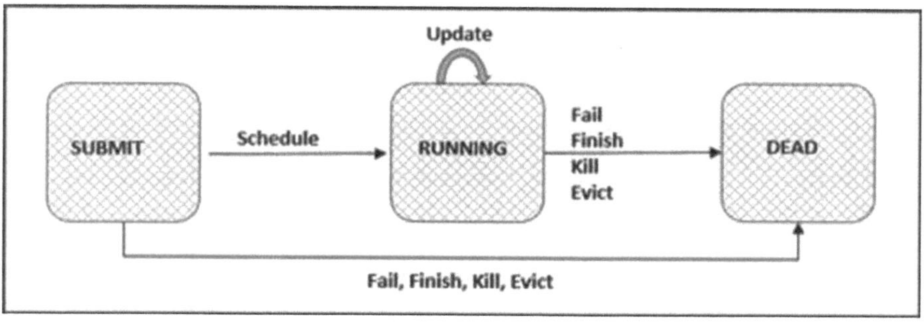

FIGURE 11.6 Event Types and Transitions.

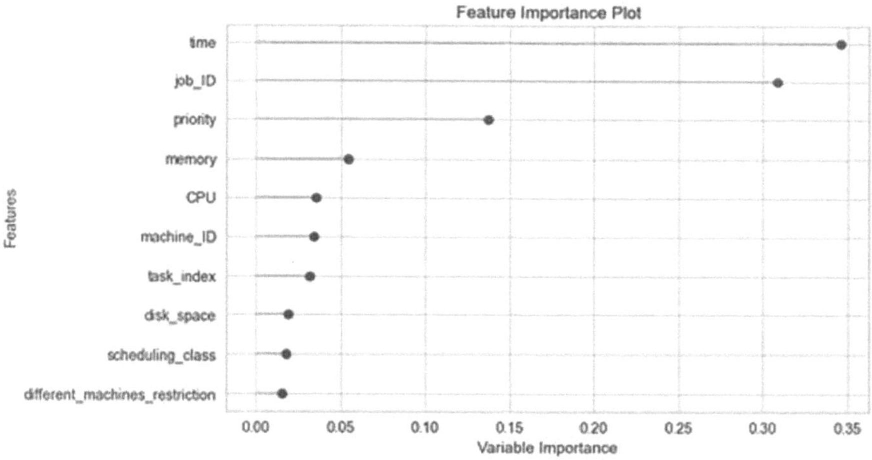

FIGURE 11.7 Feature Importance Rate.

- Evict (2): This signifies that a job has been scheduled. It often occurs when a higher-priority job is submitted.
- Fail (3): This indicates that a job has been scheduled due to a failure or error.
- Finish (4): This represents the successful completion of a job.
- Kill (5): This indicates that a job has been canceled either by the user or a driver program.

In the data set, there are 26231 "Schedule", 15000 "Evict", 3006 "Fail", 14551 "Finish" and 18216 "Kill" events. We created a feature importance plot to assess the priority of ten input values, as illustrated in Figure 11.7. This plot visually represents the significance of each input variable based on their respective importance scores. When examining Figure 11.3, it is evident that, fundamentally, three input values play a significant role. These are Time, Job_ID, and the assigned priority. On the other hand, features such as scheduling_class, disk_space, and different_machine_restriction, which are determined for each incoming transaction, seem to be less critical.

11.5.2 Result of ML Algorithms

In this study, the PyCaret library is employed to create AutoML models and compare the results of 16 different methods using performance metrics explained in the previous chapter. The PyCaret library applies models utilizing the Scikit-learn (Sklearn) library. The range of models includes decision tree-based ML algorithms such as the "decision tree classifier," "random forest classifier," and "extra trees classifier." Additionally, boosting algorithms such as "extreme gradient boosting," "light gradient boosting machine," "catboost classifier," "gradient boosting classifier," and "Ada Boost Classifier" are utilized. Regression-based algorithms, such as the "ridge classifier" and "logistic regression," are also part of the analysis. Apart from these algorithms, other models encompass the "K neighbors' classifier," "linear discriminant analysis," "quadratic discriminant analysis," "naïve bayes," "dummy classifier," and SVM-linear kernel. Upon initially classifying the data set using the PyCaret, AutoML library, it is observed that the decision tree classifier yielded the highest accuracy. The AUC, recall, F1, kappa, MCC, and TT values for the other ML algorithms are provided in Table 11.3 for comprehensive comparison.

Based on Table 11.3, decision trees, random forests, extra trees, XGBoost, and LightGBM appear to perform quite well in terms of accuracy, AUC, recall, precision,

TABLE 11.3
Results from PyCaret AutoML

Model	Accuracy	AUC	Recall	Prec.	F1	Kappa	MCC	TT
Decision Tree	0.9674	0.979	0.967	0.968	0.967	0.957	0.957	0.091
Random Forest	0.9546	0.996	0.955	0.957	0.954	0.94	0.941	0.965
Extra Trees	0.9532	0.993	0.953	0.955	0.953	0.938	0.939	0.539
Extreme Gradient Boosting	0.9495	0.994	0.95	0.952	0.949	0.933	0.934	0.803
Light Gradient Boosting MachineW	0.9423	0.993	0.942	0.945	0.942	0.924	0.925	0.972
CatBoost Classifier	0.9294	0.99	0.929	0.932	0.929	0.907	0.908	10.806
Gradient Boosting	0.8822	0.974	0.882	0.885	0.882	0.844	0.845	6.622
K Neighbors	0.8273	0.946	0.827	0.828	0.822	0.772	0.774	0.735
Ada Boost	0.5365	0.683	0.537	0.535	0.522	0.377	0.382	0.536
Linear Discriminant	0.4842	0.737	0.484	0.49	0.481	0.309	0.312	0.057
Quadratic Discriminant	0.4722	0.71	0.472	0.679	0.437	0.342	0.416	0.06
Ridge Classifier	0.4646	0	0.465	0.437	0.436	0.264	0.272	0.059
Naïve Bayes	0.4189	0.633	0.419	0.469	0.346	0.268	0.344	0.048
Logistic Regression	0.3495	0.574	0.35	0.262	0.283	0.07	0.08	0.912
Dummy Classifier	0.3406	0.5	0.341	0.116	0.173	0	0	0.045
SVM Linear	0.299	0	0.299	0.2	0.182	0.033	0.062	1.826

F1-score, and other metrics. Models such as CatBoost and gradient boosting classifier also perform relatively well but may take more time to train or test based on the last column (TT), especially CatBoost, with a value of 10.806. On the other hand, models such as SVM with a linear kernel, logistic regression, dummy classifier, naïve bayes, and ridge classifier seem to have lower accuracy and other performance metrics, making them less effective for this particular dataset or task. In our subsequent analyses, ensemble methods have been applied using the digital twin algorithm as the foundational basis.

In a classification issue, the mistake or misclassification rate for several classes is represented by the "Class Prediction Error" curve shown in Figure 11.8a. According to the Class Prediction Error chart, the decision tree classifier predicts correctly,

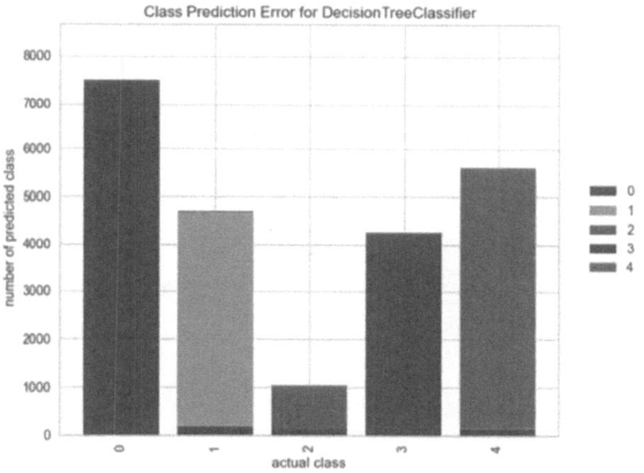

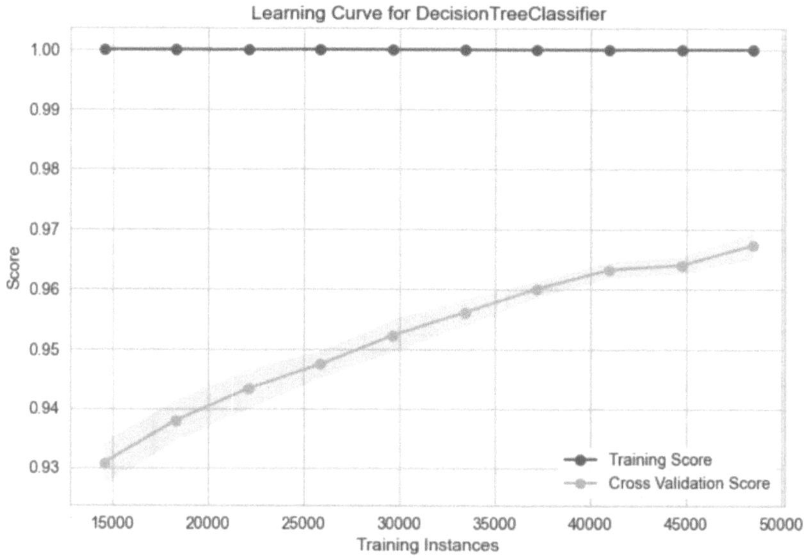

FIGURE 11.8 (a) Class Prediction Error; (b) Confusion Matrix; (c) ROC Curve; and (d) Learning Curve.

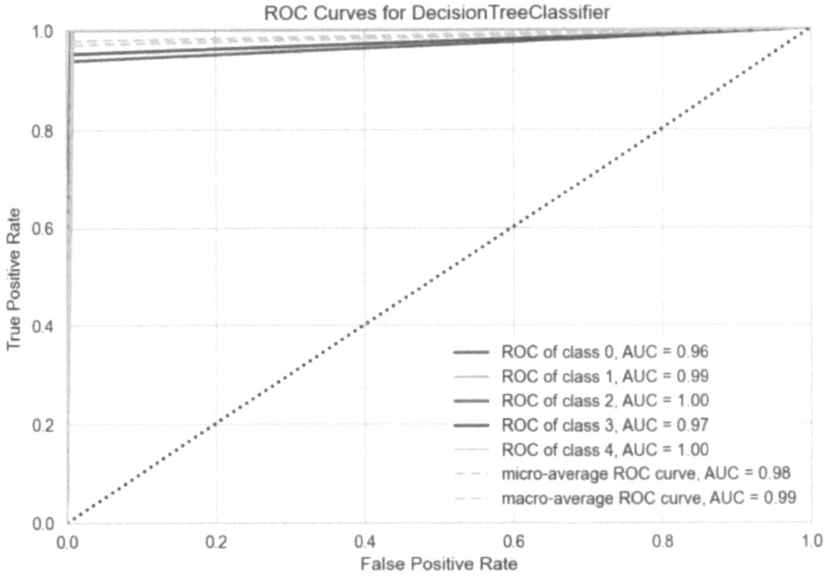

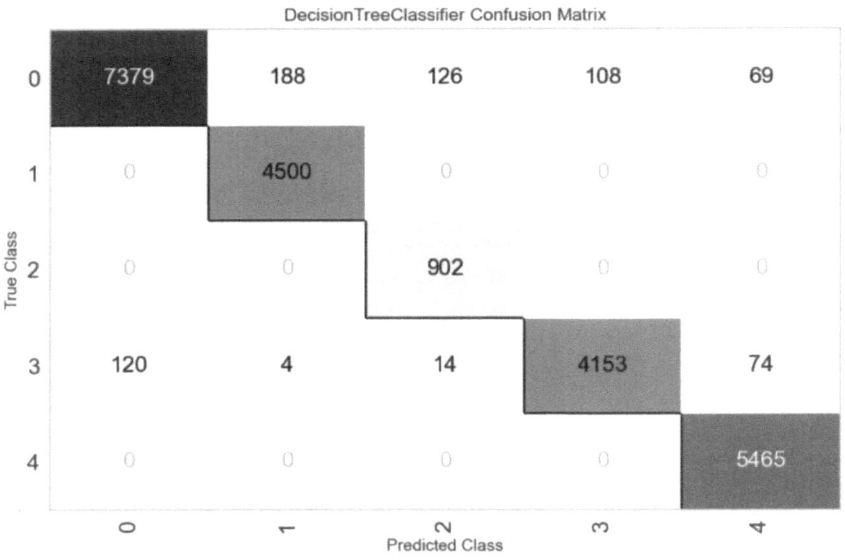

FIGURE 11.8 (Continued)

although there are also a few mispredictions. The confusion matrix in Figure 11.8b indicates that the model correctly classified classes 2 (evict), 3 (fail), and 5 (kill) with 100% accuracy. However, errors occurred in classes 1 (schedule) and 4 (finish). The AUC provides an overall performance metric that encompasses all potential categorization criteria. AUC values between 0.5 and 1 indicate random guessing and flawless categorization, respectively. Figure 11.8c shows that all AUC values are close to 1.00. The learning curve, depicted in Figure 11.8d, illustrates how a model's performance changes as the size of the training set increases for both the training and validation datasets. The figures indicate that both training and validation errors tend to stabilize at a low value.

11.5.3 Result of Advanced Ensemble Methods

In this section, results will be obtained from the four different ensemble models mentioned in this chapter, and the best model architectures will be presented. Lastly, the ensemble models will be compared in terms of accuracy. Firstly, the performance metrics for a decision tree model are shown in Table 11.4 across ten folds or ten distinct data subsets that were used for cross-validation.

Across all ten folds, the digital twin model shows consistent performance, as seen by the close values in each metric. In summary, the model appears to perform consistently well across all folds of the cross-validation, as evidenced by the closely grouped metrics. The mean values provide an overall summary of the model's performance across all folds.

The decision tree classifier in Sklearn, a well-known Python ML package, explains the arguments and their default values is performed. These parameters enable you to fine-tune the model for optimal performance depending on your unique dataset and needs, prevent overfitting, and regulate the decision tree's complexity. For example, the minimal number of samples needed to divide an internal node is indicated by the value of min_samples_split, which is set to 2.

11.5.3.1 Bagging Methods

The bagging method's performance, as represented by metrics such as accuracy, AUC, recall, precision, F1-score, kappa, and MCC, appears to be consistently high across all folds. However, it cannot be said that the Bagging method provides an improvement.

TABLE 11.4
Results from the Decision Tree Algorithm

	Accuracy	AUC	Recall	Prec.	F1	Kappa	MCC
Fold							
0	0.967	0.9789	0.967	0.9677	0.9669	0.9563	0.9565
1	0.9701	0.9812	0.9701	0.9711	0.9701	0.9605	0.9608
2	0.9638	0.977	0.9638	0.965	0.9637	0.9521	0.9525
3	0.9677	0.9795	0.9677	0.9685	0.9677	0.9573	0.9575
4	0.9677	0.9796	0.9677	0.9686	0.9676	0.9573	0.9576
5	0.9681	0.9798	0.9681	0.9691	0.9681	0.9578	0.958
6	0.9692	0.9803	0.9692	0.9699	0.9691	0.9592	0.9594
7	0.9653	0.9782	0.9653	0.9665	0.9653	0.9541	0.9545
8	0.9642	0.9776	0.9642	0.9655	0.9641	0.9527	0.9531
9	0.9709	0.9816	0.9709	0.9717	0.9708	0.9615	0.9617
Mean	**0.9674**	**0.9794**	**0.9674**	**0.9684**	**0.9673**	**0.9569**	**0.9572**

TABLE 11.5
Results from the Bagging Method

	Accuracy	AUC	Recall	Prec.	F1	Kappa	MCC
Fold							
0	0.9644	0.9929	0.9644	0.9653	0.9643	0.9529	0.9532
1	0.9677	0.9944	0.9677	0.9689	0.9677	0.9573	0.9577
2	0.9633	0.9936	0.9633	0.9647	0.9632	0.9514	0.9518
3	0.9672	0.993	0.9672	0.9682	0.9672	0.9565	0.9568
4	0.9625	0.9927	0.9625	0.9637	0.9624	0.9505	0.9509
5	0.9646	0.9918	0.9646	0.9659	0.9646	0.9531	0.9534
6	0.967	0.9932	0.967	0.968	0.9669	0.9563	0.9566
7	0.9635	0.9929	0.9635	0.9648	0.9634	0.9517	0.9521
8	0.9664	0.9936	0.9664	0.9677	0.9664	0.9556	0.956
9	0.9666	0.9932	0.9666	0.9678	0.9665	0.9559	0.9562
Mean	**0.9653**	**0.9931**	**0.9653**	**0.9665**	**0.9653**	**0.9541**	**0.9545**

Table 11.5 presents the results of applying the bagging method on an ML model, for a classification task, across multiple folds of a cross-validation process. The last row, labeled "MEAN," provides the average performance metrics across all the folds. It gives an overall summary of how the bagging method performed across the entire cross-validation process.

The optimized parameters for a BaggingClassifier in Sklearn are utilized. This ensemble meta-estimator fits base classifiers on random subsets of the original dataset and then aggregates their predictions to form a final prediction, either by voting or averaging. In summary, this BaggingClassifier uses a DecisionTreeClassifier as its base estimator, creating an ensemble of ten decision trees (as n_estimators=10), with observations and features sampled with replacement due to default settings.

11.5.3.2 Boosting Methods

The outcomes of a boosting technique applied to an ML model over several folds of a cross-validation procedure are shown in Table 11.6. Based on the stated performance indicators, the findings show that the boosting strategy applied to the ML model has consistently shown positive outcomes across multiple folds. The boosting strategy performed somewhat better than the other strategies, but not significantly.

The AdaBoostClassifier settings in Sklearn are performed. With a decision tree serving as the basis estimator, the AdaBoostClassifier is set up to employ the SAMME.R method. With a learning rate of 1.0, it will construct an ensemble of ten decision trees, each contributing to the forecast. By configuring a particular random state, the setup guarantees the reproducibility of the findings.

11.5.3.3 Blending Methods

This section shows the results of running an ML model through many folds using a blending strategy. Table 11.7 demonstrates that the blending technique employed in

TABLE 11.6
Results from the Boosting Method

	Accuracy	AUC	Recall	Prec.	F1	Kappa	MCC
Fold							
0	0.9668	0.9787	0.9668	0.9674	0.9667	0.956	0.9563
1	0.9692	0.9806	0.9692	0.9703	0.9692	0.9593	0.9595
2	0.9651	0.9779	0.9651	0.9662	0.965	0.9539	0.9542
3	0.9685	0.98	0.9685	0.9693	0.9684	0.9583	0.9585
4	0.9685	0.9802	0.9685	0.9694	0.9684	0.9583	0.9586
5	0.9673	0.9793	0.9673	0.9684	0.9673	0.9568	0.957
6	0.9699	0.9808	0.9699	0.9707	0.9699	0.9602	0.9604
7	0.9657	0.9784	0.9657	0.9668	0.9656	0.9546	0.9549
8	0.9653	0.9784	0.9653	0.9666	0.9652	0.9541	0.9546
9	0.9701	0.9812	0.9701	0.971	0.97	0.9605	0.9608
Mean	**0.9676**	**0.9795**	**0.9676**	**0.9686**	**0.9676**	**0.9572**	**0.9575**

TABLE 11.7
Results from the Blending Method

	Accuracy	AUC	Recall	Prec.	F1	Kappa	MCC
Fold							
0	0.9603	0.9967	0.9603	0.9616	0.9601	0.9475	0.9481
1	0.9655	0.997	0.9655	0.9668	0.9654	0.9544	0.9548
2	0.9623	0.9959	0.9623	0.9635	0.9622	0.9502	0.9506
3	0.9622	0.9959	0.9622	0.9631	0.962	0.95	0.9503
4	0.9618	0.9954	0.9618	0.9629	0.9616	0.9495	0.95
5	0.9636	0.9959	0.9636	0.9648	0.9635	0.9519	0.9523
6	0.966	0.9962	0.966	0.967	0.9659	0.9551	0.9554
7	0.9616	0.9967	0.9616	0.9631	0.9614	0.9493	0.9499
8	0.9603	0.9966	0.9603	0.9619	0.9601	0.9475	0.9482
9	0.9659	0.9965	0.9659	0.9669	0.9657	0.9549	0.9553
Mean	**0.9630**	**0.9963**	**0.963**	**0.9642**	**0.9628**	**0.951**	**0.9515**

the ML model has consistently delivered favorable outcomes over several folds of the cross-validation process, as seen by the performance metrics provided.

A VotingClassifier in Sklearn, an ensemble technique that aggregates several ML models and employs voting or averaging to generate predictions, is used. This VotingClassifier combines the predictions of two classifiers: random forest and decision tree. It will base its final prediction on the average probability from both classifiers using the "soft" voting approach. To potentially increase total performance, the design offers flexibility in merging predictions from several classifiers and permits parallel processing.

TABLE 11.8
Results from the Stacking Method

	Accuracy	AUC	Recall	Prec.	F1	Kappa	MCC
Fold							
0	0.3419	0.5737	0.3419	0.255	0.2753	0.0574	0.0659
1	0.3584	0.5753	0.3584	0.2717	0.2931	0.083	0.0945
2	0.3568	0.5807	0.3568	0.2692	0.2891	0.0787	0.0905
3	0.3599	0.5819	0.3599	0.2725	0.2915	0.083	0.0955
4	0.3429	0.563	0.3429	0.2533	0.277	0.0621	0.0702
5	0.3419	0.5725	0.3419	0.2545	0.2763	0.0587	0.0671
6	0.3419	0.5695	0.3419	0.2515	0.2762	0.0605	0.0686
7	0.3562	0.5729	0.3562	0.2683	0.2909	0.0809	0.0917
8	0.3434	0.5665	0.3434	0.2586	0.2788	0.0617	0.0701
9	0.3518	0.579	0.3518	0.2615	0.283	0.0719	0.0825
Mean	0.3495	0.5735	0.3495	0.2616	0.2831	0.0698	0.0797

11.5.3.4 Stacking Methods

Performance metrics are displayed for a stacking approach applied to an ML model across several folds. According to Table 11.8, the performance of a stacking method applied to an ML model is shown. The results provide insights into how well the model performs based on various evaluation metrics across different subsets of the data. The accuracy values are not adequate when compared to other advanced ensemble models.

An ensemble ML approach called StackingClassifier in Sklearn is utilized. Predictions from a random forest classifier and a decision tree classifier are combined in this StackingClassifier. The final prediction is then produced by feeding the predictions into a logistic regression model (final_estimator) once the base models have yielded predictions. The stacking procedure aims to increase prediction performance by utilizing many ML models.

11.6 CONCLUSION

Digital twin technology focuses on creating detailed digital replicas of components within manufacturing systems. One key goal of these digital replicas is to identify potential faults before they cause problems, a process known as predictive maintenance. To make these digital twin models even more effective, researchers are looking at combining them with AI algorithms. In this particular study, the researchers dive deeper into AI, especially focusing on a technique called ensemble models within ML. The idea is to see how these advanced AI methods can improve digital twin technology, especially in pinpointing and classifying faults.

While many experts and researchers have started to see the value of digital twin models for diagnosing issues in manufacturing settings, and plenty of studies showcase its applications, there is still a gap. Not many studies have delved into how AI techniques can specifically improve digital twin models, particularly in terms of

TABLE 11.9
Results from the Ensemble Method

	Accuracy	AUC	Recall	Prec.	F1	Kappa	MCC
Decision Tree	0.9674	0.9794	0.9674	0.9684	0.9673	0.9569	0.9572
Bagging Method	0.9653	0.9931	0.9653	0.9665	0.9653	0.9541	0.9545
Boosting Method	**0.9676**	**0.9795**	**0.9676**	**0.9686**	**0.9676**	**0.9572**	**0.9575**
Blending Method	0.963	0.9963	0.963	0.9642	0.9628	0.951	0.9515
Stacking Method	0.3495	0.5735	0.3495	0.2616	0.2831	0.0698	0.0797

identifying and categorizing faults. In this research, our primary emphasis was on fault classification within digital twin applications. Traditionally, fault prediction has relied on classical AI or ML algorithms. Diverging from prior studies attempting to bridge this research gap, our work incorporates ensemble techniques into the fault prediction mechanism within digital twin frameworks, showcasing potential enhancements in performance metrics.

Initially, the decision tree algorithm stood out, exhibiting superior results across seven distinct performance metrics when utilizing the PyCaret AutoML libraries. This led us to implement and subsequently compare the outcomes of four advanced ensemble methods: bagging, boosting, blending, and stacking models. Performance comparisons between several advanced ensemble ML approaches are shown in Table 11.9. The decision tree, bagging, boosting, and blending approaches exhibit comparatively positive and stable performance on several measures. On the other hand, the stacking method seems to have far lower accuracy and other performance indicators, suggesting that it may not be the ideal option for this specific job or dataset.

The fundamental contribution of this study lies in making the digital twin concept more applicable in the industry with a reduced coding infrastructure, utilizing ensemble models with AutoML libraries. Thus, digital twin technology will not only become accessible to researchers but also to various sectors within the industrial domain.

In the provided code snippets within the study, one can observe the architectures of the ensemble models that produce the results. It is recommended that future research focus on further optimization of these architectures through hyperparameter optimization. Additionally, utilizing different AutoML libraries, such as AutoKeras, H2O, or TPOT, may potentially yield better results.

REFERENCES

[1] B. He and K. J. Bai, "Digital twin-based sustainable intelligent manufacturing: A review," *Adv Manuf*, vol. 9, no. 1, Mar. 2021, doi: 10.1007/s40436-020-00302-5.

[2] J. Wu, Y. Dang, H. Jia, X. Liu, and Z. Lv, "Prediction of energy consumption in Digital Twins of intelligent factory by artificial intelligence," in *ICT-PEP 2021 – International Conference on Technology and Policy in Energy and Electric Power: Emerging Energy Sustainability, Smart Grid, and Microgrid Technologies for Future Power*

System, Proceedings, Institute of Electrical and Electronics Engineers Inc., 2021, pp. 354–359, doi: 10.1109/ICT-PEP53949.2021.9601051.

[3] T. Y. Melesse, V. Di Pasquale, and S. Riemma, "Digital twin models in industrial operations: A systematic literature review," in *Procedia Manufacturing*, Elsevier B.V., 2020, pp. 267–272, doi: 10.1016/j.promfg.2020.02.084.

[4] M. M. Rathore, S. A. Shah, D. Shukla, E. Bentafat, and S. Bakiras, "The role of AI, machine learning, and big data in digital twinning: A systematic literature review, challenges, and opportunities," *IEEE Access*, vol. 9. 2021, doi: 10.1109/ACCESS.2021.3060863.

[5] T. Gurgenc, F. Ucar, D. Korkmaz, C. Ozel, and Y. Ortac, "A study on the extreme learning machine based prediction of machining times of the cycloidal gears in CNC milling machines," *Prod Eng*, vol. 13, no. 6, pp. 635–647, Dec. 2019, doi: 10.1007/s11740-019-00923-1.

[6] D. Lehner, S. Sint, M. Vierhauser, W. Narzt, and M. Wimmer, "AML4DT: A model-driven framework for developing and maintaining digital twins with AutomationML," in *IEEE International Conference on Emerging Technologies and Factory Automation, ETFA*, 2021, doi: 10.1109/ETFA45728.2021.9613376.

[7] V. Tihanyi *et al.*, "Towards cooperative perception services for its: Digital twin in the automotive edge cloud," *Energies (Basel)*, vol. 14, no. 18, 2021, doi: 10.3390/en14185930.

[8] A. M. Madni, C. C. Madni, and S. D. Lucero, "Leveraging digital twin technology in model-based systems engineering," *Systems*, vol. 7, no. 1, 2019, doi: 10.3390/systems7010007.

[9] A. Rasheed, O. San, and T. Kvamsdal, "Digital twin: Values, challenges and enablers from a modeling perspective," *IEEE Access*, vol. 8, 2020, doi: 10.1109/ACCESS.2020.2970143.

[10] M. J. Kaur, V. P. Mishra, and P. Maheshwari, "The convergence of digital twin, IoT, and machine learning: Transforming data into action," in *Internet of Things*, 2020, doi: 10.1007/978-3-030-18732-3_1.

[11] M. A. Ganaie, M. Hu, A. K. Malik, M. Tanveer, and P. N. Suganthan, "Ensemble deep learning: A review," *Eng Appl Artif Intelligence*, vol. 115, 2022. https://doi.org/10.1016/j.engappai.2022.105151

[12] A. Mohammed and R. Kora, "A comprehensive review on ensemble deep learning: Opportunities and challenges," *J King Saud Univ – Comput Inf Sci*, vol. 35, no. 2, 2023, doi: 10.1016/j.jksuci.2023.01.014.

[13] M. Tanveer *et al.*, "Ensemble deep learning in speech signal tasks: A review," *Neurocomputing*, vol. 550, 2023, doi: 10.1016/j.neucom.2023.126436.

[14] J. Lee, H. Davari, J. Singh, and V. Pandhare, "Industrial Artificial Intelligence for industry 4.0-based manufacturing systems," *Manuf Lett*, vol. 18, pp. 20–23, 2018, doi: 10.1016/j.mfglet.2018.09.002.

[15] R. S. Peres, X. Jia, J. Lee, K. Sun, A. W. Colombo, and J. Barata, "Industrial artificial intelligence in Industry 4.0 -Systematic review, challenges and outlook," *IEEE Access*, pp. 220121–220139, 2020, doi: 10.1109/ACCESS.2020.3042874.

[16] G. Lugaresi and A. Matta, "Automated manufacturing system discovery and digital twin generation," *J Manuf Syst*, vol. 59, no. February, pp. 51–66, 2021, doi: 10.1016/j.jmsy.2021.01.005.

[17] W. Zhang, Z. Liu, and Z. Liao, "Digital twin inspired intelligent bearing fault diagnosis method based on adaptive correlation filtering and improved SAE classification model," *Math Probl Eng*, vol. 2022, 2022, doi: 10.1155/2022/8767974.

[18] K. Guo, X. Wan, L. Liu, Z. Gao, and M. Yang, "Fault diagnosis of intelligent production line based on digital twin and improved random forest," *Appl Sci (Switzerland)*, vol. 11, no. 16, 2021, doi: 10.3390/app11167733.
[19] D. Gao *et al.*, "Intelligent instrument fault diagnosis and prediction system based on digital twin technology," *J Phys Conf Ser*, 2021, doi: 10.1088/1742-6596/1983/1/012106.
[20] Y. Xu, Y. Sun, X. Liu, and Y. Zheng, "A digital-twin-assisted fault diagnosis using deep transfer learning," *IEEE Access*, vol. 7, 2019, doi: 10.1109/ACCESS.2018.2890566.
[21] Y. Wen, M. Fashiar Rahman, H. Xu, and T. L. B. Tseng, "Recent advances and trends of predictive maintenance from data-driven machine prognostics perspective," *Measurement (Lond)*, vol. 187, 2022, doi: 10.1016/j.measurement.2021.110276.
[22] R. K. Phanden, P. Sharma, and A. Dubey, "6 A review on simulation in digital twin for aerospace, manufacturing and robotics," in *Materials Today: Proceedings*, Elsevier Ltd, 2020, pp. 174–178, doi: 10.1016/j.matpr.2020.06.446.
[23] L. Pan, X. Guo, Y. Luan, and H. Wang, "Design and realization of cutting simulation function of digital twin system of CNC machine tool," *Procedia Comput Sci*, 2021, doi: 10.1016/j.procs.2021.02.057.
[24] C. H. dos Santos, J. A. B. Montevechi, J. A. de Queiroz, R. de Carvalho Miranda, and F. Leal, "11 Decision support in productive processes through DES and ABS in the Digital Twin era: A systematic literature review," *Int J Prod Res*. Taylor and Francis Ltd., 2021, doi: 10.1080/00207543.2021.1898691.
[25] W. Luo, T. Hu, C. Zhang, and Y. Wei, "Digital twin for CNC machine tool: Modeling and using strategy," *J Ambient Intell Humaniz Comput*, vol. 10, no. 3, 2019, doi: 10.1007/s12652-018-0946-5.
[26] J. C. P. Cheng, W. Chen, K. Chen, and Q. Wang, "Data-driven predictive maintenance planning framework for MEP components based on BIM and IoT using machine learning algorithms," *Autom Constr*, vol. 112, Apr. 2020, doi: 10.1016/j.autcon.2020.103087.
[27] S. Ayvaz and K. Alpay, "Predictive maintenance system for production lines in manufacturing: A machine learning approach using IoT data in real-time," *Expert Syst Appl*, vol. 173, Jul. 2021, doi: 10.1016/j.eswa.2021.114598.
[28] H. Singh, "Big data, industry 4.0 and cyber-physical systems integration: A smart industry context," *Materials Today: Proceedings*, 2021, doi: 10.1016/j.matpr.2020.07.170.
[29] W. J. Lee, H. Wu, H. Yun, H. Kim, M. B. G. Jun, and J. W. Sutherland, "Predictive maintenance of machine tool systems using artificial intelligence techniques applied to machine condition data," in *Procedia CIRP*, Elsevier B.V., 2019, pp. 506–511, doi: 10.1016/j.procir.2018.12.019.
[30] A. Theissler, J. Pérez-Velázquez, M. Kettelgerdes, and G. Elger, "Predictive maintenance enabled by machine learning: Use cases and challenges in the automotive industry," *Reliab Eng Syst Saf*, vol. 215, Nov. 2021, doi: 10.1016/j.ress.2021.107864.
[31] C. Kammerer, M. Gaust, M. Küstner, P. Starke, R. Radtke, and A. Jesser, "Motor classification with machine learning methods for predictive maintenance," *IFAC-PapersOnLine*, vol. 54, no. 1, pp. 1059–1064, 2021, doi: 10.1016/j.ifacol.2021.08.126.
[32] H. A. Gohel, H. Upadhyay, L. Lagos, K. Cooper, and A. Sanzetenea, "Predictive maintenance architecture development for nuclear infrastructure using machine learning," *Nucl Eng Technol*, vol. 52, no. 7, pp. 1436–1442, Jul. 2020, doi: 10.1016/j.net.2019.12.029.

[33] Q. Zhang, Z. Yang, J. Duan, Z. Liu, and J. Qin, "Three-dimensional visualization interactive system for digital twin workshop," *J Southeast Univ (Engl Ed)*, vol. 37, no. 2, 2021, doi: 10.3969/j.issn.1003-7985.2021.02.003.

[34] Z. Han, Y. Li, M. Yang, Q. Yuan, L. Ba, and E. Xu, "Digital twin-driven 3D visualization monitoring and traceability system for general parts in continuous casting machine," *J Adv Mech Design Syst Manufact*, vol. 14, no. 7, 2020, doi: 10.1299/jamdsm.2020jamdsm0100.

[35] Y. Wang, W. Ren, Y. Li, and C. Zhang, "Complex product manufacturing and operation and maintenance integration based on digital twin," *Int J Adv Manufact Technol*, 2021, doi: 10.1007/s00170-021-07350-6.

[36] R. B. Roy *et al.*, "Digital twin: current scenario and a case study on a manufacturing process," *Int J Adv Manufact Technol*, vol. 107, no. 9–10, pp. 3691–3714, 2020, doi: 10.1007/s00170-020-05306-w.

[37] S. Liu, S. Lu, J. Li, X. Sun, Y. Lu, and J. Bao, "Machining process-oriented monitoring method based on digital twin via augmented reality," *Int J Adv Manufact Technol*, vol. 113, no. 11–12, pp. 3491–3508, 2021, doi: 10.1007/s00170-021-06838-5.

[38] M. R. Cesur and Cesur Elif, "Prototype of a Digital Twin Model: A Case Study on Cartesian Axes CNC Workbench," in *1st International Congress on Artificial Intelligence and Data Science*, 2021, pp. 217–222.

[39] N. Tzanis, N. Andriopoulos, A. Magklaras, E. Mylonas, M. Birbas, and A. Birbas, "A Hybrid Cyber Physical Digital Twin Approach for Smart Grid Fault Prediction," in *2020 IEEE Conference on Industrial Cyberphysical Systems (ICPS)*, IEEE, Jun. 2020, pp. 393–397, doi: 10.1109/ICPS48405.2020.9274723.

[40] Z. Liu, W. Chen, C. Zhang, C. Yang, and H. Chu, "Data super-network fault prediction model and maintenance strategy for mechanical product based on Digital Twin," *IEEE Access*, vol. 7, pp. 177284–177296, 2019, doi: 10.1109/ACCESS.2019.2957202.

[41] Y. Wang, F. Tao, M. Zhang, L. Wang, and Y. Zuo, "Digital twin enhanced fault prediction for the autoclave with insufficient data," *J Manuf Syst*, vol. 60, pp. 350–359, Jul. 2021, doi: 10.1016/j.jmsy.2021.05.015.

[42] Q. Qiao, J. Wang, L. Ye, and R. X. Gao, "Digital twin for machining tool condition prediction," *Procedia CIRP*, vol. 81, pp. 1388–1393, 2019, doi: 10.1016/j.procir.2019.04.049.

12 Digital Twins for an Irrigation System

R. Parvathi, Rajkumar Jhapte, M. Sucharitha, Soorya M. Nair and V. T. Krishnaprasath

12.1 INTRODUCTION

With global climate change and population growth, agricultural production faces many challenges. In order to improve agricultural production efficiency, reduce agricultural production costs, and achieve sustainable development, our country is actively developing agricultural digital twin technology [1]–[3].

At present, most farmland irrigation is still in the manual control and management stage. Data such as farmland equipment and crops are collected manually or entered manually on a regular basis. The amount of data is small, and the real-time performance is poor. It is difficult for users to intuitively and timely obtain the environmental status and status of farmland. The growth status of crops; in addition, most farmland irrigation equipment has manual switches, lacking mature and reliable control strategies, making it difficult to achieve automatic control [4]–[6]. Although some farmlands adopt relatively advanced Internet of Things (IoT) technology to achieve real-time collection of farmland environment and crop soil moisture (SM) data, users can also receive corresponding data in a timely manner. However, the function of the system in this method is limited to real-time monitoring of data, which is too simple, and functions such as intelligent decision-making and control are still weak.

The application of digital twins in the agricultural field mainly includes the following aspects:

- Agricultural production optimization
- Agricultural machinery and equipment management
- Agricultural product quality and safety supervision
- Integration of the agricultural industry chain

12.1.1 Agricultural Production Optimization

Digital twin technology can provide real-time and accurate data support for agricultural production, helping farmers to better understand crop growth conditions, thereby achieving refined agricultural production management. By analyzing the key parameters during crop growth, such as light, temperature, and humidity, among

DOI: 10.1201/9781003465904-15

others, farmers can adjust their planting plans, fertilization, watering and other measures to improve crop yield and quality [7].

12.1.2 Agricultural Machinery and Equipment Management

Digital twin technology can monitor the operating status of agricultural machinery and equipment in real time, and provide farmers with equipment maintenance, fault diagnosis and other services. Indicators, such as equipment failure rate and maintenance costs, can be analyzed by collecting equipment operation data, and targeted equipment maintenance suggestions can be provided to farmers [8].

12.1.3 Agricultural Product Quality and Safety Supervision

Digital twin technology can support agricultural product quality and safety supervision. By monitoring key indicators in the production process of agricultural products in real time, such as the number of pesticides and fertilizers used, agricultural product contamination and quality and safety issues can be effectively prevented. In addition, digital twin technology can also provide data support for agricultural product quality and safety supervision, as well as provide a scientific basis for regulatory decision-making [9].

12.1.4 Integration of Agricultural Industry Chain

Digital twin technology can provide data support for and promote collaborative innovation in all aspects of the agricultural industry chain [10]. Through digital twin technology, we can realize the digitization and networking of agricultural production, agricultural product processing, and sales, and improve the overall competitiveness of the industrial chain [11].

Given this context, the literature highlights the emergence of the concept of the smart farm or smart farming, which indicates how a farm can benefit from the digital and technological transformations that are taking place in agriculture.

12.2 LITERATURE REVIEW

In order to resolve these difficulties [12], the development goals of the Smart-FarmNet platform are achieving integration between different devices produced by different manufacturers, ensuring the scalability of large-scale farm systems and achieving a level that farmers themselves can use to analyze the characteristics of their farms. The authors also show that certain platforms can serve as a basis for creating agricultural platforms, including Unidosts, Xively, ThingSpeak, Open.Sen.se, SensorCloud, Amazon IoT, and IBM IoT.

The author [13] developed a platform called FarmBeats based on IoT technology, which can transmit data between sensors, cameras, and drones through the network. The system proposed by the authors has been used for six months on two farms in the United States (US). The authors highlight four key points of the platform: large-scale transmission, the ability of base stations to identify the climate in which they are located, compressing the information contained in images to transmit them to the cloud, and flight planning algorithms for agricultural drones.

According to [14], when developing the platform, they mainly focused on four aspects: usability in the event of electrical and connectivity problems; the ability to support large amounts of data and sensors; cloud connectivity that enables farmers to access information even outside the farm; and adequate availability of data despite issues related to gaps in data transmission and storage.

The author [15] developed a drip irrigation system based on different nodes. Each node is divided into a processing unit, monitoring unit, power supply unit, and communication unit. In this work, two different nodes were developed: a final node that calculates reference evapotranspiration using the Penman-Monteith method [16] and another responsible for detecting excess moisture on the terrain, determining precipitation rates, planning fertigation, and controlling battery size vitality.

The authors [16] developed an IoT platform that can collect soil condition parameters, such as pH, humidity, texture, color, air temperature, and light intensity. Therefore, the authors propose a low-cost integrated device capable of collecting these different parameters and sending them to farmers over the Internet for analysis and visualization.

The authors [17] developed an IoT platform using LoRaWAN transmission technology, capable of collecting data from multiple sensors connected to Arduino hardware containing LoRaWAN modules. The data is transmitted to the LoRaWAN gateway, which transmits it to the platform, so that it can be processed and then used in applications and services.

The authors [18] developed a system that utilizes IoT, GPS, and artificial intelligence (AI) to determine agricultural implementation coverage. The proposed system uses the k-nearest neighbor (KNN) method with or without the fast nearest neighbor (FLAN) library.

It should be pointed out that most IoT articles only provide small prototypes or platform architectures. Most articles about prototypes demonstrate electronic circuits and show that the proposed systems work, but they do not put the prototypes out in the field to collect real data on a given crop. Articles emphasizing IoT platform architecture provide operational diagrams but do not apply the platform to actual farms.

12.3 FARMLAND IRRIGATION BASED ON DIGITAL TWINS

An overview of the digital twin system is provided below:

- A smart decision-making system for farmland irrigation is based on digital twins and is characterized by including a physical layer, a data collection layer, a twin model layer, a functional layer and an application layer.
- The data collection module collects field sensor data using different protocols and supports analog, digital, and RS485 data access. The data collection module communicates with the remote communication module via the LoRa wireless network. The remote communication module packages the data from each data

collector. Then, it processes the data into message queuing telemetry transport (MQTT) protocol data. This is transmitted through 4G/5G and Ethernet or Wi-Fi and cloud platforms to enable data transmission.
- The twin data fusion performs data fusion on a temporal scale and a spatial scale. For the temporal scale, the distribution diagram method is used to eliminate inadvertent errors in the data on temperature and humidity. On the spatial scale, a weighted fusion algorithm fuses temperature and humidity, pipeline pressure, and flow data to obtain accurate parameters. The fused data is then inputted into the digital twin model.
- The specific steps of the distribution map method are provided in section 4 (S1-S4).
- The construction of a digital twin model includes the steps that are provided in section 3.1 (T1-T4).
- The visual real-time monitoring platform gathers the environmental parameter information collected by the weather station, along with the data on the soil's temperature, humidity, and moisture. The acquired information is then passed on to the precision irrigation decision-making system.
- The intelligent irrigation decision-making system collects data from weather stations, crop soil temperature and humidity, wind speed, sunshine hours, and pipeline flow. The irrigation decision algorithm determines the required amount of water for the crops, which is then combined with the pipeline flow to obtain the actual irrigation time needed, forming an irrigation decision control strategy.
- The irrigation decision-making algorithm is provided in section 3.2 (A1-A3).
- The irrigation control strategy uses an SM prediction model to forecast SM at the next time point. The SM is assessed to determine whether it is in the most suitable growth interval for the irrigation strategy. The principal component analysis (PCA) is used to reduce the dimensionality of the collected datasets, and the radial basis function (RBF) prediction model of SM is created using the datasets before and after dimensionality reduction. The newrbe function is called to design a path. For the base network, the expansion speed spread is set to 30.
- The method for establishing the RBF SM content prediction model is provided in Section 3.2 (B1-B8).

In Figures 12.1 to 12.6, a smart decision-making system for farmland irrigation based on digital twins of the current method includes a physical layer, a data collection layer, a twin model layer, a functional layer, and an application layer (see Figure 12.1 and Figure 12.2).

Specifically, as shown in Figure 12.2, the physical layer comprises a water source, filter, water pump, pressure sensor, flow sensor, pressure-reducing valve, solenoid valve, sprinkler nozzle, SM sensor, and field weather station. After the water from the source is processed by the filter, it is pumped out by the water pump, and transferred to each irrigation field. The water is then evenly sprayed by the sprinkler nozzle during irrigation operations. Additionally, a water pump controller is integrated within the

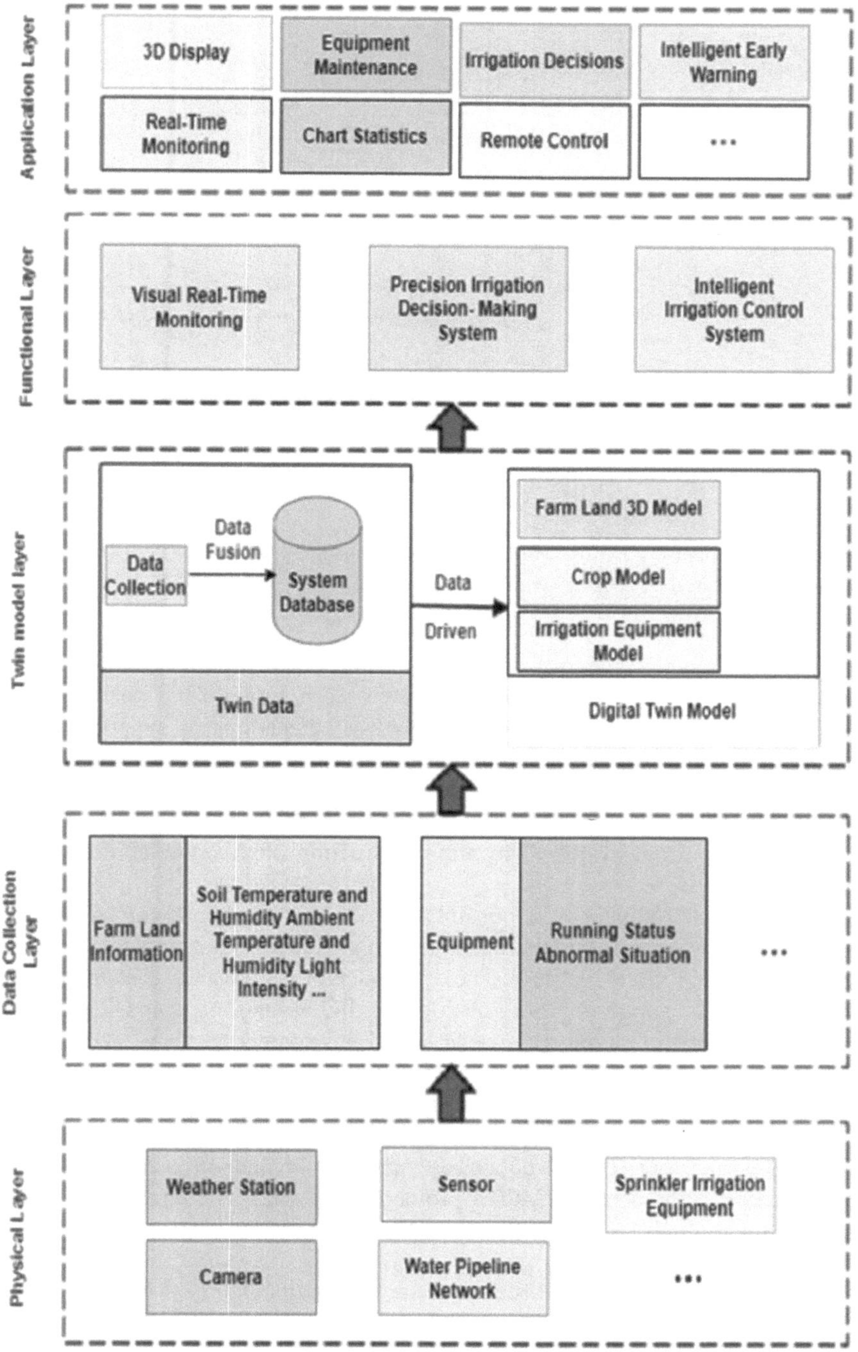

FIGURE 12.1 A Physical System Structure Diagram of a Farmland Irrigation Intelligent Decision-Making System Based on Digital Twins.

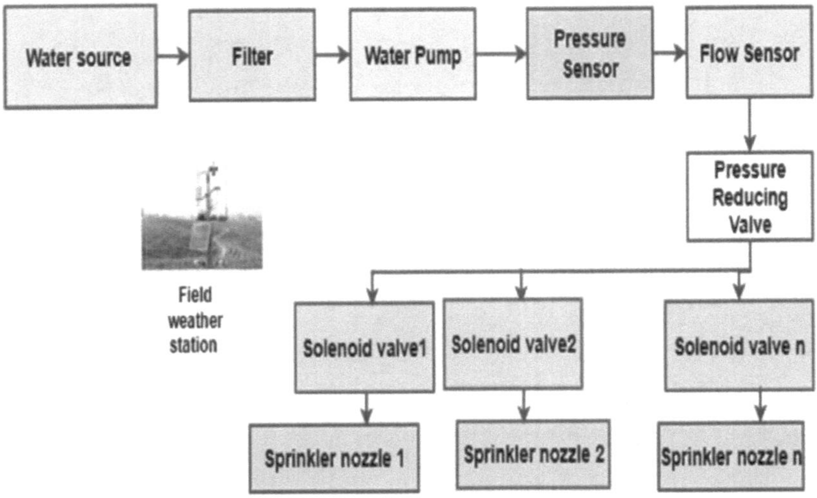

FIGURE 12.2 An Overall Framework Diagram of a Farmland Irrigation Intelligent Decision-Making System Based on Digital Twins.

water pump to remotely control its operation. Sensors such as vibration, temperature, flow, pressure, and water level sensors are used to monitor its operating status. Pressure, flow, and SM sensors, as well as field weather stations, all support the LoRa network communication.

A flow chart of digital twin construction is shown in Figure 12.3. 3D Scene modeling is performed using the solid modeling tool MeshLab. After modeling, optimization is carried out, followed by the construction of the simulation using the digital twin.

The data collection layer is responsible for collecting farmland information and monitoring the equipment's operating status. This includes a data collector, an edge gateway, and an IoT cloud platform. The data collector gathers field sensor data from different protocols, supporting both analog and digital quantities, and RS485 data, which enables real-time collection of farmland environmental data, SM data, and equipment operating statuses. This includes soil temperature and humidity, ambient temperature and humidity, light intensity, equipment operating parameters and other data types. The data collector and the edge gateway communicate using the LoRa wireless network. The edge gateway packages the data of each data collector. Then, it processes the data into MQTT protocol data, which is communicated to the data collector through 4G/5G, Ethernet or Wi-Fi. The cloud platform handles data transmission.

The twin model layer is used for twin data fusion and the construction of digital twin models. Twin data fusion occurs on both temporal and spatial scales. On the temporal scale, the distribution diagram method eliminates inadvertent errors in the data for parameters such as temperature and humidity. On the spatial scale, a weighted

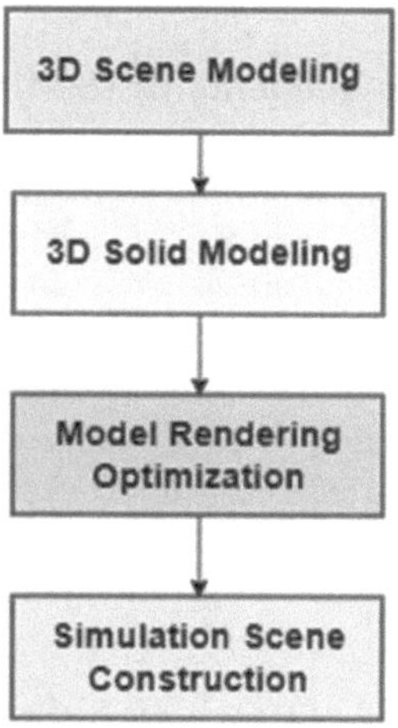

FIGURE 12.3 A Flow Chart of Digital Twin Construction.

fusion algorithm is applied to fuse temperature, humidity, pipeline pressure, flow, and other data to obtain accurate parameters. The fused data is then inputted into the digital twin model. The specific steps of the distribution diagram method are as follows:

Step 1: Sort the n measurement results, measured by m sensors at different times in the farmland from smallest to largest to obtain the measurement sequence: $X = \left(x_1, x_2, \ldots, x_n\right)$

Step 2: When n is an even number, define the median value x_N as $x_N = \frac{x_{\frac{n}{2}+1} + x_{\frac{n}{2}}}{2}$

When n is an odd number, define the median value x_N as: $x_N = \frac{x_{n+1}}{2}$

Step 3: The upper quartile, F, is the median of the $[x_N, x_n]$ interval, and the lower quartile, F, is the median of the $[x_1, x_N]$ interval. The difference between the quartiles, or the degree of dispersion, is defined as $dF = F\,upper - F\,lower$

Step 4: Set the effective judgment interval as $\left[F\,lower - dF, F\,upper + dF\right]$. If the data falls within this interval, then it is considered valid. If the data is outside the interval, then it is regarded as an inadvertent error and will be eliminated.

Using field sensors to collect SM as an example, two sensors continuously collected eight SM data points at intervals of 15 minutes, as shown in Table 12.1.

To calculate the median value, with the degree $dF = 0.8$, the effective judgment interval is [38.7, 41.1]. Therefore, the three data points of 38.4, 44.3, and 37.3 are eliminated.

TABLE 12.1
Eight SM Data at a Frequency of 15 Minute Intervals

Sensor	SM Data 1	SM Data 2	SM Data 3	SM Data 4
1	38.4	39.5	44.3	39.6
2	37.3	40.6	40.3	40.0

The variance (δ2) value for each sensor is obtained from previously collected data. For the two sensors, the mean square error $\delta = 0.172$ for the five deleted data points. The weighted fusion formula, $W1 = 0.4584$, $W2 = 0.5416$, and the fused humidity data $S = 39.956$ were used. This data fusion operation is completed with high accuracy.

12.3.1 Construction of a Digital Twin Model

Further, the specific steps to construct a digital twin model are:

Step T1: Utilize unmanned aerial vehicle (UAV) oblique photographs of the farmland and employ DJI Map software tools to combine global digital elevation data (DEM) to synthesize and process the pictures. This will enable you to realize the construction of a three-dimensional (3D) scene of the farmland.

Step T2: 3D modeling tools can help you to establish models of multiple physical entities (e.g. sensors or equipment) within the intelligent farmland scenes. Use finite element analysis to determine the structural, geometric, and material parameters, as well as the states and boundary conditions of the 3D model.

Step T3: According to the 3D model obtained in Step T2, use the 3ds Max rendering tool to render the model, apply materials, and optimize its edges.

Step T4: Import the model from Step T3 into the 3D scene using the TerraExplorer visualization management software. Assign physical, geometric, behavioral, and regular dynamic attributes to the digital twin model, which will enable it to dynamically reflect each aspect of the physical irrigation equipment. This will achieve a visual representation of digital twins and a visual display of smart farmland application scenarios in a virtual space.

The functional layer includes a visual real-time monitoring platform, a precision irrigation decision-making system, and an intelligent irrigation control system. The purpose of the functional layer is to monitor farmland information data and equipment operating statuses in real time through a digital twin visualization front-end. The system embeds the PCA-RBF algorithm model to predict SM and make intelligent decisions on the amount of irrigation and its timing. The instructions generated by these decisions can be input to the microcontroller (MCU) through the 4G/5G wireless transmission module, allowing for the automatic control of irrigation equipment. Additionally, information exchanges between the physical farmland and the digital twin platform are facilitated, with on-site data, equipment status, alarm information, and decision-making results presented in the form of data and charts. Users can also input different control commands as needed to remotely adjust the operating status of the equipment.

The visual real-time monitoring platform collects the environmental parameter information from weather stations and the soil's temperature, humidity, and moisture levels, transferring this data to the precision irrigation decision-making system. The intelligent irrigation decision-making system uses data from weather stations, crop soil temperature and humidity, wind speed, sunshine hours, and pipeline flow. The irrigation decision-making algorithm calculates the amount of irrigation that is required for the crops and combines this with the pipeline flow to determine the actual irrigation time that is needed to form an irrigation control strategy. The intelligent irrigation control system then issues instructions to remotely control the irrigation equipment based on the irrigation time that the irrigation decision-making system calculated.

12.3.2 The Irrigation Decision-Making Algorithm Using Digital Twins

Further, the irrigation decision-making algorithm is based on the reference crop transpiration amount ET_0, combined with the water demand coefficient (Kc) of the crop growth stage, to obtain the actual water demand *ETc* of the crop. It then calculates the total irrigation amount based on the irrigation amount per unit area and the total land area, combined with the SM prediction results for the next moment. This enables intelligent decision-making regarding the current irrigation amount and irrigation duration. Specifically, the process includes the following steps:

Step A1: First, calculate reference crop water requirement RT_0 according to the Penman-Monteith modified formula (12.1):

$$RT_0 = \frac{0.408\Delta\left(R_n - H\right) + \sigma \dfrac{900}{T+273} \vartheta_2 \left(W_s - W_a\right)}{\Delta + \sigma\left(1 + 0.34\vartheta_2\right)} \quad (12.1)$$

In this step:

- RT_0 is the crop evapotranspiration, mm/d.
- T is the average temperature at a distance of two meters above the ground.
- Δ refers to the slope of the curve of plant air pressure, kPa/°C.
- R_n is the solar net radiation, MJ/(m²·d).
- H refers to the soil heat flux, MJ/(m²·d).
- σ is psychrometric (monitored dry and wet) constant, kPa/°C.
- W_s is saturated water vapor pressure, kPa.
- W_a refers to the actual water vapor pressure, kPa.
- ϑ_2 is the average wind speed at a height of two meters from the ground, m/s.

Step A2: Using the water storage coefficient (Kc), which is based on the crop growth stage (and follows the Food and Agriculture Organization's recommended values), calculate the actual water demand (ETc) of the crop.

$$ET_c = K_c RT_0 \tag{12.2}$$

A3: Calculate the irrigation amount per unit area based on the air temperature, humidity, wind speed, and sunshine hours measured two meters above the ground. Then, calculate the total irrigation amount by considering the irrigation amount per unit area and land area. Finally, the irrigation duration is calculated based on the flow rate of the irrigation pipes.
A4: Send the irrigation amount and irrigation time information to the system control module remotely.

Further, the hardware structure diagram of the intelligent irrigation control system is shown in Figure 12.4.

The power module supplies power to the MCU and the 4G/5G wireless communication module. The MCU controls the relay switch according to the process, and then drives the solenoid valve to work (Figure 12.5).

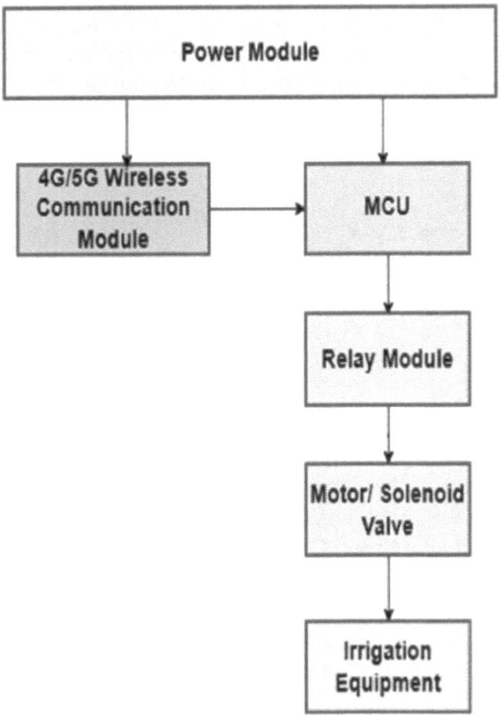

FIGURE 12.4 System Control Module Structure Diagram Using Digital Twins.

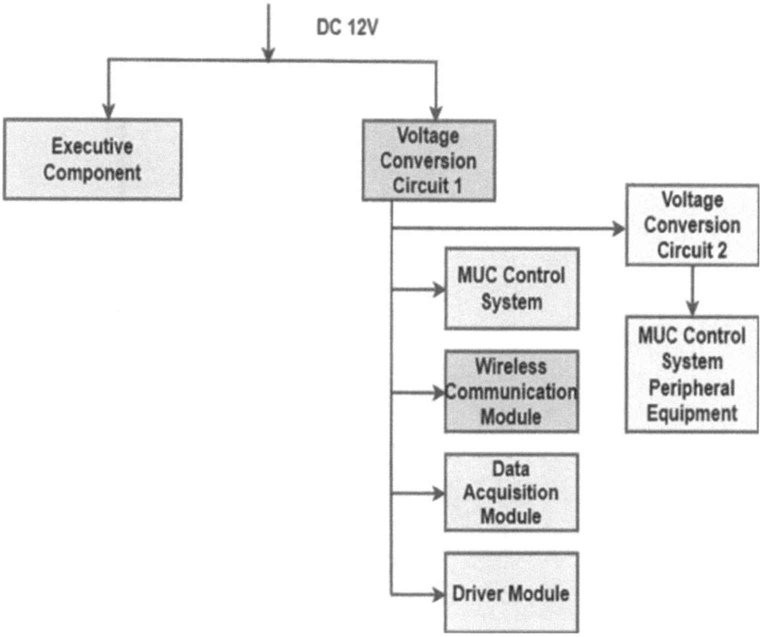

FIGURE 12.5 A System Power Supply Circuit Structure Diagram Using Digital Twins.

The irrigation control strategy flow is shown in Figure 12.6, with the MCU as the core. After receiving instructions wirelessly, the relay module controls the motor speed or the solenoid valve switch to initiate irrigation and determines whether the predicted irrigation level is reached. The system automatically adjusts the water supply by managing the on/off status of the irrigation equipment.

The irrigation control strategy uses an SM prediction model to forecast the SM at the next time point. The irrigation decision is made by comparing the predicted SM to assess whether it falls within the most suitable growth interval. PCA is applied to the collected datasets for dimensionality reduction, and the datasets before and after reduction are used to establish the RBF prediction model for SM. The newrbe function is used to design a radial basis network (RBN), with the expansion speed spread set to 30.

12.3.3 The Method of the RBF SM Content Prediction Model

It is established as follows:

B1: Determine the weight coefficient W1 from the input layer to the radial base layer (RBL). This weight coefficient is $W1 = RT$.

B2: Solve the offset b1 from the input layer to the RBL using the formula $b_1 = sqrt\left(-log\left(0.5\right)\right) / spread$.

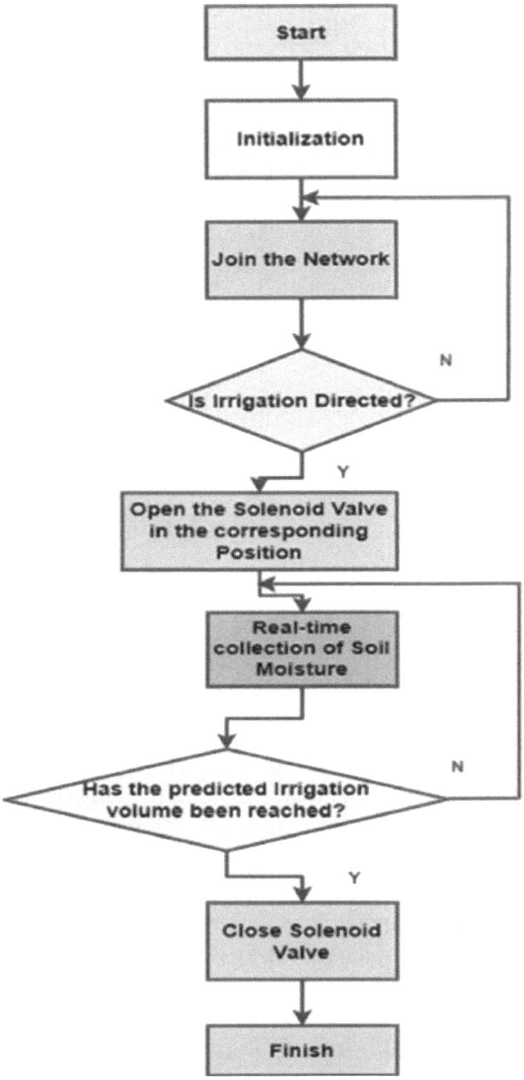

FIGURE 12.6 A Flow Chart of an Irrigation Control Strategy Using Digital Twins.

B3: Calculate the distance between each sample and the radial basis neuron using the formula $dist(x, W1) = \|x - W1\|$.
B4: Calculate parameter n using the formula $n = \|x - W1\| * b1$
B5: Repeat steps B1 to B4, and then proceed to B6.
B6: Calculate the output result A{1} of the first layer using the transfer function.
B7: Solve for the second layer's weight coefficient W2 and bias b2 using the formula $[W\{2,1\}b\{2\}]*[A\{1\}; ones(1,Q)] = T$
B8: End.

Where:

- $R = [Tc,\ Tair,\ RH,\ Rn,\ V]$ is the input.
- $T = [\theta]$ is the output.
- Q represents the input vectors.
- A{1} is the output of the first layer.
- b1 and b2 are the deviations from the input layer to the RBL and from the RBL to the output layer, respectively.
- W1 is the weight of the RBL, and W2 is the weight from the RBL to the output layer.

The system application layer creates an interface in the user interface (UI) to enable 3D display, equipment maintenance, irrigation decision-making, intelligent early warning, real-time monitoring, chart statistics, and remote control. In order to complete the design of the human-computer interaction interface, ThingJS software is used to design and arrange the visual interface. It connects twin data through the software's built-in interface, supporting Ajax, JSONP, WebSocket, and other data docking formats. In addition, various visual charts and buttons can be added to the 3D scene, corresponding to applications such as 3D display, equipment maintenance, irrigation decision-making, intelligent early warning, real-time monitoring, chart statistics, and remote control.

The system power supply circuit structure diagram is shown in Figure 12.5. A 12V power supply is used as the input power supply and then split into two channels. One channel controls the solenoid valve, with a relay enabling its on-off logic; the other channel is stabilized and converted to 5V to power the wireless network communication model, sensor module, and relay module. This 5V channel is further converted to 3.3V to supply power to other peripheral devices in the main control chip.

12.4 CONCLUSION

The work highlights the emergence of the concept of a smart farm or smart farming, which illustrates how a farm can benefit from the digital and technological transformations that are taking place in agriculture. The intelligent irrigation control system, using digital twin technology, issues instructions for remote control of irrigation based on the actual irrigation time obtained by the irrigation decision-making system. The equipment executes irrigation actions. The invention enables visual monitoring, intelligent early warning, and irrigation functions for intelligent farmland scenes in virtual space through a digital twin.

REFERENCES

1. B. Kavitha, A. Jose Anand, B. Rammya, R. Dinesh Kumar, K. Kalanithy, and J. N. Mathan. 2023. Implementation of Effective Rainfall Forecast Model using Machine Learning. *4th International Conference on Smart Electronics and Communication (ICOSEC)*, pp. 1655–1660.

2. T. Anuradha, A. Suganya, R. Muthalagu, M. Narender, and A. Jose Anand. 2023. Evaluation of a Wireless Sensor Network for the Detection of Forest Fires. *2nd International Conference on Edge Computing and Applications (ICECAA)*, pp. 1021–1027.
3. A. Jose Anand, S. Sandhiya, A. Sweety Sharon, and R. Vinisha. 2023. Disease Detection of Various Plant Leaf Using Image Processing Technique. *Eighth International Conference on Science Technology Engineering and Mathematics (ICONSTEM)*, pp. 1–5.
4. A. Shaik, N. Badruzaman, D. Gajendran, M. Geethalakshmi, and Jose Anand. 2022. Organic Farming in Drainage System with Advanced Automation through Robotics and IoT. *2022 International Conference on Data Science, Agents & Artificial Intelligence (ICDSAAI)*, pp. 592–597.
5. M. S. Sundari, A. Jose Anand, and T. S. Nagarajan. 2022. Smart Aerial Imaging Solution in Precision Agriculture. *International Conference on Data Science, Agents & Artificial Intelligence (ICDSAAI)*, pp. 1–3.
6. G. Villani, P. Castaldi, A. Toscano, C. Stanghellini, T. Salmon Cinotti et al. 2018. Soil Water Balance Model CRITERIA-ID in SWAMP Project: Proof of Concept. *Proceedings of the 2018 23rd Conference of Open Innovations Association (FRUCT)*, pp. 2–3.
7. R. Gomes Alves, R. Filev Maia, and F. Lima. 2023. Development of a Digital Twin for Smart Farming: Irrigation Management System for Water Saving. *Journal of Cleaner Production*, 388: 135920. https://doi.org/10.1016/j.jclepro.2023.135920.
8. P. Angin, M. Hossein Anisi, F. Goksel, C. Gursoy, and A. Buy ukgulcu, 2020. AgriLoRa: A Digital Twin Framework for Smart Agriculture. *Journal of Wireless Mobile Networks, Ubiquitous Computing, and Dependable Applications (JoWUA)*, 11(4): 77–96.
9. J. R. Tsay, C.-T. Lu, and T.-C. Tu. 2019. *Application of Common Information Plat form to Foster Data-Driven Agriculture in Taiwan*. Food Agricultural Policy Plat form Article.
10. P. Verboven, T. Defraeye, A. K. Datta, and B. Nicolai. 2020. Digital Twins of Food Process Operations: The Next Step for Food Process Models?. *Current Opinion in Food Science*, 35: 79–87.
11. T. Defraeye, G. Tagliavini, W. Wu, K. Prawiranto, S. Schudel, M. A. Kerisima, P. Verboven, and A. Bühlmann. 2019. Digital Twins Probe into Food Cooling and Biochemical Quality Changes for Reducing Losses in Refrigerated Supply Chains, Resources. *Conservation and Recycling*, 149: 778–794.
12. P. Prakash Jayaraman, Yavari, A., Georgakopoulos, D., Morshed, A., and Zaslavsky, A.. 2016. Internet of Things Platform for Smart Farming: Experiences and Lessons Learnt. *Sensors*, 16 (11): 1884.
13. D. Vasisht, Z. Kapetanovic, J. Won, X. Jin, R. Chandra, A. Kapoor, S.N. Sinha, and M. Sudarshan. 2017. FarmBeats: An IoT Platform for Data-Driven Agriculture. *14TH USENIX Symposium on Networked Systems Design and Implementation (NSDI 17)*, Boston, MA: USENIX Association, pp. 515–529.
14. I. Mohanraj, V. Gokul, R. Ezhilarasie and A. Umamakeswari. 2017. Intelligent Drip Irrigation and Fertigation using Wireless Sensor Networks. *2017 IEEE Technological Innovations in ICT for Agriculture and Rural Development (TIAR)*, Chennai, India, 36–41.
15. G. A. Richarda, L. S. Pereira, D. Raes, and M. Smith. 1998. Crop Evapotranspiration: Guidelines for Computing Crop Water Requirements. *Food and Agriculture Organization for the United Nations*, 56: 1998.

16. A. Oliveira-Jr, C. Resende, A. Pereira, P. Madureira, J. Gonçalves, R. Moutinho, F. Soares, and W. Moreira. 2020. IoT Sensing Platform as a Driver for Digital Farming in Rural Africa. Sensors, 20(12): 3511.
17. Y. Ma, and J. Chen. 2018. Toward Intelligent Agriculture Service Platform with Lora-Based Wireless Sensor Network. *2018 IEEE International Conference on Applied System Invention (ICASI)*, pp. 204–207.
18. M. Waleed, T.-W. Um, T. Kamal, A. Khan, and A. Iqbal. 2020. Determining the Precise Work Area of Agriculture Machinery Using Internet of Things and Artificial Intelligence. *Applied Sciences*, 10(10): 3365.

13 Diagnostics, Treatment and Patient Care in the Age of Digital Twins

A Game-Changer in Healthcare

S. Muthukaruppasamy, G. Arun Sampaul Thomas, J. Nandha Gopal, A. Jose Anand and E. Parimalasundar

13.1 INTRODUCTION

At its foundation, digital twin technology involves the creation of virtual clones of actual things, such as products, systems, processes, or whole environments. The digital counterpart is built using real-time data and provides a dynamic portrayal of its physical counterpart. It is a powerful instrument for in-depth investigating, modeling, and monitoring of a physical object's functioning [1].

The interior functions are as follows:

A. Data Gathering and Unification: A digital twin is constructed based on data. Sensors, the Internet of Things (IoT) devices, and other data sources collect real-time data from the physical object. This information is then incorporated into a digital platform.
B. Simulation and Modeling-based Data Processing: The data is processed by sophisticated algorithms to generate a model that accurately represents the physical object. This model is dynamic and continually updated with fresh data to ensure it remains an accurate portrayal.
C. Continuous Monitoring: The digital twin receives live data from its physical counterpart continually. This includes details about its current status, circumstance, and performance. This real-time input enables quick modifications or solutions.
D. Analytics Predictive: One of the most useful features of digital twin technology is its ability to make predictions. It can predict how things will behave and perform in the future by looking at past data and using machine learning (ML) techniques [2].

DOI: 10.1201/9781003465904-16

E. Feedback Mechanism: The information obtained from the digital twin could be used to optimize the operational effectiveness of the physical entity. It creates a feedback mechanism where improvements are continuously incorporated.

However, it is essential that we approach digital twin technology with caution as we enter this new era. A significant leap is required to transition from its use in mechanical systems to applications in human health. Digital twins present their own special set of difficulties in healthcare. After reviewing the fundamentals, it is encouraged to reconsider digital twins within the healthcare context by delving into their advantages, disadvantages, and the significance of thoroughly comprehending their implementation in this critical domain [3].

13.1.1 Applications in Various Sectors

The sectors of production and Industry 4.0 are being transformed using digital twins, which allow for virtual testing and optimization of production processes. This technology reduces downtime and improves product quality. Blockchains can be applied in the following sectors:

A. Healthcare: Digital twins of patients could improve specific therapies, facilitating more efficient and accurate care [23].
B. Smart Cities: Digital twins of urban areas may assist in the distribution of resources, environmental responsibility, and planning for cities.
C. Aviation and automotive: Digital twins enable corporations to monitor performance, forecast maintenance requirements, and optimize fuel efficiency.
D. Utilities and Energy: Digital twins of power plants and utility grids help to optimize energy creation, transfer, and use.

Digital twin technology is poised to transform many fields by offering new perspectives, the ability to work more efficiently, and improve outcomes. As this technology continues to evolve, we can expect even more cutting-edge applications that push the limits of what is possible in the digital world. The future seems to hold vast potential for digital twins. Digital twin technology works by generating a virtual replica, or "twin," of a tangible object in the digital domain. The virtual representation is created by collecting data in real time from sensors and the IoT components, and other data sources linked to the physical item [5].

13.2 RETROSPECTING DIGITAL TWIN FOR HEALTHCARE

The merits of the digital twin concept, derived from the realm of machinery and intricate procedures, are not universally applicable to the healthcare sector. Understanding the human factor, the need for consistent utilization of various multifaceted data, the intricacy of determining causality from correlations, and the importance of health events are all critical consideration. The phrase "digital twins" originated during NASA's missions. NASA scientists recognized the need for a comprehensive terrestrial replica (referred to as a "twin") of the spacecraft to diagnose issues and model reactions to different scenarios. As the technical capabilities grew, these physical reproductions gradually transformed into digital counterparts [6]. A digital twin is essentially a dynamic virtual model that accurately reflects a tangible object or operation in real time. This is facilitated by the continuous stream of information from the IoT and sensors connected to the physical entity. Digital twins have proven to be immensely useful in fields such as manufacturing and aviation. They assist engineers in monitoring equipment, forecasting faults, and simulating various situations—all without interfering with the real machinery.

With the ongoing digital revolution in healthcare, the notion of a digital twin has naturally become relevant in this field. The concept of reproducing a patient's entire health record digitally is intriguing. This includes their genetic composition, medical background, and lifestyle, accompanied by real-time biometric information (to the minute). An all-encompassing digital replica could enable healthcare practitioners to anticipate health concerns before they reach a critical or diagnosable stage. This technology has the potential to completely transform the customization of preventive and treatment strategies, ultimately leading to improved patient outcomes and an enhanced quality of life for all individuals [7]. The use of digital twins in healthcare promises an innovative opportunity. Nevertheless, while the concept has been successful in mechanized and system-focused fields, its implementation in the complex and individualized realm of healthcare is not as simple. Now, let's focus on the central point of this study—the constraints and potential misapplication of the digital twin analogy in healthcare [8].

13.2.1 The Equivocal Parable of Digital Twins

The concept is very intriguing, particularly within the realm of healthcare. Nevertheless, it is important to scrutinize whether this phrase, which was first created for mechanical systems, can be applied to people's health. An engine in an aircraft or a manufacturing line is an example of a mechanical or physical process that is mostly deterministic and predictable. The state of such a system is determined by specified rules. These systems are specifically engineered to function in a definite manner, and unless there are any involuntary malfunctions or external disruptions, they perform accordingly [9]. Compiling all health-related factors, such as life status, environmental, societal, and other events, into one complete personal record can make it easier to create a thoroughly individual model. When applied to an individual, this model has the potential to guide the development of personalized treatment and preventive approaches. An immediate occurrence or an interval is the normal mode of communication for human beings when it comes to occurrances. These occurrences, which often comprise different aspects of varying importance, can influence our health and well-being significantly. However, they do not seamlessly integrate with the continuous flow of data that a digital twin model system is specifically engineered to process. Instead of focusing on timestamps, people are skilled at remembering and linking events to one another.

In addition, establishing causation from data streams is a more complex task. In industrial settings, certain occurrences tend to be linked to variations in the functioning of machines. However, the task is made more challenging when it involves human health. If two health occurrences take place in quick succession, then it does not necessarily mean that one of them caused the other. Inferring causality from health data requires meticulous study and, often, a deeper understanding of the underlying biology and behavior. When considering the digital twin metaphor in the context of healthcare, it is necessary to proceed with caution given these complexities. Simply inputting biometric data into a digital model is not enough; rather, it is essential to master an understanding of the connections between the events that are recognized in different data streams. To achieve this, it is vital to infer relevant information about events from common sources, such as geographical, environmental, and social factors, and to extract meaningful insights from this information, including those related to fatality. This is a highly challenging operation. Incorrect assumptions, as well as possibly inaccurate health diagnoses, projections, and actions, could result from misinterpretation and oversimplification [15].

13.2.2 Machines vs Humans: Two Distinct Realms in Modeling

In the case of digital twins, the crucial contrast between people and computers goes beyond their fundamental natures and includes the inherent uncertainties associated with each. The actions of mechanical systems and machines are predictable because they are created by human engineers and adhere to accepted rules and principles. These systems are generally well understood and have predictable properties. By contrast, people are quite different. Human physiology and health are incredibly complicated, influenced by a multitude of interrelated variables, including lifestyle choices, emotions, surroundings, and genetic predispositions. Despite the significant advances in medical research, our knowledge remains largely stochastic and is always changing.

Consider diabetes as an example. Its development and control depend on several factors, such as diet, exercise, stress, and sleep. The interaction between these components may greatly differ between people and even within the same person under different circumstances. Moreover, our understanding of these connections is perpetually progressing. Given these complexities, an unchanged database may not be most appropriate solution for storing health data. Treating data as a history of personal health occurrences could be a more effective method. A stressful workday, a night of restless sleep, a missed workout, or a high-cholesterol meal all contribute to a person's health narrative. This record could enhance our comprehension of the timing, interactions, and effects of these occurrences on general health. Furthermore, when new information arises, the chronicle-based method allows for a dynamic reinterpretation of historical health data. An occurrence that initially seems inconsequential could later be viewed as a significant turning point in one person's health journey. Previously concealed patterns could become apparent, offering fresh perspectives into control and prevention strategies. Although the digital twin concept works well for machinery, applying it to humans requires a fundamentally distinct and more sophisticated approach. One way to make digital twins more useful and adaptable in healthcare is to use the concept of a personal health chronicle. This strategy evolves as our knowledge of health continues to expand [11].

A digital twin may not only help healthcare providers but also serves as a valuable tool for navigating your personal health. Implementing digital twins in healthcare requires a subtle and nuanced strategy. It is crucial to develop and implement this notion with caution and accountability, considering its limitations, and adapting it to suit the complex and diverse nature of human health. While the digital twin is powerful, it remains a metaphor that is borrowed from the realm of machinery systems. It should function as a tool to assess and forecast health conditions, directed by individual health systems that are built using data. People's health information comprises a combination of subjective information, factual observations, constant and intermittent sensor data, as well as social and atmospheric data. An optimal system integrates all sources, recognizing health-related occurrences that do not always coincide with uninterrupted data streams. Moreover, when interacting with other individuals, health issues are often discussed in relation to specific occurrences rather than the exact time they occurred. This highlights the need to view health information as distinct incidents and to organize all personal information chronologically.

Caution should be exercised when applying causal relationships. Establishing causation in health data poses substantial challenges. Correlation does not automatically indicate casualty, and comprehending causal connections typically requires a more profound knowledge of the underlying biology and the lifestyle of the patient. Compassion in healthcare, human intuition, and judgment remain vital despite technological advancements. Private information may be a valuable resource for understanding an individual's emotional sensitivity, enabling compassionate interactions. Ensuring data privacy and security should be of utmost importance. Due to the delicate nature of personal information, it is essential to provide strong tools for the management, safeguarding, and responsible use of this data. Continuous learning is important to stay updated with the latest study and advances in healthcare, as digital twin ideas change and gain popularity. As new information and ideas emerge, adjustments should be made [12].

Understanding what a digital twin is and how it is used in patient care is crucial. Openness and honesty can help to set realistic goals, build trust, and ensure that patients are actively involved in their own care. By following guidelines, digital twins will be used to their fullest potential in healthcare, taking advantage of their benefits while minimizing risks. The term "digital twin" refers to a computerized copy of a real thing, method, or system. It encompasses various elements, such as patient data, medical equipment, healthcare centers, and even entire healthcare environments. These virtual replicas are created using data gathered from different sources [13]. One key aspect of digital twins is their ability to change in real time. Digital twins are not static models or pictures; they evolve and adapt constantly based on real-world data. This real-time feature enables healthcare workers to monitor digital twins and use them to make better decisions. Additionally, digital twins enable the simulation and analysis of different healthcare scenarios. This capability provides useful information about patients' conditions, treatment options, and system operations, empowering healthcare professionals to improve tactics and results [14].

13.2.3 Types of Digital Twins in Healthcare

There are different kinds of digital twins used in healthcare, and each one has its own purpose. A few of the most popular types are listed below.

i. *Patient Digital Twins:* These digital twins are accurate representations of actual patients and their conditions. They use information from smart tech health records and tracking tools. Personalized treatment plans and tracking from far away are made possible by the digital twins of patients.
ii. *Bio Medical Device Digital Twins:* Health device digital twins are copies of medical tools and machines, like magnetic resonance imaging (MRI) machines and ventilators. They help with predictive maintenance, which ensures that the equipment works well, and that patients are safe.
iii. *Healthcare Facility Digital Twins:* The healthcare centers with digital twins enable entire hospitals or offices to be seen virtually. They help in planning how to respond to emergencies, allocating resources efficiently, and using layouts most effectively [15].
iv. *Digital Twins for Medicine and Pharmaceutical Research:* Digital twins simulate how chemicals behave, enabling researchers to predict how well and safely new medications may function.
v. *Ecosystem Digital Twins:* The entire healthcare supply chain, which includes vendors, consumers, suppliers, and individuals, is modeled by these digital twins. They support discovering opportunities to reduce costs and minimize processes [16].

The virtual representation becomes worthless in the absence of precise and pertinent information. Important data sources and inputs for digital twins in the healthcare industry comprise the following.

A. *Electronic Health Records*: The fundamental elements of patient digital twins are the patients' medical histories, diagnosis, and treatment plans.
B. *Wearable Devices and Sensors*: Real-time patient status updates are provided through data from sensors, cardiovascular monitors, and wearables.
C. *Imaging and Diagnostics*: By offering comprehensive information, medical imaging techniques, such as computed tomography (CT) and MRI scans, support medical device and patient digital twins.
D. *IoT Devices*: Smart beds and medicine dispensers are examples of IoT technologies in healthcare that provide useful data for optimizing healthcare facilities.
E. *Electronic Monitoring Systems*: Healthcare institutions use monitoring systems to gather information on patient vital signs, equipment conditions, and other topics. Understanding these fundamental ideas is the basis for

effective digital twins in healthcare, as well as the many kinds and sources of data. In later parts of this extensive manual, we will examine the necessary procedures for using digital twins' capabilities in order to revolutionize healthcare systems and enhance patient care.

13.3 THE ROLE OF DIGITAL TWINS IN HEALTHCARE: A USE CASE

Digital twin integration in the healthcare industry requires meticulous planning and a thorough understanding of the objectives of healthcare institutions. It needs to examine the essential stages of establishing healthcare objectives, collecting and ensuring data accuracy, and data protection concerns. Before delving into the technical parts of digital twin deployment, setting unambiguous goals for healthcare organizations is crucial. The following must be taken into account:

A. Enhancing Medical Attention: Learn how using digital twins enhances the treatment of patients. Can they help with remote surveillance, tailored treatment plans, and early detection?
B. Effectiveness of Operations: Examine how hospitals may use digital twins to streamline healthcare procedures, save expenses, and better allocate resources.
C. Creativity and Investigation: Consider how digital twins can aid in drug discovery, clinical studies, and the development of innovative solutions for healthcare.
D. Making Decisions Based on Data: Define how digital twins may be useful, and how healthcare workers can make better decisions by providing actionable information.

13.3.1 Information Collection and Quality Control

After determining objectives, gathering data is a crucial next step. The efficacy of digital twins is directly impacted by the data quality. There are a few tips for ensuring that data is properly gathered and preserved, using the following facts.

A. Sources of Data: Determine the sources of all relevant data, such as electronic health records (EHRs), wearable devices, sensors, and other healthcare systems.
B. Integration of Data: Use data combination technologies to gather and combine information from numerous sources. Ascertain this integration is smooth and real-time.
C. Quality Assurance of Data: Establish data quality control systems to verify and sanitize incoming data. Incorrect or inconsistent data could result in erroneous digital twins.
D. Storage of Data: Select scalable and safe data storage options that can manage the amount and diversity of medical data. Cloud-based solutions are suggested for flexibility.

E. Compliance and Data Privacy Consideration: Stringent data privacy policies apply to the healthcare sector.
F. Data Encryption: To safeguard patient data both during transmission and storage, establish strong encryption methods.
G. Access Controls: Strict access restrictions should be established and maintained to ensure that only authorized individuals can access patient data.
H. Audit Logs: Keep detailed audit records to track who retrieves patient information and when they do so. This will assist with conformity and security.
I. Guidelines for Data Preservation: To prevent needless data preservation, establish data retention and destruction procedures that comply with applicable requirements.
J. Patient Consent: Protect patients' privacy rights by obtaining informed permission for digital twin data collection and usage.

Healthcare data privacy laws are mandatory and may have serious implications. Legal and regulatory professionals must design and execute digital twin deployments to ensure effective implementation in healthcare organizations, it is crucial to establish precise healthcare goals, gather data of exceptional quality, and handle regulatory and data protection concerns. These steps will provide a strong foundation for the implementation process. In the next section, we will delve into the technical elements of developing and using digital twins to accomplish healthcare objectives [17].

The use of digital twins in the health sector necessitates a series of technological procedures and considerations to guarantee the development of precise and interactive virtual simulations. This section examines the essential elements of constructing digital twins, including modeling and simulation, developing seamless integration of real-time data, and harnessing IoT and sensor innovations to boost functionality.

13.3.2 Constructing Digital Twins: Modeling and Concepts

Creating digital twins starts with constructing precise systems and simulations that accurately mirror present-world things and procedures. The recommended strategy is to tackle this crucial step:

A. Model Selection: It selects suitable modeling methodologies, such as 3D modeling, or data-driven modeling, based on the specific characteristics of the digital twin.
B. Data Combination: It integrates the data gathered from different sources, ensuring consistency with the selected modeling technique. This data forms the basis for the behavior of the digital twin.
C. Truthfulness and Realism: It targets a significant level of verisimilitude and faithfulness in systems to replicate current world behavior precisely. Utilizing historical data for validation can help to accomplish this objective.
D. Continuous Updating: It develops procedures to continuously update and calibrate models that guarantee that digital twins accurately and faithfully represent real-world changes.

E. Digital Twins Rely on Up-to-date Data: It maintains the dynamic and flexible characteristics of digital twins by ensuring immediate and seamless integration and data connection.
F. Data Pipeline: It creates a data pipeline that efficiently collects, analyzes, and transfers live data from many sources, such as sensors, wearables, EHRs, and others, directly into the digital twins.
G. Cooperation: It ensures compatibility among various systems and data formats to facilitate smooth data transmission.
H. Synchronization of Data: It employs synchronization procedures to ensure coherence between the digital twins and their physical equivalents.
I. Decrease in Latency: It reduces the time it takes to transmit data to facilitate immediate decision-making and fast responses in healthcare procedures.

13.3.3 Making Use of Sensor and IoT Innovations

The IoT and sensors are essential for improving the capabilities of digital twins. The following is a guide on how to efficiently use these technologies.

A. Sensor Selection: Select appropriate sensors for specific medical applications. For example, wearables can be used to monitor a patient's vital signs, while temperature sensors can be employed to oversee equipment.
B. IoT Platform Integration: Employ IoT systems that effectively gather, automate, and transmit information from the sensors to the digital twins.
C. Cutting-Edge Computing: This technology can manage data aggregation from various sources, reducing delays and enabling real-time analytics.
D. Security and Privacy: Prioritize security measures to safeguard the data produced by IoT sensors and devices. It is also essential to comply with the laws that protect data confidentiality.
E. Scalability: This is a pivotal aspect in crafting the infrastructure for IoT and sensor systems, as it ensures that the system can adapt seamlessly to escalating data volumes and expanding device deployments. The key lies in constructing resilient models that enable real-time data integration through leveraging IoT and sensor technologies. This strategic approach empowers digital twins to stay synchronized with the real world, offering healthcare professionals invaluable insights and reliable support.

Embracing digital twins in healthcare yields significant advantages, chiefly in furnishing valuable insights and fortifying decision-making processes. This segment delves into the strategic utilization of digital twins, elucidating their potency in a utilizing data analytics, predictive analytics, and crafting clinical decision support systems [18]–[19].

13.3.4 Utilizing Predictive Data Analytics to Enhance Informed Decision-Making

Digital twins produce a wealth of data, presenting a valuable resource for informed decision-making in healthcare. Effectively leveraging data analytics involves the following strategic steps.

A. Visualizing Data for Clarity and Insight: Use data visualization tools to transform complex healthcare data into clear graphs, charts, and dashboards.
B. Ongoing Monitoring in Real Time: Utilize intuitive dashboards to continuously monitor patient health in real time, device performance, and facility operations, ensuring timely responsiveness.
C. Identifying Irregularities Through Anomaly Detection: Deploy ML algorithms to identify anomalies within data, promptly alerting healthcare professionals to potential issues and deviations.
D. Tracking and Assessing Performance Indicators: Establish and monitor key performance indicators (KPIs) to accurately assess the effectiveness of healthcare processes and interventions.
E. Analyzing Historical Data for Insights: Conduct thorough analyses of historical data to discern patterns and trends, serving as a foundation to develop predictive modeling.
F. Deploying ML Models for insightful analysis: Construct robust models that are capable of forecasting patient health outcomes, predicting equipment failures, or anticipating the occurrence of disease outbreaks.
G. Implementing Early Warning Systems for Proactive Monitoring: Establish early warning systems that are designed to inform healthcare providers about potential patient health deterioration or equipment malfunctions, allowing for timely intervention before issues escalate.
H. Strategically Planning Resources for Optimal Operational Efficiency: Utilize predictive analytics to forecast and evaluate patient admission rates and determine resource needs. This facilitates optimized staff management and effective allocating of resources for enhanced operational efficiency. Integrating digital twins into clinical decision support systems provides invaluable assistance to healthcare professionals, aiding them in making well-informed decisions.

The development of effective clinical decision support systems involves the following key considerations.

A. Implementing Guidelines Rooted in Evidence for Informed Decision-Making: Align the clinical decision support system with the recognized best practices by integrating evidence-based recommendations and protocols.
B. Utilizing Current Patient Information: Leverage digital twins to incorporate real-time patient data, ensuring that the system provides up-to-date information for clinical decision-making.
C. Warnings and Suggestions: Configure the system to produce alerts and suggestions based on past data and patient situations, enhancing its responsiveness and relevance.
D. Customization: Customize recommendations for each patient by considering their specific needs and medical history, thereby improving the relevance and effectiveness of the support system.

Ethical considerations, such as securing patient consent and protecting patient privacy, are incorporated into developing and implementing clinical decision support systems and processes. The accuracy and efficiency of clinical decision support systems are significantly increased by digital twins, leading to improved patient outcomes and more streamlined healthcare delivery. This comprehensive strategy includes clinical decision support systems, data analytics, and predictive analytics, all within the framework of digital twins. Healthcare organizations can improve patient care, streamline operations, and proactively address emerging healthcare issues [20].

13.3.5 Ensuring Adoption, Protecting Privacy, and Securing

As healthcare organizations embrace digital twins to change patient care and streamline operations, prioritizing data security and privacy is crucial for effectively deploying this transformative technology. This section explores methods for protecting patient data, implementing efficient adoption and training programs, and resolving typical challenges.

A. Data Security Measures: Prioritize data security measures by having access limits, encryption techniques, and frequent audits in place. Ensure compliance with industry standards and regulations to protect sensitive patient information from unauthorized access and breaches.
B. Privacy Safeguards: Implement comprehensive privacy safeguards, including anonymization techniques and strict access controls to uphold patient confidentiality. Develop and enforce policies that comply with privacy regulations, fostering trust among patients and stakeholders in handling their health data securely.
C. Adoption Strategies: Develop tailored training programs for healthcare professionals and staff to enhance their understanding and proficiency in using digital twins. Foster a culture of continuous learning and adaptation, encouraging the seamless integration of this technology into daily workflows.
D. Stakeholder Engagement: Engage key stakeholders, including healthcare providers, administrators, and IT personnel in the decision-making process. Solicit feedback to address concerns and promote a collaborative approach that ensures a smoother adoption of digital twins across various departments.
E. Usability and User Experience: Prioritize the usability and user experience of digital twins by designing intuitive interfaces and functionalities. Streamline workflows to minimize disruption, making it easier for healthcare professionals to incorporate digital twins into their routine practices.
F. Addressing Implementation Challenges: Anticipate and proactively address common challenges associated with implementing digital twins, such as technical glitches, interoperability issues, and resistance to change. Establish a robust support system to troubleshoot issues promptly and provide ongoing assistance.
G. Ongoing Evaluation and Enhancement: Establish ongoing monitoring systems to evaluate the security and functionality of the digital twin. Frequently assess and refine adoption strategies based on feedback and evolving technology standards to ensure sustained success.

By adhering to these strategies, healthcare organizations can fortify the security and privacy of patient data while fostering a successful adoption of digital twins. This comprehensive approach contributes to the seamless integration of transformative technologies into the healthcare landscape, ultimately enhancing patient care and operational efficiency [21].

13.3.6 Addressing Challenges in Implementation

The implementation of healthcare digital twins may encounter several challenges, but these can be overcome using thoughtful strategic planning, as listed below.

A. Integration Difficulty: Collaborate with the IT team to address the complexities of integrating digital twins with existing healthcare systems and operations. Establish clear communication channels to simplify the integration process.
B. Data Quality Issues: Implement a robust monitoring system to continuously assess and enhance data quality. This ensures that the digital twins deliver accurate insights and reliable support for healthcare decision-making.
C. Resource Constraints: Alleviate resource constraints by allocating sufficient personnel and technological resources to aid in the smooth deployment and ongoing maintenance of digital twins. Prioritize scalability to accommodate future growth.
D. Change Management: Develop a comprehensive change management plan to help healthcare staff adapt smoothly and transition to new processes and technologies. Provide training and support to foster a positive reception to the changes.
E. Compatibility: Invest in interoperability technology and standards to guarantee effective communication between digital twins and other healthcare systems. Ensure compatibility to enhance the overall efficiency of the healthcare ecosystem.
F. Expense Control: Create a realistic budget that considers the initial implementation costs and the recurring expenses for updates and maintenance. Prioritize cost-effectiveness without compromising the quality and sustainability of the digital twins initiative.

By actively addressing these challenges through collaborative efforts, continuous improvement, and thoughtful resource allocation, healthcare organizations can successfully implement digital twins, unlocking their potential to revolutionize patient care and streamline healthcare operations.

13.4 ADVANCING DIGITAL TWINS: TRANSFORMING PATIENT CARE AND MEDICAL RESEARCH

A digital twin is a virtual replica of a physical object, process, or system. Within the field of medicine, this idea entails the production of an exact digital copy that includes the anatomy of a patient's biology, health history, and heredity. This virtual representation is continually updated, incorporating real-time data and providing a dynamic and comprehensive view of the patient's health using the following strategies [22].

13.4.1 Transformative Effects on Diagnostics

The integration of AI and ML has ushered in a transformative era for diagnostics, improving the accuracy and speed of medical assessments. These advanced technologies not only streamline the diagnostic process but also contribute to early detection and personalized treatment strategies, revolutionizing the healthcare landscape as listed below.

A. Customized Diagnostics for Individualized Care: Digital twins enable healthcare practitioners to mimic and analyze a patient's exact condition, facilitating personalized diagnoses. The degree of personalization at this level improves the accuracy of diagnoses, resulting in more precise and customized treatment strategies.
B. Prompt Identification of Diseases in their Initial Stages: Utilizing digital twins for continuous monitoring and data analysis enables the early diagnosis of diseases. Digital twins are capable of detecting minute changes in a patient's well-being, so that illnesses in their first phases can be identified. This, in turn, allows for timely interventions when they are most impactful.
C. Optimizing Drug Testing and Treatment Strategies: Digital twins are essential for replicating a patient's responses to different drugs. This assists healthcare practitioners in selecting the most appropriate treatments and adjusting dosages for the most optimal outcome. This individualized method for medication testing and optimization enhances the efficacy and customization of therapeutic treatments.

13.4.2 Revolutionizing the Patient Experience

A. The advent of digital health technologies and patient-centric applications is revolutionizing the patient experience, fostering greater engagement and empowerment. From telehealth services that provide convenient access to healthcare professionals to personalized mobile apps delivering real-time health information, these innovations are reshaping how individuals interact with and navigate their healthcare journeys. These are listed as follows.
B. *Fostering Active Patient Engagement:* Patient involvement is improved by providing people with access to the data of their digital twin. Patients could actively monitor their progress, understand treatment choices, and make well-informed decisions in partnership with their healthcare professionals.
C. *Empowering Remote Patient Monitoring:* By using digital twins, healthcare professionals may monitor patients remotely, thereby surveying a patient's health. This capability ensures timely intervention in case of any emerging health issues.

D. *Integrating Digital Twins into Telemedicine Platform:* Digital twins seamlessly integrate into Telemedicine systems, which provide healthcare practitioners with a thorough overview of a patient's health during virtual consultations. This integration enhances the efficiency and effectiveness of remote healthcare delivery.

The significance of digital twins in healthcare is monumental, presenting a transformative strategy for delivering personalized care. Several factors contribute to their growing importance in the healthcare industry, including compliance with regulatory standards and a shift toward preventive healthcare. Regulatory standards play a pivotal role in ensuring patient safety and regulatory compliance. Digital twins virtually represent a patient's health condition, empowering healthcare providers to meet these standards effectively. Digital twins offer healthcare professionals a holistic perspective on a patient's health, enabling the delivery of customized treatments tailored to specific requirements [23].

The surge in digital twin technology is fueled by the growing emphasis on proactive healthcare. Unlike traditional healthcare focused on addressing illnesses once they manifest, digital twins enable a proactive stance by identifying risk factors, forecasting health outcomes, and empowering patients to make informed choices about their well-being. Harnessing data from digital twins enables healthcare providers to assist patients in embracing preventive measures and adopting healthier lifestyles. The dynamic healthcare landscape is influenced by factors such as an aging population, increased chronic illnesses and rising healthcare costs, which demand innovative healthcare delivery methods. Digital twins provide a robust solution by facilitating precise diagnostics, tailored treatments, and streamlined healthcare administration. They promise to transform healthcare, improve patient outcomes, enhance the quality of care, and optimize resource allocation.

13.4.3 Advancing Digital Twins: In Silico Data Development

A. The progress of digital twin technology relies on harnessing in silico data, involving computer simulations rooted in real-world patient data. In silico data empowers researchers to simulate the dynamics of biological systems, encompassing cells, organs, and complete organisms. This data facilitates the generation of virtual patient models, enabling the exploration of diverse treatment options, the forecasting of patient reactions, and the detection of biomarkers crucial for the innovation of novel therapies and diagnostics, as outlined below.
B. Innovative Medical Device Design: This technology designs and evaluates medical devices before production. Virtual prototypes of these devices are created through digital twins, enabling engineers to identify potential problems and optimize safety and efficiency. This approach minimizes the risk of complications while ensuring patient safety.
C. Continuous Patient Monitoring through Digital Twins: Digital twin technology generates virtual representations of patients, enabling continuous real-time

monitoring and analysis. Healthcare providers can observe patients' vital signs, identify irregularities, and address emerging concerns pre-emptively. Remote patient monitoring that is facilitated by digital twins leads to more efficient and cost-effective healthcare delivery.

D. Tailoring Treatments with Digital Twins for Personalized Medicine: Digital twins simplify the generation of individualized models based on a patient's unique physiology. This personalized approach enhances the precision of diagnoses and allows for treatment regimens tailored to each patient's individual needs. Identifying the most effective treatments for every patient helps to reduce expenses and enhance patient outcomes.
E. Enhancing Healthcare Training through Simulations: Digital twins establish virtual training *settings* for healthcare professionals, providing a secure and regulated space for honing procedural and treatment skills. This empowers providers to elevate their competencies, diminish the likelihood of complications, and ultimately secure improved patient results.

13.4.4 Benefits and Applications of Digital Twins in Healthcare

Digital twins, a cutting-edge technology that creates virtual replicas of physical objects or processes, have emerged as a transformative force in healthcare. They offer an innovative approach with a myriad of benefits, from enhancing medical diagnostics and treatment planning to facilitating personalized patient care through precise simulations of individual physiological conditions.

A. Enhancing Patient Outcomes through Digital Twins: Customized modeling using digital twin technology enables healthcare providers to deliver precise diagnoses and tailor patient treatments, improving patient outcomes and minimizing healthcare expenses.
B. Reducing Healthcare Costs with Digital Twins: Digital twin technology in healthcare reduces costs by enhancing medical device safety and efficacy, mitigating complications, and facilitating remote monitoring. More efficient and economical healthcare delivery reduces healthcare expenditure.
C. Improving Efficiency in Healthcare Delivery through Digital Twins: Digital twins pave the way for distant patient monitoring, empowering healthcare providers to oversee patients' well-being from afar. It enhances the efficiency of healthcare delivery and personalized medicine, reducing the need for trial-and-error methods and ensuring superior patient outcomes.

Digital twins can revolutionize the healthcare sector by streamlining medical device design, elevating patient monitoring, facilitating personalized medicine, and enhancing healthcare professional training. Their utilization yields advantages, including better patient outcomes, cost savings, and more effective care delivery, ultimately resulting in superior healthcare experiences for both healthcare providers and patients.

A. Precision Medicine: Digital twins create personalized models based on individual patient data, enabling healthcare providers to tailor treatments and interventions to each patient's specific needs and characteristics.

B. Medical Device Design and Testing: Digital twins are employed in the design and testing of medical devices, providing a virtual environment for engineers to simulate and evaluate the performance of devices before they are physically produced. This enhances the effectiveness and safety of medical equipment and reduces potential complications.
C. Remote Patient Monitoring: Digital twins facilitate the creation of virtual representations of patients, enabling continuous real-time monitoring. Healthcare providers can remotely monitor vital signs, detect anomalies, and intervene proactively, leading to more efficient and cost-effective healthcare delivery.
D. Simulation Training for Healthcare Professionals: Virtual environments created by digital twins serve as platforms for simulation training for healthcare professionals. This also minimizes complications and enhances patients' overall outcomes by enabling practitioners to refine their abilities in a safe and regulated environment.
E. Disease Modeling and Prediction: Digital twins simulate the dynamics of biological systems, including cells, organs, and organisms. This capability enables researchers to model and predict disease progression, identify potential biomarkers, and explore innovative therapeutic approaches.
F. Treatment Optimization: By simulating patient responses to various treatments, digital twins help healthcare providers to optimize treatment plans. This includes predicting the effectiveness of medications and adjusting dosages for better outcomes.
G. Healthcare Facility Design and Planning: Digital twins play a role in designing and planning healthcare facilities. Architects and planners can create virtual models of healthcare spaces, which optimize layouts for efficiency and patient care.
H. Continuous Quality Improvement: Digital twins contribute to continuous quality improvement in healthcare by providing real-time data for analysis. This data can identify areas for improvement, streamline workflows, and enhance overall healthcare delivery.
I. Drug Discovery and Development: Digital twins aid in drug discovery by simulating the effects of potential medications on virtual patient models. It accelerates the drug development process by predicting drug efficacy and possible side effects before entering clinical trials.
J. Patient Education and Engagement: Digital twins empower patients by providing them with access to their virtual health models. They foster patient education, engagement, and collaboration in decision-making processes regarding their healthcare.

The diverse applications of digital twin technology in healthcare highlight its potential to revolutionize the industry, offering innovative solutions for personalized medicine, efficient healthcare delivery, and improved patient outcomes.

13.4.5 Digital Twins for Diagnosis and Treatment Decision Support

Digital twins facilitate the creation of virtual duplicates of patients or specific medical conditions, enabling healthcare practitioners to model various scenarios and evaluate the efficacy of diverse treatment alternatives. This innovation allows for tailored and precise suggestions based on individual patient information and medical studies, resulting in heightened diagnostic precision, optimized treatment strategies, and superior patient outcomes.

A. Medical Device Design and Optimization: Digital twin technology provides significant advantages in designing and optimizing medical devices and equipment. Manufacturers can use virtual replicas to simulate real-world situations, identify potential concerns, and make essential enhancements before physical production. This methodology reduces costs, shortens development timelines, and enhances the overall performance of devices. Digital twins also enable predictive maintenance through constant monitoring, enabling manufacturers to scrutinize real-time data and pre-emptively detect defects or malfunctions, ensuring patient safety, prolonged device longevity, and uninterrupted healthcare services.
B. Hospital & Facility Management: Digital twin technology enhances the operational efficiency of healthcare facilities by optimizing resource allocation and patient pathways, leading to improved services through the effective use of existing resources. Applications of digital twins extend to energy planning, resource logistics, administrative processes, and facility maintenance, promoting sustainability and streamlining operations.
C. Remote Monitoring & Telemedicine: Digital twins transform patient monitoring through wearable devices that capture real-time health data and transmit it to a virtual patient replica. Swift analysis and interpretation by digital twin technology yield valuable insights for healthcare practitioners. Ongoing monitoring enables customized interventions, early identification of irregularities, and remote patient surveillance, empowering healthcare providers to offer optimized care and enhance health outcomes.
D. In-depth Medical Research: Digital twin technology enables economical and secure simulation of scenarios for medical research. Researchers can recognize patterns of illness, simulate the impact of therapies, and pinpoint domains for additional investigation in live subjects. By comparing a substantial cohort of digital twins, researchers can assess diverse treatment choices for individuals with similar characteristics and discover biological markers for disorders, as well as the progression of medical understanding.

The versatile applications of digital twin technology in healthcare underscore its potential to revolutionize diagnostics, treatment decisions, medical device design, facility management, remote monitoring, telemedicine, and medical research. These advancements contribute to more personalized, efficient, and effective healthcare services.

13.4.6 Regulatory Framework and the FDA's Perspective on Digital Twins

The regulatory landscape for digital twins is evolving, and the United States Food and Drug Administration (FDA) has taken an open-minded approach to this emerging domain. The FDA recognizes the crucial role of modeling and simulation in its strategic objective, especially when assessing medical equipment. The FDA's Office of Science and Engineering Laboratories (OSEL) has committed a significant number of resources to transform computational modeling from a critical scientific tool into a regulatory asset. This change is motivated by the possible cost savings in evaluating medical devices and modeling their behavior in scenarios that may be too risky for humans to use or are more effectively assessed through simulation [15].

In a noteworthy initiative, the FDA has entered a collaboration with Siemens to investigate the application of digitized twin technology in the development of medical devices. The FDA has allocated nearly $2 million in funding to Siemens for a pilot program to demonstrate how digital twins can enhance product quality, development processes, and the commercialization of medical devices. These initiatives highlight the FDA's pledge to explore the possibilities of digital twins in healthcare and its recognition of the benefits they offer in improving regulatory procedures and driving innovation in the field of medical devices [19].

13.5 CASE STUDY ADVANCEMENTS OF DIGITAL TWINS IN HEALTHCARE SYSTEMS

The integration of digital twins in the healthcare system has ushered in a new era of personalized and efficient patient care. Case studies exemplify the transformative impact of digital twins, demonstrating their prowess in improving diagnostics, treatment strategies, and overall healthcare outcomes. These digital replicas of patients enable real-time monitoring, simulation, and analysis, enabling healthcare professionals to tailor interventions based on individual characteristics. From predicting disease progression to optimizing treatment plans, digital twins empower medical practitioners with invaluable insights, enhancing the precision and efficacy of healthcare delivery. The amalgamation of cutting-edge technologies and healthcare expertise in these case studies showcases a promising trajectory for the widespread adoption of digital twins, heralding a future where healthcare is increasingly tailored to each patient's unique needs.

13.5.1 Practical Instances of Digital Twins in the Future

In the rapidly advancing landscape of technology, the practical implementation of digital twins is poised to revolutionize various sectors in the future. From optimizing industrial processes and predicting equipment failures to simulating complex urban environments for smart city planning, the deployment of digital twins promises unprecedented levels of efficiency and innovation across diverse fields, as listed below.

A. Optimizing Cancer Therapy: Digital twins are being created by experts to customize cancer therapy. These computerized models forecast the tumor response to various treatments, allowing for more specialized and efficient care.
B. Managing Cardiovascular Well-being: Heart digital twins are used to monitor and maintain cardiac health. They can prevent and treat cardiovascular illnesses by simulating the effects of medicine and lifestyle modifications on heart function.

C. Managing Treatment for Diabetes: Digital twins offer a chance to find the best treatment plans for diabetes. Through simulations of a patient's routine schedule, they can plan how to control their insulin, calculate doses and schedules, and recommend optimal diets.
D. Quick Diagnoses: Patients can receive quicker, more precise diagnoses, which reduce their anxiety and allow for earlier treatment.
E. Enhanced Results of Treatment: Tailored therapeutic interventions have the potential to yield improved outcomes and an elevated standard of living.
F. Patient Empowerment: By gaining access to the data and insights of their digital twins, patients will be granted increased autonomy in managing their health.

All of these can be achieved using the following main constraints:

A. Security and Privacy of Data: Safeguarding sensitive information requires rigorous safety precautions, which are essential for protecting patient data.
B. Data synthesis: Massive volumes of data from several sources are needed for digital twins. There are technological obstacles in ensuring correctness and smooth integration.
C. Implement Synthetic Data: Synthetic data may be utilized, as training digital twins requires a large amount of data. One significant problem is to ensure that synthetic data is impartial and fair.
D. Regulatory structures: To allow for the use of digital twins in diagnosis and treatment, rules about healthcare may need to change [22].

On the other hand, humans are complex biological mechanisms, who are greatly affected by environmental, genetic, and lifestyle elements. Living systems have intricate structures and behaviors that are not totally predictable, but rather governed by probabilistic interactions among several factors. Despite having the same genetic composition, even identical twins may have significant disparities in their health characteristics because of changes in their surroundings, behaviors, and life encounters. Therefore, it may not be entirely realistic to attempt to apply the notion of a 'digital twin' to the intricate and diverse realm of human health, as it was originally intended to depict predictable and rule-based equipment or processes. The human body does not function uniformly on every occasion, as a machine would. Organisms are adaptable and constantly changing, reacting to many internal and external stimuli. Essentially, we cannot use metaphors from the machine world to oversimplify the complicated realm of human health. It is essential to reconsider the idea and develop a new one that can offer valuable information and make sound predictions about healthcare. This model would start with genes and then incorporate culture, surroundings, and mental states, as well as the aging process, which has a significant impact.

Another crucial distinction between the equipment and patient health pertains to the inherent characteristics of the information being handled. The digital twin idea principally relies on the constant flow of sensor data. This strategy is very effective in an industrial context, as machines consistently provide information about their condition. Preservation professionals examine the information gathered in digital twins and implement suitable measures. Machines cannot express their emotions or convey the factors that impact them. In human health, the information is intricate, including several modes of data collection, which are not simply obtained via sensors. Humans provide crucial but subjective knowledge on occurrences in their health and existence. Healthcare professionals could supply instinctive event information that is crucial for comprehending the person.

While activities of daily living (ADL) are helpful, we believe they may be enhanced and improved. Activities can describe events, and characterize, explain, and guide lifestyles. Figure 13.1 illustrates that a person's daily activity, such as eating or working, only provides information about their competence at completing a task. Events are a popular idea in daily life that summarize activities and qualities into meaningful entities. We believe events offer better abstractions for aligning human states with everyday activities, semantic context, and physiological signals, as well as modeling health using learning techniques. In this article, we use daily living events, which are semantically rich in spatial-temporal, informational, experiential, structural, and causal aspects, as the top data stream to associate multimodal data [24].

13.5.2 Case Study on Hosp.App—Hospital Task Management System

This application will help the client to create a novel mobile system to facilitate effortless communication and collaboration among hospital employees within patient-specific groups. This system enables streamlined interactions, file sharing, prescription management, and task execution for individual patients within these groups [25].

Key features of the application are defined as follows:

A. Messages: List groups tied to patients and categories such as cleaning, food, doctor, medical record, etc.
B. Chat Room/Group: Members can exchange text messages, convert speech-to-text, attach photos and files, call online members, create tasks, and fulfill them.
C. Tasks: Tasks are automatically assigned to the member on shift and remain "in progress" until marked as "completed."
D. Challenges: Differentiate between text messages and tasks clearly, making tasks stand out, and organizing tasks chronologically with filters.

13.5.3 Case Study on IsDocIn Mobile Application for Appointment Booking

The goal was to create a mobile application that simplifies the process of booking appointments with doctors. The app enables doctors to establish profiles and receive notifications when someone schedules an appointment with them [26].

Key features of the application are defined as follows:

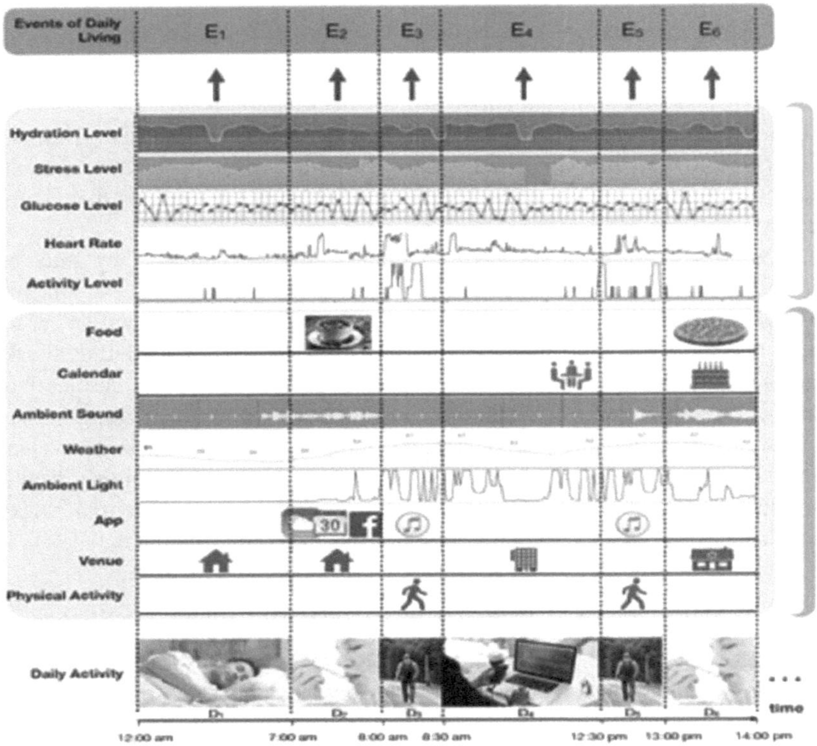

FIGURE 13.1 Human Healthcare Chronicle.

A. User Stories: Detailed stories provided by the customer guided the prototype development.
B. Consultations: Collaborate with the client and their back-end team to create a user-friendly interface.
C. Challenges: Ensure compliance with medical procedures while maintaining a clear and easy-to-use interface.

Considering the swift progress of digital twin technology, many organizations and industries are enthusiastic about adopting this ground-breaking method to digitize their operations. Nevertheless, to embark on this transformative path, they necessitate a dependable technology partner to guide them through every phase of the process. Stfalcon is a reputable service provider, delivering essential expertise, software solutions, and integration services to ensure the smooth and effective adoption of digital twin technology. The central objective is to protect valuable data, simulations, and real-world entities depicted by the digital twin, assuring the highest levels of security and privacy.

13.6 CONCLUSION

The transformation that promises extraordinary insights, efficiency, and optimization across various industries is being led by the technology known as digital twins, which is at the vanguard of the revolution. We anticipate seeing even more imaginative applications that push the boundaries of what is possible in the digital arena as it continues to develop for the foreseeable future. The future has found a virtual counterpart in digital twin technology, which serves as a powerful bridge between the digital and real worlds. It functions as a revolutionary instrument for creativity, optimization, and real-time decision-making across a wide range of areas. A healthy dose of both caution and optimism is required to successfully navigate the future of healthcare with technology. When it comes to understanding the chronicles of a person's life, the collection of massive longitudinal data holds immense promise; nonetheless, its limitations must be acknowledged and respected, particularly in the complex subject of human health. As we continue along this path, there is a chance not only to use technology for the benefit of health but also to redefine and improve the standards of healthcare delivery. There is no straightforward narrative regarding the future of digital twins in the healthcare industry, yet if we take the necessary steps, then we can mold it into a story of triumph. To successfully adopt a digital twin in the healthcare industry, a comprehensive strategy is required—one that prioritizes data protection, privacy, training, and overcoming barriers. By doing so, healthcare organizations will be able to make full use of digital twins, enabling them to enhance patient care, speed up operations, encourage innovation, and maintain the highest possible privacy and data security requirements. The use of this game-changing technology represents a significant step forward in terms of both outcomes for patients and the delivery of healthcare. In the future, digital twins will significantly impact diagnostics and radically transform how patients interact with medical professionals.

REFERENCES

[1] www.jashminsingh17.medium.com/what-is-digital-twin-technology-and-how-does-it-work-3b01d759cbd4

[2] Rashmi R Nair., Senthamizh Selvi, R., Harshaa Vartini, G ., Jayashri, B., & Suruthi, S., & Jose Anand A. (2023). Maturity Level Classification of Moringa Leaf using DNN Algorithm. In 2023 4th IEEE International Conference on Smart Electronics and Communication (ICOSEC) (pp. 1667–1672), Tiruchy, India https://ieeexplore.ieee.org/document/10276048

[3] Kabat, W., Schertzer, A., De La Torre, D., Taraboulos, P., & Co. (2023). The power of Digital Twins in healthcare: Revolutionizing patient care. *KSDT CPA*. www.linkedin.com/pulse/revolutionizing-patient-care-how-digital-kwicf?trk=public_post_main-feed-card_feed-article-content

[4] Mourya, R., Gadekar, L., & Aher, S. (2023). Exploring the Revolution in Healthcare Systems through the Applications of Digital Twin Technology. *ScienceDirect, 49(2): 370–382*. www.sciencedirect.com/science/article/pii/S2949723X23000065

[5] Anuradha, T., Suganya, A., Muthalagu, R., Narender, M., & Jose, Anand A. (2023). Evaluation of a Wireless Sensor Network for the Detection of Forest Fires. In 2023

2nd IEEE International Conference on Edge Computing and Applications (ICECAA) (pp. 1021–1027), Namakkal, India https://ieeexplore.ieee.org/document/10212390
[6] Allam, Z., Eltumi, A., & Hassan, M. (2023). Digital twin in healthcare: Recent updates and challenges. *PMC – NCBI.* www.ncbi.nlm.nih.gov/pmc/articles/PMC9830576/
[7] Zheng, Y., Jin, J., Xu, W., Yang, S., & Ai, W. (2023). Digital twin for healthcare systems. *PMCNCBI.* www.ncbi.nlm.nih.gov/pmc/articles/PMC9830576/
[8] Wang, F., Wang, F., Zhang, R., Zhang, A., & Zhang, Y. (2022). A Review of Digital Twins in Healthcare: Architecture, Applications, and Challenges. *IEEE Access, 10(3): 8351–8365.* www.ieeexplore.ieee.org/document/10099646/
[9] Zhang, H., Dastbaz, A., Fan, Z., & Li, H. (2022). Digital Twin Technology in Healthcare: Enabling Precision Medicine. *IEEE Journal of Biomedical and Health Informatics, 26(12): 2907–2918.* www.ieeexplore.ieee.org/document/9255249
[10] Meguerdichian, S., Cardoso, J., Santos, J., & Cunha, J. P. (2022). Application of Digital Twins in Healthcare: A Systematic Review. *Journal of Healthcare Engineering, 13(1): 39–52.* www.igi-global.com/book/digital-twins-healthcare/298637
[11] www.medium.com/@jain49/rethinking-digital-twin-for-health-71ae07dbb44
[12] McKinsey & Company. (2023). Digital Twins in Healthcare: Transforming Care Delivery. www.mckinsey.com/industries/industrials-and-electronics/our-insights/digital-twins-the-key-to-smart-product-development
[13] Accenture. (2023). The Rise of the Healthcare Digital Twin. www.accenture.com/us-en/insights/health/mirrored-world
[14] GE Healthcare. (2023). How Digital Twins Are Revolutionizing Healthcare. www.gehccommandcenter.com/press-coverage/21-ways-medical-digital-twins-will-transform-healthcare
[15] Philips Healthcare. (2023). The Power of Digital Twins in Precision Medicine. www.ncbi.nlm.nih.gov/pmc/articles/PMC9830576/
[16] Li, W., Li, H., Wu, Z., & Wang, S. (2023, October). Digital Twins for Healthcare: Opportunities and Challenges. In 2023 IEEE 17th International Conference on BioInspired Systems and Computational Intelligence (BICI) (pp. 797–802). IEEE. www.ieeexplore.ieee.org/document/8792622
[17] Chen, Y., Yang, M., & Yu, X. (2023, June). The Application of Digital Twins in Healthcare. In International Conference on Computational Science and Applications (pp. 332–344). Springer, Cham. www.link.springer.com/chapter/10.1007/978-981-99-2746-3_69 1
[18] www.medium.com/@Larisa10/a-comprehensive-guide-to-implementing-digital-twins-in-healthcare-dfd832a6f72
[19] Arun Sampaul Thomas, G., & Harold Robinson, Y. (2020, February 25). IoT, Big Data, Blockchain and Machine Learning Besides Its Transmutation with Modern Technological Applications. In *Springer Book Chapter-Internet of Things and Big Data Applications- Part of the Intelligent Systems Reference Library Book Series* ISRL (Vol. 180, pp. 47–63), ISSN: 978-3-030-39118-8
[20] Arun Sampaul Thomas, G., Muthukaruppasamy, S., Saravanan, K., & Muleta, N. (2023). Revolutionizing Smart Grids with Big Data Analytics: A Case Study on Integrating Renewable Energy and Predicting Faults. In: Kumar Sharma, D., Sharma, R., Jeon, G., & Kumar, R. (eds) *Data Analytics for Smart Grids Applications—A Key to Smart City Development. Intelligent Systems Reference Library*, (Vol. 247). Springer, Cham. https://doi.org/10.1007/978-3-031-46092-0_11.
[21] Silva, D. A., Costa, C. A., Rodrigues, J. J., & Oliveira, J. L. (2022). Digital Twins in Healthcare: Towards a New Era of Patient Care. *International Journal of Environmental Research and Public Health, 19(8): 5003.* www.mdpi.com/2075-4426/12/8/1255

[22] www.medium.com/@gianlucaradesich/digital-twins-in-healthcare-revolutionizing-diagnostics-and-transforming-the-patient-journey-4b11866f9425
[23] www.stfalconcom.medium.com/digital-twins-in-healthcare-improving-patient-care-medical-research-49671045d4e8
[24] Kahneman, Daniel, Krueger, Alan B., Schkade, David A., Schwarz, Norbert, & Stone, Arthur A. (2004). A survey method for characterizing daily life experience:The day reconstruction method. *Science*, *306(5702)*: 1776–1780.
[25] www.stfalcon.com/en/portfolio/hospapp
[26] www.stfalcon.com/en/portfolio/isdocin-app

Part 4

Case Study and Challenges

14 Digital Twins in Industry 5.0

Some Challenges and Future Directions

Anupama and Geeta Rani

14.1 INTRODUCTION

Civilization is in the midst of a digital transition. Issues such as Industry 4.0, the Internet of Things (IoT), and big data, to name just a few, are becoming more prevalent, particularly in business contexts. Digital twins were first used almost a decade ago and are a useful piece of technology similar in every way to its true version. They are capable of being used for simulations, forecasting defects in things, merchandise, and so on. In 2002, Michael Grieves [20] openly presented the digital twin notion as a conceptual framework underpinning "product life cycle management" at a conference in Michigan. Digital or virtual twins are the consequence of ongoing innovation in product development and engineering efforts. As more and more of intellectual resources are consistently diverted toward their usage, the prospects of digital twins seem almost endless. As a result, digital twins are continually acquiring new abilities and competencies, enabling them to come up with the insights required to improve goods and operations. Although digital twins are currently in practice across numerous sectors, the fast-increasing digital twin industry suggests that the market for digital twins will keep on growing for the foreseeable future. Globally, the digital twin market was expected to surpass USD 73.5 billion by 2027 [14]. Many sectors have witnessed significant improvements such as Industry 4.0, physical Internet, cyber-manufacturing, Made in China 2025, and cloud manufacturing. Because of advances in digital technology, IoT, and big data, Industry 4.0 has experienced a transformation [18], [34]–[45]. Industry 4.0 is a manufacturing transformation spurred by breakthroughs in digitization, artificial intelligence (AI), the IoT, and data analytics [2], [4], [19]. The digitalization and networking of machines, systems, and processes mark this industrial revolution, leading to the rise of advanced manufacturing facilities and smart production systems. Traditional industrial systems are evolving into networked cyber-physical systems capable of communicating, analyzing data, and making autonomous decisions. Because of Industry 4.0, the safekeeping of every bit of information in digital form, and the incorporation of sensors into business settings, digital twin deployments became practical, revitalizing the notion [5], [16]. Fortunately, with the introduction of Industry 4.0, digital twins have found use in a variety of industries, including manufacturing, logistics, product development, service,

DOI: 10.1201/9781003465904-18

and maintenance [22], [59]. Furthermore, with the development of sophisticated capabilities for simulation and a large rise in processing resources, practical testing in virtual settings became conceivable. As a result of these technological improvements, firms such as IBM, Siemens, and GE began adopting an effective digital twin as a tool for themselves as well as for their customers [37] [27]–[28].

14.2 BACKDROP AND CONCEPTUAL FRAMEWORK

The core of contemporary industry is innovative production and the rise of Industry 4.0. The move from labor-intensive techniques to mechanized manufacturing characterized the beginning of the Industrial Revolution (Industry 1.0).

During the second wave of industrialization, mass manufacturing and assembly lines emerged(Industry 2.0). The use of computers and automation in industrial processes were introduced during the Third Industrial Revolution (Industry 3.0). The progress in the adoption of digital technologies, including the IoT, AI, machine learning (ML), big data analytics, robots, additive manufacturing, and digital twins, is referred to as Industry 4.0 [30]–[31]], [38]. The interconnection and real-time data interchange among machines, systems, and humans distinguishes Industry 4.0, enabling extremely adaptable and effective manufacturing processes. Industry 5.0, commonly referred to as the Fifth Industrial Revolution (5IR), is a novel and developing era of industrialization in which people collaborate with modern technology and AI-powered robots to improve workplace procedures. This is accompanied by a more a human-centered orientation [25] , [43]–[44], [47], better resilience, and a stronger emphasis on sustainability [3], [51]–[52] (Figure 14.1).

As Industry 4.0 evolves, the significant developments and advances listed above are defining the foreseeable future of manufacturing as well as production, resulting in improved productivity, adaptability, and sustainable operations [1]. In the industrial and production industries, these advancements facilitate more automation, enhanced productivity, and enhanced interaction among machines, systems, and humans. The digital twins are an innovation that stems from the "Fourth Industrial Revolution" concept. Javaid and Haleem [24] stated that 5IR was the manufacturing area's next evolutionary path, intending to employ human intelligence to interact with smart and technologically advanced robots. The present 5IR approach combines the distinctive creativity of human professionals with strong, intelligent, and precise robots. To organize business and trade, 5IR would depend on data, equipment, and AI. The

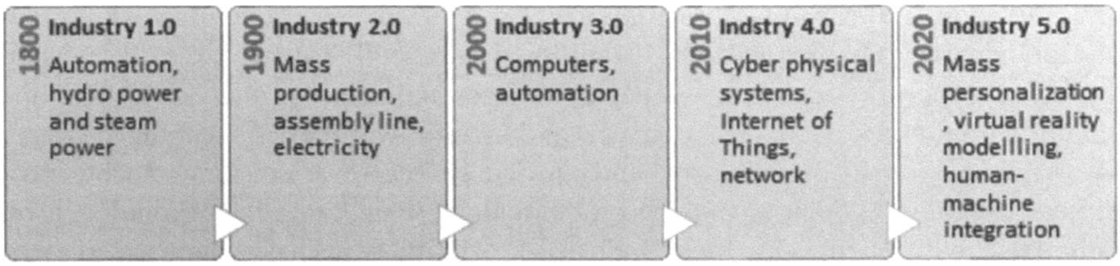

FIGURE 14.1 Industrial Revolution from Industry 1.0 to 5.0.

aforementioned elements were entirely memory-based, analogous to the way human brain works. Memory, in reality, provided intelligence in AI by giving data to execute algorithms as well as the context behind behaviors and responses [31], [35]], [54]–[55], [62], [65]–[66].

A digital twin notion emerged from the aviation industry, where it was first put to use in order to track and optimize the operation of intricate systems [60]. Michael Grieves [20] originated the concept and model underpinning the phrase "digital twin" in his book "Virtually Perfect".

A digital twin's theoretical framework has three core elements: the physical thing, the digital representation, and the link that exists between the two [21] (Figure 14.2); and has five imperative elements lacking which, digital twin's distinctiveness vanishes [1], [16].

1. IoT devices – to gather sensor data from various component parts of physical assets and peripheral devices. High-fidelity connectivity among IoT devices is required for precise and rapid exchange of data [8].
2. Data – information acquired from various IoT elements and software; it is essential for keeping tabs on the system, ensure proper behavior, and offer stimulus in relation to the ML approach. Facilities for big data evaluation and archiving are required for obtaining usable insights from data [16].
3. ML – in extreme instances, for forecasts plus feedback, in addition to identifying appropriate ways to mitigate. A combined optimizing mechanism for the digital twin's component parts is required.
4. Information as well as data transit protection-Security methods for information exchange and verification, as well as authorization processes, are required [16].
5. Assessment of digital twin performance. Assessment measures (e.g., accuracy, robustness, robustness, costs), as well as assessment methodologies and tests, are required.

Digital twins combine IoT, AI, cloud computing, smart visualization tools, and other related technologies to create a one-of-a-kind synthetic replica of a real thing that observes and replicates its physical condition and behavior. Digital twins connect with actual manufacturing equipment in the real world, boosting the digitalization powers of computer-aided design. Because of this, digital twin solutions can provide better replicas with greater authentic, stochastic and comprehensive measures of uncertainty.

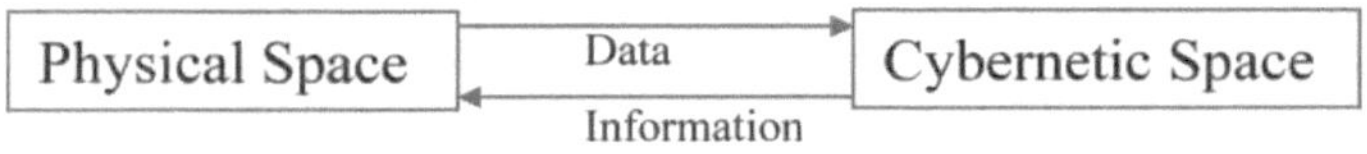

FIGURE 14.2 Adaption of Grieves's 3D Digital Twin Model.

14.3 LITERATURE REVIEW

Considering the digital twin's critical relevance, developing digital twin systems is still an uphill battle for organizations of all sizes. Sharma et al. [57] have examined pertinent digital twin studies as well as industrial works, emphasizing the major digital twin attributes, present practices in several fields, and effective digital twin implementations in order to extrapolate the key digital twin elements and properties, and figure out the present constraints and causes behind the backlog in the generalization and implementation of digital twins, such as two-way syncing is extremely difficult for large businesses since it necessitates resources along with a high-stream IoT network; digital twin requirements are still uncommon, transforming; dependence on the problem and the domain.; security issues across shared information; the inadequate digital twin performance metrics; and the digital twin's dependency on other rapidly emerging technologies [7]. The application of ML to the IoT and improvements in big data have resulted in considerable gains in digital twin components, such as continuous surveillance and exact prediction [57].

Fuller et al. [18] have explored applications, difficulties, and valid technologies related to digital twins, IoT/Industrial IoT and data analytics. They have also attempted to establish a connection between IIoT, IoT and data analytics, and digital twin technologies, as well as the issues connected with digital twins-infrastructure for data analytics and IoT; accessibility of quality data; privacy, trust and risk; modeling [13].

Stephanie [60] states that digital twins require a massive computing scale. The expertise and finances required to run these distributed, massively scalable compute and storage infrastructures are out of scope for except for a few companies.

Cooper et al. [11] have identified challenges in applying digital twins in industrial settings, such as vertical integration disrupts hierarchical levels; making usage of pre-existing information and & splitting down silos; huge testing prior to consumer release.

Wagg et al. [67] offer an assessment of the modern facilities for virtual twins in the practical field of engineering dynamics. Furthermore, numerous open research concerns and technology obstacles were presented, covering workflow, joints, risk management, and credibility assessment [31], [35], [40], [46], [67].

In the study conducted by Lei et al. [32] to discover urbanized digital twin issues, fourteen technical along with nine non-technical issues has been identified and mapped to stages of the life cycle of digital twin. The utmost serious issues tend to be those of interoperability (for instance, divergent semantic guidelines) and practical utility (e.g., an absence of economic models). The findings revealed that technical and non-technical difficulties pose varying degrees of difficulty for urban digital twins and geographic spatial domains. Some obstacles, such as definitions and goals, are basic concerns that must be addressed right away. In a similar vein, numerous panelists see the practical benefits associated with digital twins in cities as among the most important matters. For example, who are the intended users, what are their needs, and what are the goals of digital twins in cities. Data-related concerns tend to be the most prevalent among all categories of stakeholders. At the same time, various groups regularly emphasize teamwork and authority.

Anderton [8] highlighted digital modeling methodologies, the notion of integrity as a measure of resemblance between the two worlds, and the benefits of the digital twin in general, which are leading to valuing assets, keeping connected to consumers, and improving agility and speed. The writers also made it more relevant to industrial circumstances encountered by control specialists and ops. The contributor also discussed the digital twin in the backdrop of digital and Industry 4.0, as well as the manner in which it facilitates a human-machine victorious pairing. The case studies provided addressed modeling numerous variables that led to distinct advantages.

Gulewicz [24] has conducted a study to find out the extent of familiarity with virtual twins among industry representatives in addition to pinpoint the main obstacles in putting the technology into action within enterprises. The study concentrated on market responsiveness of the stated technology, as well as the way companies make use of ideas from staff to enhance their companies and the variables that impact the efficacy of processes.

Amthiou et al. [1] stated that digital twin capabilities for any item may be realized throughout its existence, from conception to final disposal, solving the problems of Industry 4.0 through remote surveillance and proactive upkeep.

Eleftheriou et al. [15] in their study stated that while digital twins operate alongside a variety of other technologies, including IoT, AI and data analytics, each of them somewhat shares the same difficulties such as updated infrastructure, high quality data, privacy and security concerns as well as standardized modeling.

In their study, Ghatti et al. [20] identified a number of significant obstacles that must be tackled so as to effectively utilize the potential for digital twins for medical applications, such as efficient computing, security and confidentiality, seamless integration, ethical standards and norms, and stakeholder participation along with instruction for professionals and patients.

The road to effective digital twin implementation does not come without obstacles. There are four significant areas of difficulty that are limiting digital twin implementation-a lack of a solid vision, managerial support, and appropriate governance; a particular expertise mix; insufficient internal digital architecture; and platforms for collaboration, identified in a study conducted by Capgemini research institute authors (DigitalTwins_adding-intelligence-to-the-real-world.pdf).

The majority of digital twin concepts merely include geometrical models. Despite more recent studies focusing on tangible models, there persists a gap in behavioral [12] and rule modeling. Data transfer experiences delays and aberrations. Owing to the massive amount of data that must be communicated concurrently, existing data transfer systems are incapable of meeting the increasing need for outstanding precision along with rapid pace. Existing data analysis algorithms and processes require enhancements in speed as well as precision. There are numerous digital twin platforms for various purposes, particularly in the multifaceted equipment surveillance and IoT domains. Yet, due to a variety of transmission standards and service expectations, the platform's progress time is insufficient to establish an integrated digital twin framework [19].

A successful simulation necessitates accuracy, precision, collection of data, and synchronicity [12], [53]. Some stated that real-time dialogue, intricacy, reliability, integration, and technical underpinnings are all challenges in the practical application of digital twins. The difficulty of running the digital twin in real time stems not only from the necessity to re-run the optimization method, but moreover from the need to migrate the digital twin via one edge server onto the next [6]. A proper integration of elements with a local intellect defines a system's operation and performance [9]. In view of numerous formats, regulations, and standards, present technologies cannot be combined or applied concurrently for a particular purpose [36]. Furthermore, the digital twin's confidentiality and safety should be thoroughly investigated [48]. In a different study proposed by [20], certain issues are identified with the adoption of digital twin technology. The most prevalent issue encountered was the high cost of technology implementation, which included the technology itself, infrastructure modification, people costs (hiring new staff), and eventual costs of rectifying problems caused by the solution's incorrect operation. Another major impediment was a dearth of understanding about the technology plus widespread awareness of its application. Following that, technological hazards are identified as the primary impediment to technology implementation.

The battery digital twin's virtual physical-essentials as described by Wu et al. [68] under the scope of electric vehicles (EVs) pose an architectural engineering problem on account that the majority of the elements employed are multi physics structures integrating combinations of mechanical, electronic, thermal, as well as another field [39], [45].

Kukushkin et al. [31] have identified over 100 subjects, with the most well-liked and quickly expanding being "digital twins of industrial robots, production lines, and products". It was also observed that fascination in basic investigations regarding the digital twin, as well as publications analyzing the digital twin's function in Industry 4.0, is steadily waning.

Yao et al. [70] conducted a study on "systematic review of digital twin technology and applications" and find out that at the moment, comprehensive and aligned virtual modeling techniques and tools that can be beneficial throughout the digital lifespan are insufficient; systems that evaluate the degree of development of existing digital twins and clarify the direction of digital twin construction to guide upgrading and optimization are still lacking; the emerging research orientation of "practice first, concept later," which is frequently associated with information technology, big data, AI, and IoT in the development process, lacking of a suitable theoretical foundation in research [8].

Xie et al. [69] aimed at contributing to better comprehension of incorporating the application of digital twin for Modular Integrated Construction (MiC) by evaluating its prospects and constraints. According to the findings of the research, a digital twin is frequently explored for fusion with other forms of digital technology. A theoretical foundation is established for using digital twins in MiC module deployment in the backdrop of Hong Kong as a typical scenario of a high-density city. The implementation of digital twins in MiC is expected to provide exciting prospects with substantial advantages, such as improved logistics and construction management by employing digital twins for tracking on-site activities throughout the module installation phase. Nonetheless, obstacles are highlighted in not just the technology for sensing and cyber system, but additionally the social-political framework that supports creativity and governance.

Madubuike et.al [37] evaluated the advancement and deployment of virtual twins in the construction sector and compared it to its usage in other sectors while evaluating the advantages of digital twins to the construction sector that involves anticipating a facility's life cycle efficiency and providing economic benefit by optimizing potential future circumstances. This lifespan projection enables the adoption of a condition-based maintenance (CBM) system, in which real-time data is used to prioritize and optimize maintenance resources. The research showed the building industry's limitations in digital twin adoption [37].

14.4 CHALLENGES AND FUTURE DIRECTIONS OF DIGITAL TWINS

Despite the fact that digital twins have received substantial attention from both academia and business, this section concentrates on digital twin difficulties and solutions.

1. Information Technology Infrastructure: The digital twin's functioning entails a setup that empowers the effective implementation of IoT as well as data analytics; they will support a digital twin's smooth functioning [7], [34]. The digital twin is unlikely to realize its aims unless it has a thoughtful and well-connected IT infrastructure [26], [32]. The notion of graphics processing units (GPUs) as a sort of service will be critical in overcoming this difficulty. In this approach, it is envisioned that large leading firms such as Google, Amazon, and Microsoft would supply innovative on-demand services comparable to typical cloud-based apps, overcoming the barrier of expensive pricing.
2. Informative data: Superior quality substantial enterprises find two-way synchronization particularly challenging; it demands resources and a high-stream IoT connectivity. If the data is weak and unreliable, the digital twin might perform poorly since it is operating on data that is poor and missing. The quantity and the quality of IoT signals are critical for digital twin data [7], [32], [35], [46], [61]. To determine the relevant data is gathered and utilized for optimal usage of a digital twin, device setup and evaluation must be done and also to prevent power failures at software and sensor issues, a high degree of connection and durable solutions must be devised.
3. Compatibility with current software in a production lifecycle: Different industries utilize different software for different functions including product management, operations, and inventory control. Dealing with the difficult problem of digital twin's compatibility with these might cause implementations to be delayed [7].
4. Add-Ons: There are specific costs, materials, and examinations associated with implementing digital twins [7], [46], [61]. Digital twins can be expensive if a project has a short lifespan because it takes time to execute and reap the benefits of. Consequently, industries that employ digital twins for an extended period of time will have to continue making investments in this

investigation, which might result in increased costs. Combining real-time technologies, massive data resources, a collaborative optimization problem formulation process, and interoperability across several components can be time-consuming and provide unintended diversions while implementing digital twins [19]– [20].

5. Update and versioning: These are critical steps in maintaining digital twins, and they address three distinct challenges. One possible issue is maintaining version management. However, dealing with multiple model versions is problematic. Users find it difficult to discern differences between versions, which impairs their interpretation of alterations to the physical world. In some cases, such as mimicking traffic scenarios in a metropolitan digital twin, it is vital to handle data as quickly as feasible. Otherwise, delay in detecting changes may reduce the analysis's accuracy.
6. Privacy and Security: In an industrial setting, it is clear that the security and privacy associated with digital twins are a concern due to the enormous extent of data they consume and the potential threat this exposes to critical system-related information [7], [32], [41], [46], [69]. To solve this difficulty, the primary technology platforms for digital twins – data analytics and IoT – must adhere to current security and privacy rules. Considering the security and privacy of digital twins' data helps to address trust difficulties regarding digital twins. Techniques for dealing with these concerns nowadays involve the use of block chain technologies to assure data privacy and integrity, as well as trust and transparency in a variety of use cases [48], [61]. However, the literature on digital twin security continues to be in its early stages.
7. Trust: It discusses concerns from both sides of the company and the user. A method for overcoming trust issues is model validation. It is critical to ensure that digital twins work as intended in order to maintain user confidence. The facilitating technology will provide more visibility into the actions that guarantee privacy and security practices are followed throughout development, hence solving trust concerns [7], [45]–[46], [61].
8. Outlooks: The requirement for robust underpinnings for the IoT framework, as well as a better knowledge of data necessary for analytics, will ensure enterprises embrace digital twin technology. It is also difficult to resist the notion that the digital twin ought to be utilized simply for current developments. The benefits and drawbacks of digital twins must be examined in order to take proper action while designing digital twin systems. There are certain special issues associated with the designing and construction of the digital twin [49].
9. Technologies for environmental interfacing. The present digital twin has no connection to the outside world. The system that describes how real objects connect to their surroundings has yet to be entirely incorporated in the digital representations [17], [32], [69]. Several investigations have been conducted to investigate the method by which tangible objects connect to their surroundings in reality. However, there is still a pressing demand to establish an equivalent

digital expression approach, which will result in a more effective and precise prognosis in the near future digital twin.

10. Modeling Standardization: Since there exists no uniform technique for modeling, an additional problem arises that of modeling such systems [42], [56], [61]. The parameters of data analysis in the factory include "performance, capacity, energy consumption, quality, cost, and efficiency". Each dimension has several linkages, making data collection and modeling complex [33]. From the digital twin's initial concept to its computerized simulation, a methodical approach, regardless physics-based or designed-based, is required. Standardized methodologies encourage user and domain awareness despite the fact that they enable flow of data throughout stages of digital twin development and deployment.
11. Modeling of Domains: Another problem resulting from the requirement for consistent utilization is making certain that information pertaining to domain use is conveyed across all of the development and operational phases of digital twin models [42]. This empowers interoperability thru areas such as IoT and data analytics make it possible to effectively exploit the digital twin in the years to come.
12. Massive computing scale. The expertise and finances required to run these distributed, massively scalable compute and storage infrastructures are out of scope for except for a few companies.
13. Digital Twin and AI projects: Future research will undoubtedly focus on boosting up smaller proven digital twins and AI endeavors [12].
14. Energy usage in manufacturing courses: To optimize revenues, these industries are supposed to lower their energy consumption while maintaining total production [46], [71]. Recognizing how energy is used at each stage of the process is critical for optimizing expensive activities and subsequently lowering energy consumption, which will lead to greater profits throughout the way. Intelligent energy use control paired with energy-efficient technology may cut energy exhaustion by 50% compared to improved procedure, which can lower it by as much as 20%.

Through exploiting the benefits of a digital twin, careful navigation around the hurdles and attentive management will result in an eventual return on investment.

14.5 FUTURE DIRECTIONS

Automated production and maintenance of equipment are two areas where digital twin technology offers a bright future. It is progressively gaining traction in both the military and commercial sectors, encompassing robots, aircraft, new energy, and others [31]. Each of the stated industries underway to investigated the technological systems, core technologies, and practical possibilities of digital twins. The foreseeable growth pattern of digital twins is expected to take following paths:

A. The incorporation of related technology: In the years to come, digital twin technologies will cover a greater spectrum of sectors. Although digital twin technology acceptance has been progressively expanding in the past few years, a survey of extant works suggests that digital twin's growth even now lies in its premature phases. Additionally, the majority of published articles concentrate on theoretical principles and paradigms instead of specific case studies proving the practical implementation and advantages of digital twins.
B. Continual developments in commercial applications: As the technology evolves, investigators will need to perform extensive case studies in a variety of industrial settings to figure out the potential rewards and real-world effects of digital twin solutions. These instances can provide useful insights into each industry's specific issues and requirements by analyzing varied sectors such as manufacturing, health care, transport, energy, and agriculture, enabling the creation of personalized digital twin solutions [10]. As more solid proof of the positive impacts of digital twins becomes apparent, we may anticipate rising investments in R&D and placing more importance on embedding digital twins into current processes and workflows. Finally, the continuing development and refinement of digital twin technology will enable organizations to optimize their operations, improve decision-making, and foster innovation across a wide range of sectors.
C. IoT protocols that are needed to operate digital twins: As digital twins count extensively on IoT mechanism for data capture and sharing, actual time synchronization, and surveillance, recognizing which IoT norms are best-suited for these tasks will increase digital twin acceptance and facilitate widespread adoption less difficult. The similar necessity for a data transmission standard is recognized by [24].
D. Collaboration among subject matter experts to expand the usage of digital twins across industries: Interacting with subject matter experts will make the process simpler to deploy and embrace digital twins across fresh domains. Once a competent domain-specific layout is complete, coders and programmers can oversee the digital twin framework's realization.
E. Worldwide Applications: At the moment, only local or country-level applications exist. Worldwide applications could have a considerably greater impact on bigger infrastructures that include supply chain networks and logistics.
F. Establishing a data-sharing network and pursuing inter-organizational collaboration: Specifying the data information hierarchy in the initial stage based on the level of responsiveness and accessibility, establishes the foundation to construct a data-sharing network among agencies, researchers, and user communities, with each stakeholder's obligations clearly defined. Subsequently in order to handle the ownership of data in action, it is necessary to collaborate both outside and internally, involving discussion among numerous stakeholders. With such initiatives, it has the potential to realize new data gathering in communities and areas, which might aid in quality

of data upgrading and fulfill objectives across various accessible geospatial databases.

G. Workflows implementation: Network theory formalizations tend to be the most useful tool in efficiently implementing workflows. Furthermore, how the workflow might move via the various aspects of the digital twin for this a viable option appears to be the concept of a knowledge network (which might be created from an initial ontology) [50].

H. Error diagnosis: The building of more complicated systems increases the danger of errors. A little breakdown amid production can result in irreversible repercussions. As a result, mistake identification is critical in smart building. The error diagnosis standard has evolved from a solely data analysis to an immersive experience using digital twins.

I. Applicability in Healthcare: Digital twin healthcare (DTH) can be applied to patient care, hospital administration, and architectural design. Before planning and executing real changes, such as staffing schedules and bed scheduling, a variety of possibilities can be evaluated virtually with DTH. Medical personnel can only plan their actions based on their expertise in the field and an initial evaluation if DTH is not available; however, using a digital pair will lower hazards and save money [23], [29], [63].

J. Digital models of humans: Ross et al. [37] discussed Hewlett-Packard (HP) research that uses the IoT to build digital avatars—digital representations of real people. According to HP, it can set up individuals' avatars depending on dietary regimens, exercise regimens, and workouts in conjunction with clinical data to forecast wellness depending on a variety of lifestyle decisions. HP sees a far greater rapid applicability and demand in the garment business, despite the enormous scope of this application. Furthermore, HP is witnessing additional integration of this system into games. Some have [57] claimed that in terms of these technologies, we seek to speed up advancements on human dissociated tumor cells (DTCs) by emphasizing the distinctiveness of individuals and diving deeper into individual's inner selves [8]. Furthermore, human DTC must be expanded beyond AI and engineering into domains such as biology and medicine as well as ethics, philosophy, and other humanities for instance, behavioral economics [64].

K. Cloud computing and edge computing: Cloud computing infrastructure is accessible and versatile, allowing for more effective and affordable digital twin systems. Edge computing enables more efficient data processing at the point at which it is generated, hence minimizing the volume of data required to be transmitted, as well as lowering latency [13], [59].

L. Sophisticated digital twin models: There is a need to build more reliable and complex digital twin simulations, as well as addressing compatibility and integration concerns, as well as further investigate the possibilities of digital twin technology in developing fields such as online medical care and customized healthcare [12]–[13].

M. Uniformity with real-world data: The creation of a data structure that includes every bit of what is needed and is consistent with real-life data such as physical, geometry, behavioral, and rule information [19].
N. Factors influencing digital twin implementation: Further research suggested by [20] would incorporate firms that use digital twins, discovering and validating the elements (organizational, financial) that favor and influence implementation. Furthermore, it is likely to validate the usage of workforce proposals by refined businesses, the association of the aforementioned phenomenon with the degree of inventiveness of businesses, and the likelihood of classifying it as a facet influencing the execution and growth of digital twin technology.

14.6 DISCUSSIONS AND CONCLUSION

The digital twin has the potential to add significant worth to businesses, generate new income streams, and assist them in answering crucial strategic issues. Corporations perhaps capable of begin their endeavors to develop a digital twin with lesser investment in capital and faster time to worth than ever before considering to advances in technology, adaptability, dexterity, and reduced cost. A digital twin has various uses throughout the lifespan of a product as well as may respond to queries in real time that could not be addressed before, giving sorts of value that were previously unthinkable. Digital twin has elementary and imperative elements. The fundamental elements that a digital twin cannot function without are "physical asset, digital asset and information flow between the physical and digital asset (this could be 1-way or 2-way/objective)". The critical or imperative components enhance the digital twin's features, making it a comprehensive instrument for imitation, analytics and real-time monitoring [12]. Lacking these, the distinctiveness of the digital twin is lost. It has been discovered that 5IR integrates and strengthens the key aspects of Industry 4.0. In the context of technology, 5IR wishes to capitalize on the prospects of advanced digitalization, big data, and AI, emphasizing the crucial part these technologies may play in solving novel and vital requirements in the manufacturing, societal and environmental landscapes [6]. This entails leveraging data and AI in manufacturing to gradually boost production adaptability and strengthen the value chain [10]. The primary causes of the delays are identified in this work are-firms were not yet exposed to the concept, or other companies have heard of the concept but lack the necessary facilities to apply it; issues about shared information's safety; implementation cost; modeling in digital space that accurately reflect items; impact on environment; attempts and constraints to standardization; data collection and processing; two-way, real-time communication between our virtual and physical worlds is difficult to achieve [58]. This study has compiled the most thorough list of obstacles to the design and deployment of digital twins in practice. One of the most important findings is that the issues described in the source material and survey are largely consistent. Recent articles reveal that technological obstacles are addressed in a deeper manner and often take precedence over social and legal ones, however some (infrastructure, computer equipment, and data storage) are regarded as irrelevant in realistic performance [9].

The most recent studies on approaches in ML and big data may help to improve the success rate of digital twins and lessen the complexity associated with their execution. The study looked at the possible application of digital twin in 5IR, the key issues now addressing the use of digital twin technology, and the potential advantage of giving directions for study toward the continuing evolution of intelligence in the arena of industry in the wake of Industry X.0 [33]. In addition, there is a requirement to widen our perspectives while recognizing potential issues that digital twins as a technology may encounter or generate. Understanding the comprehensive perspective of the digital twin—what counts as a digital twin, its physical appearance, its benefits, how it may be put into effect and its challenges—is critical if we are to realize the technology to its fullest [58].

REFERENCES

1. H. Amthiou, M. Ariouaand and T. Benbarrad. 2023. *ITM Web of Conferences* 52:01002. https://doi.org/10.1051/itmconf/20235201002
2. M. Attaran, S. Attaran and B.G. Çelik.2023. The impact of digital twins on the evolution of intelligent manufacturing and Industry 4.0. *Advances in Computational Intelligence*, 3(3). https://doi.org/10.1007/s43674-023-00058-y
3. A. Barari, M. de Sales Guerra Tsuzuki, Y. Cohen and M. Macchi. 2021. Intelligent manufacturing systems towards industry 4.0 era. *Journal of Intelligent Manufacturing*, 32(7):1793–1796.
4. B.R. Barricelli, E. Casiraghi and D. Fogli. 2019. A survey on Digital Twin: Definitions,characteristics, applications, and design implications. *IEEE Access*, 7:167653–167671. https://doi.org/10.1109/access.2019.2953499
5. A. Benveniste, E. Fabre and S. Haar. 2003. Markov nets: Probabilistic models for distributed and concurrent systems. *IEEE Transactions on Automatic Control*, 48(11):1936–1950.
6. R.N. Bolton, J.R. McColl-Kennedy, L. Cheung, A. Gallan, C. Orsingher, L. Witell and M. Zaki. 2018. Customer experience challenges: Bringing together digital, physical and social realms. *JOSM*, 29(5):776–808.
7. D.M. Botín-Sanabria, A. Mihăiţă, R.E. Peimbert-García, M.A. Ramírez-Moreno, R.A. Ramírez-Mendoza and J. De J Lozoya-Santos. 2022b. Digital Twin technology challenges and applications: A comprehensive review. *Remote Sensing*, 14(6):1335. https://doi.org/10.3390/rs14061335
8. C.B. Anderton. Challenges and benefits of implementing a Digital Twin in composites manufacturing. *White Paper*. www.Challenges-and-Benefits-of-Implementing-a-Digital-Twin-in-Composites-Manufacturing.pdf
9. O. Chukhno, N. Chukhno, G. Araniti, C. Campolo, A. Iera and A. Molinaro.2020. Optimal placement of social Digital Twins in edge IoT networks. *Sensors*, 20:6181.
10. C. Cimino, E. Negri and L. Fumagalli.2019. Review of Digital Twin applications in manufacturing. *Computers in Industry*, 113:103130. https://doi.org/10.1016/j.compind.2019.103130
11. J.P. Cooper, S. Jackson, S. Kamojjala, G. Owens, K. Szana and S. Tomić.2022. Demystifying Digital Twins: Definitions, applications, and benefits. *J AWWA*, 114(5):58–65. https://doi.org/10.1002/awwa.1922

12. U. Dahmen and J. Rossmann. 2018, October 1–3. Experimentable Digital Twins for a Modeling and Simulation-based Engineering Approach. In Proceedings of the 2018 IEEE International Systems Engineering Symposium (ISSE). Rome, Italy.
13. D. Gurdur Broo and J. Schooling. 2023. Digital twins in infrastructure: definitions, current practices, challenges and strategies. *International Journal of Construction Management*, 23(7):1254–1263. https://doi.org/10.1080/15623599.2021.1966980
14. DUBLIN. February 17, 2023 /PRNewswire/ -- The "Global Digital Twin Market by Usage Type, Enterprise, Application, Industry By Region: Global Opportunity Analysis and Information Industry Forecast, 2022–2027".
15. O.T. Eleftheriou and C. Anagnostopoulos. 2022. Digital twins: A brief overview of applications, challenges and enabling technologies in the last decade. *Digital Twin*, 2:2. https://doi.org/10.12688/digitaltwin.17581.1
16. H. Eriksson, J. Sun, V. Tarandi and L. Harrie. 2020. Comparison of versioning methods to improve the information flow in the planning and building processes. *Transactions in GIS*, 25(1):134–163. https://doi.org/10.1111/tgis.12672
17. J. Feng, L. Ma, T. Broyd, W. Chen and H. Luo.2022. Digital Twin enabled sustainable urban road planning. *Sustainable Cities and Society*, *78*:103645. https://doi.org/10.1016/j.scs.2021.103645
18. A. Fuller, Z. Fan, C. Day and C. Barlow.2020. Digital Twin: Enabling technologies, challenges and open research. *IEEE Access*, 8:108952–108971. https://doi.org/10.1109/access.2020.2998358
19. T. Gabor, L. Belzner, M. Kiermeier, M.T. Beckand A. Neitz.17–22 July 2016. A simulation-based architecture for smart cyber-physical systems. In Proceedings of the 2016 IEEE International Conference on Autonomic Computing (ICAC), Wuerzburg, Germany, pp. 374–379.
20. S. Ghatti, L.A. Yurish, H. Shen, K.S. Rheuban, K.B. Enfield, N.R. Facteau, G. Engel and K. Dowdell. 2023. Digital Twins in healthcare: A survey of current methods. *Archives of Clinical and Biomedical Research*, 07(03). https://doi.org/10.26502/acbr.50170352
21. M. Grieves. 2014. Digital twin: Manufacturing excellence through virtual factory replication. *White Paper*, 1:1–7.
22. M. Grieves and J. Vickers.2017. Digital Twin: Mitigating Unpredictable, Undesirable Emergent Behavior in Complex systems. In *Transdisciplinary Perspectives on Complex Systems (Issue August)*. Springer International Publishing, pp. 85–113.
23. M.W. Grieves.2019. Virtually Intelligent Product Systems: Digital and physical twins, in: *Complex Systems Engineering: Theory and Practice*, American Institute of Aeronautics and Astronautics; pp. 175–200.
24. M. Gulewicz. 2022. Digital twin technology — Awareness, implementation problems and benefits. *Engineering Management in Production and Services*, 14(1):63–77. https://doi.org/10.2478/emj-2022-0006
25. R. Harrison, D. Vera and B. Ahmad. 2021. A connective framework to support the lifecycle of cyber–physical production systems. *Proceeding of IEEE*, 109:568–581.
26. L. Hou, S. Wu, K. Zhang, Y. Tan and X. Wang. 2020. Literature review of Digital Twins applications in construction workforce safety. *Applied Sciences*, 11(1):339. https://doi.org/10.3390/app11010339
27. W. Hu, T. Zhang, X. Deng, Z. Liu and J. Tan. 2021. Digital twin: A state-of-the-art review of its enabling technologies, applications and challenges. *Journal of Intelligent Manufacturing and Special Equipment*, 2(1):1–34. https://doi.org/10.1108/jimse-12-2020-010

28. S. Huang, B. Wang, Li X, P. Zheng, D. Mourtzis and L. Wang. 2022. Industry 5.0 and Society 5.0—Comparison, complementation and co-evolution. *Jorurnal of Manufacture Systems*, 64:424–428.
29. M. Javaid and A. Haleem. 2020. Critical components of Industry 5.0 towards a successful adoption in the field of manufacturing. *Journal of Industrial Integration Management*, 5(03):327–348.
30. K.J. Kuehner, R.K. Scheer and S. Straßburger. 2021. Digital Twin: Finding common ground – A meta-review. *Procedia CIRP*, 104:1227–1232. https://doi.org/10.1016/j.procir.2021.11.206
31. K. Kukushkin, Y.F. Ryabov and A. Borovkov. 2022. Digital Twins: A systematic literature review based on data analysis and topic modeling. *Data*, 7(12):173. https://doi.org/10.3390/data7120173
32. B. Lei, P. Janssen, J. Stoter and F. Biljecki. 2023. Challenges of urban digital twins: A systematic review and a Delphi expert survey. *Automation in Construction*, 147:104716. https://doi.org/10.1016/j.autcon.2022.104716
33. K. Lim, P. Zheng and C. Chen. 2020. A state-of-the-art survey of Digital Twin: Techniques, engineering product lifecycle management and business innovation perspectives. *Journal of Intelligent Manufacturing*, 31:1313–1337.
34. Y. Lu, C. Liu, K.I. Wang, H. Huang and X. Xu. 2020. Digital Twin-driven smart manufacturing: Connotation, reference model, applications and research issues. *Robotics and Computer-Integrated Manufacturing*, 61:101837. https://doi.org/10.1016/j.rcim.2019.101837
35. Z. Lv. 2023. Digital Twins in Industry 5.0. *Research,* 6:Article 0071. https://doi.org/10.34133/research.0071
36. F. Longo, A. Padovano and S. Umbrello. 2020.Value-oriented and ethical technology engineering in industry 5.0: A human-centric perspective for the design of the factory of the future. *Applied Science*, 10(12):4182.
37. O.C. Madubuike and C.J. Anumba. 2023. Digital Twin–Based health care facilities management. *Journal of Computing in Civil Engineering*, 37(2). https://doi.org/10.1061/jccee5.cpeng-4842
38. A. Masaracchia, V. Sharma, B. Canberk, O.A. Dobre and T.Q. Duong. 2022. Digital Twin for 6G: Taxonomy, research challenges, and the road ahead. *IEEE Open Journal of the Communications Society*, 3:2137–2150. https://doi.org/10.1109/ojcoms.2022.3219015
39. A. Mehdiabadi, V. Shahabi, S. Shamsinejad, M. Amiri, C. Spulbar, R. Birau. 2022. Investigating Industry 5.0 and its impact on the banking industry: Requirements, approaches and communications. *Applied Science*, 12(10):5126.
40. D. Mourtzis.2023. Digital twin inception in the Era of industrial metaverse. *Frontiers in Manufacturing Technol*ogy, 3:1155735. https://doi.org/10.3389/fmtec.2023.1155735
41. Oracle. 2018. Developing Applications with Oracle Internet of Things Cloud Service: Digital Twins. Oracle. Available online: www.docs.oracle.com/en/cloud/paas/iot-cloud/iotgs/oracle-iot-digital-twin-implementation.html
42. J. Orozco-Messana, M. Iborra-Lucas and R. Calabuig-Moreno. 2021. Neighbourhood modelling for urban sustainability assessment. *Sustainability*, 13:4654.
43. A. Parrott and L. Warshaw. 2017. Industry 4.0 and the DT. Available online: Industry 4.0 and the digital twin technology | Deloitte Insights.
44. H. Paulheim. 2017. Knowledge graph refinement: A survey of approaches and evaluation methods. *Semantic Web*, 8(3):489–508.

45. M. Perno, L. Hvam and A. Haug. 2022. Implementation of digital twins in the process industry: A systematic literature review of enablers and barriers. *Computers in Industry*, 134:103558. https://doi.org/10.1016/j.compind.2021.103558
46. E.O. Popa, M. Van Hilten, E. Oosterkamp and M. Bogaardt. 2021. The use of digital twins in healthcare: Socio-ethical benefits and socio-ethical risks. *Life Sciences, Society and Policy*, 17(1). https://doi.org/10.1186/s40504-021-00113-x
47. Q. Qi, F. Tao, T. Hu, H. Anwer, A. Liu, Y. Wei, L. Wang and A. Nee. 2021. Enabling technologies and tools for digital twin. *Journal of Manufacturing Systems*, 53:3–21.
48. R. and T. Damerau. 2019. Digital Twin, in Chatti, S. and Tolio, T. (Eds.), *CIRP Encyclopedia of Production Engineering*, Vol. 66, Springer Berlin Heidelberg, Berlin, Heidelberg, pp. 1–8. https://doi.org/10.1007/978-3-642-35950-7_16870-1
49. A. Rasheed, O. San and T. Kvamsdal. 2020. Digital Twin: Values, challenges and enablers from a modeling perspective. *IEEE Access*, 8:21980–22012. https://doi.org/10.1109/access.2020.2970143
50. B. Rodríguez, E. Sanjurjo, M. Tranchero, C. Romano and F. González. 2021. Thermal parameter and state estimation for digital twins of E-Powertrain components. *IEEE Access*, 9:97384–97400.
51. R. Rosen, G. Von Wichert, G. Lo and K.D. Bettenhausen. 2015. About the importance of autonomy and DTs for the future of manufacturing. *IFAC-Papers Online*, 48:567–572.
52. D. Ross. 2016. Digital twinning [Information Technology Virtual Reality]. *Engineering Technology*, 11(4):44–45.
53. H. Russell. 2020. Sustainable urban governance networks: Data-driven planning technologies and smart city software systems. *Geopolitics, History, and International Relations*, 12:9–15.
54. N. Schleich, L. Anwer, S. Mathieu and Wartzack. 2017. Shaping the digital twin for design and production engineering shaping the digital twin for de-sign and production engineering. *CIRP Annals-Manufacture Technol*ogy, 66 (1). http://doi.org/10.1016/j.cirp.2017.04.040
55. S. Shan, X. Wen, Y. Wei, Z. Wang and Y. Chen. 2020. Intelligent manufacturing in industry 4.0: A case study of Sany heavy industry. *Systems Research and Behaviour Science*, 37(4):679–690.
56. P. Shames and J. Skipper. 2006. Toward a framework for modelling space systems architectures. In *Space Ops 2006 Conference*: 5581.
57. A. Sharma, E.E. Kosasih, J. Zhang, A. Brintrup and A. Calinescu. 2022. Digital Twins: State of the art theory and practice, challenges, and open research questions. *Journal of Industrial Information Integration*, 30:100383. https://doi.org/10.1016/j.jii.2022.100383
58. M. Singh, E. Fuenmayor, E.P. Hinchy, Y. Qiao, N. Murray and D.M. Devine. 2021. Digital Twin: Origin to future. *Applied System Innovation*, 4(2):36. https://doi.org/10.3390/asi4020036
59. Y.H. Son, G. Kim, H.C. Kim, C. Jun and S.D. Noh. 2021. Past, present, and future research of digital twin for smart manufacturing. *Journal of Computational Design and Engineering*, 9(1):1–23. https://doi.org/10.1093
60. R. Stephanie. 2018. The Digital Twin. Available from: www.ge.com/digital/sites/default/files/download_assets/The-DigitalTwin_Compressing-Time-to-Value-for-Digital-Industrial-Companies.pd
61. S. Suhail, R. Hussain, R. Jurdak, A. Oracevic, K. Salah, C.S. Hong and R. Matulevičius. 2022. Blockchain-based Digital Twins: Research trends, issues, and

future challenges. *ACM Computing Surveys*, 54(11s):1–34. https://doi.org/10.1145/3517189

62. F. Tao, J. Cheng, Q. Qi, M. Zhang, H. Zhang, F. Sui. 2018. Digital twin-driven product design, manufacturing and service with big data. *International Journal Advanced Manufacture Technol*ogy, 94(9–12):3563–3576.
63. E.J. Topol. 2019. High-performance medicine: The convergence of human and artificial intelligence. *Nature Medicine*, 25(1):44–56. https://doi.org/10.1038/s41591-018-0300-7
64. I. Toshima, S. Kobashikawa, H. Noto, T. Kurahashi, K, Hirota and S. Ozawa. 2020. Challenges facing human digital twin computing and its future prospects. *NTT Technical Review*, 18(9):19–24. https://doi.org/10.53829/ntr202009fa2
65. M. Uschold and M. Gruninger.1996. Ontologies: Principles, methods and applications. *The Knowledge Engineering Review*, 11(2):93–136.
66. S. Vitalis, A. Labetski, G.A. Ohori, H. Ledoux and J. Stoter. 2019. A data structure to incorporate versioning in 3D city models. *ISPRS Annals of the Photogrammetry, Remote Sensing and Spatial Information Sciences*, IV-4/W8:123–130. https://doi.org/10.5194/isprs-annals-iv-4-w8-123-2019
67. D. Wagg, K. Worden, R.J. Barthorpe and P.A. Gardner. 2020. Digital Twins: State-of-the-Art and future directions for modeling and simulation in engineering dynamics applications. *ASCE-ASME Journal of Risk and Uncertainty in Engineering Systems*, 6(3). https://doi.org/10.1115/1.4046739
68. B. Wu, W. Widanage, S. Yang and X. Liu. 2021. Battery digital twins: Perspectives on the fusion of models, data and artificial intelligence for smart battery management systems. *Energy AI*, 1:100016.
69. M. Xie and W. Pan. 2020. Opportunities and challenges of Digital twin applications in modular Integrated Construction. *Proceedings of the ISARC*. https://doi.org/10.22260/isarc2020/0040
70. J. Yao, Y. Yang, X. Wang and X. Zhang. 2023. Systematic review of digital twin technology and applications. *Visual Computing for Industry, Biomedicine, and Art*, 6(1). https://doi.org/10.1186/s42492-023-00137-4
71. M. Zhou, J. Yan and D. Feng. 2019. Digital Twin framework and its application to power grid online analysis. *CSSE Journal of Power Energy Systems*, 5:391–398.

15 Researching Trends on Smart Tourism Technology Using Bibliometric Analysis

Duong Thanh Tung and Nguyen Thi Bich Ngoc

15.1 INTRODUCTION

The rapid growth of technology in recent years, particularly in the tourist business, has resulted in great opportunities, which puts forward the need to keep up with technological advancements [1]. Subsequently, it is the primary responsibility of all stakeholders in the tourism industry to stay abreast of the latest innovations and implement a variety of cutting-edge technologies, as technology not only facilitates progress but also offers several advantages and contentment to tourists. Furthermore, it is anticipated that the development of society and the popularity of smart cities will foster the advancement of smart tourism technology (STT).

As the concept of "smart people" becomes more widely recognized, the public will become more internationalized and amenable to the need for STT [2]. Resultantly, smart tourism (ST) has grown in popularity.

According to some research, an ST destination is an innovative tourist destination constructed on a state-of-the-art technological infrastructure that guarantees the sustainable development of tourist areas, accessibility for all, and the establishment of favorable conditions for visitors to engage and integrate with the surrounding environment [3]. Due to digital transformation in the tourism industry, STT was formed to use cutting-edge technology and practices to boost sustainability, competitiveness, and better resource management [4]. Research indicates that traveler experiences with smart travel technology are typically positive regarding accessibility, interactivity, and personalization [5]–[6]. These technologies include the Internet of Things (IoT), wireless communications, near-field communication (NFC), augmented reality (AR), ambient intelligence (AmI), big data, machine learning (ML), metaverse, digital twins, gamification [7]–[13], etc.

Using big data, artificial intelligence (AI), ML, and the IoT in the tourism industry has led to rapid transformation and innovation in the industry. Smart travel applications have utilized big data analytics, which enables quick data processing, visualization, and analysis to aid in daily operational decision-making and strategy development [10]. Also, applying virtual reality travel material may result in numerous positive

 DOI: 10.1201/9781003465904-19

outcomes, including customer satisfaction, intent to visit destinations and word-of-mouth recommendations. VR technology has proven to be beneficial to this industry by drawing visitors and providing distinctive experiences, particularly during the COVID-19 pandemic [14].

Gamification in tourism is a future trend that applies game design and game-thinking elements to elicit motivation and behavior changes in tourists [15]. It refers to the use of game elements and mechanics to enhance the tourist experience. Gamification strategies and methodologies have been used in-game-oriented cultural tourism applications to improve visitor experiences and encourage constant interaction with the applications [16]. They can contribute to more rewarding interactions, higher satisfaction levels, increased brand awareness, and stronger loyalty to destinations. Moreover, digital gamified applications in the tourism industry enable users to socialize and have fun in non-game environments. It has been found that gamification can improve the VR tourism experience by reducing tourism fatigue and increasing tourist satisfaction [13]. Gamification innovations in industrial tourism products have been shown to extend travel time and distance, improve the travel experience, and enhance tourists' willingness to revisit and recommend [17]–[20].

Metaverse was a breakthrough technology that unites the digital and physical realms, transforming tourist administration and marketing. Thanks to its ability to create immersive experiences in both virtual and real-world settings, users can travel through time and visit various virtual destinations [21]. The metaverse can revolutionize the travel experience and transform tourism marketing and management. Indeed, integrating the metaverse into the tourism industry offers advantages to tourists and destinations. A prime example can be seen in establishing a new marketing channel that enables destinations to enhance their credibility and forge closer bonds with their target audience [22]. Therefore, some authors emphasized the need to understand the implications of using the metaverse for different factors in the value chain as well as the wider spillover effects in the tourism industry [11]. The metaverse ecosystem for tourism is identified and conceptualized, focusing on its potential for increasing awareness, destination positioning and branding, and information coordination and administration through digital twins [21].

Digital twin technology has the potential to revolutionize sustainability in the tourism and hospitality industry, helping to increase productivity, efficiency, and profits. It reduces energy waste, maintenance costs, and time wasted during production by enabling prompt forecasts and, thus, early remedial actions [22]. For example, the digital twins of a city, which involve creating digital copies of individual elements of the urban environment on a single hardware and software platform, are linked in a cooperative system that enables them to use data generated by other digital twins. Similarly, digital twins can also offer real-time data on a range of tourism-related aspects, enabling stakeholders to track performance, spot issues, and make timely informed decisions [23]. By adopting an evolutionary approach, digital twin systems can be developed and expanded in a more manageable and scalable way, which enables them to adapt to the complexity and dynamics of urban settings [12].

From the above analysis, it is evident that the use of STT has a positive impact on satisfaction, happiness, and intent to return, along with enhancing the tourist experience. As a result, reviewing and researching trends in STT is extremely necessary and important.

By utilizing bibliometric analysis, this study attempts to provide a macroscopic overview of research publications on STT based on the Scopus and Web of Science databases. Researchers can use the study's results to ascertain the researched topics, nearby reference sources, new research trends in STT, selecting appropriate research topics, and the level of influence exerted by authors, author groups, organizations, and countries, as well as the accomplishments made by cooperative relationships between these stakeholders.

15.2 RESEARCH METHODS

Bibliometric analysis is the principal method in this study. This is the use of statistical techniques to analyze data on a specific topic, helping researchers to gain an in-depth knowledge of the issue at hand [24]. Accordingly, the researcher will methodically delve into a variety of fascinating but obscure areas of the overall picture, bringing new inspirations for research ideas and problems. This approach reduces the time needed to gather data and conduct longitudinal research by aggregating a large number of published articles within a self-determined time frame [25].

The research mainly focuses on the following contents: (1) Research related to STT; (2) The level of influence and research cooperation relationship of authors, author groups, organizations, and countries on STT; (3) Publications that have great importance and impact on the research issue; (4) Content analysis to identify research gaps and trends; (5) Appropriate journal sources for publishing research results.

The research is carried out in three steps:

Step 1, collect and analyze data using the PRISMA method [26]. Therefore, published studies on the search topic will be carefully analyzed based on a comprehensive and semantic review. This method goes through four stages: (1) Identifying articles to evaluate, (2) Screening articles to evaluate, (3) Assessing suitable conditions of the research, and (4) Completing the list of documents based on the review in Figure 15.1.

The authors collected the research sources from two academic databases, Web of Science (WOS) of Clarivate and Scopus of Elsevier. The two bibliographic databases hold the largest and most relevant influential scientific articles in the field of tourism research and are frequently chosen as a data source for bibliometric analysis [27]. This is one of the important factors that determines the quality of research data sources.

After conducting some tests on the two databases with different keywords, the keyword phrases ("Smart Tourism Technologies" OR "Smart Tourism") AND "Tourists' Experiences" OR trends (updated December 22 October 2023) and ("Digital twin" OR Metaverse") AND "Tourists' Experiences" (updated December 21, 2023) have led to the most suitable research results among the 374 related articles. This includes 30 research articles closest to the research problem, which combines two data sets searched from two databases and excludes duplicate cases. The results will be the data for bibliometric analysis in the second step, as in Figure 15.1.

Step 2, to perform bibliometric analysis, the research continues to use the Scope review method combined with three analysis tools: (1) Sciencescape, an AI tool owned

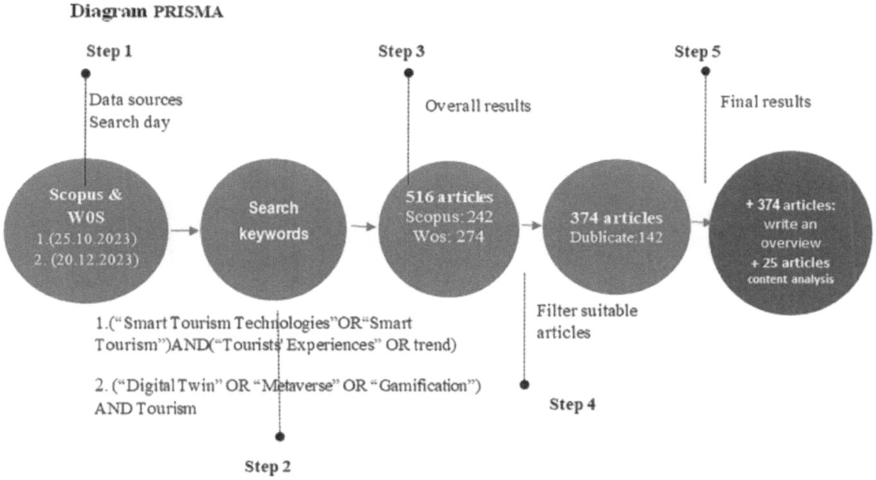

FIGURE 15.1 Data Search Process According to the PRISMA Method. (Source: Authors' Team.)

by Meta big data analysis company that supports extracting descriptive information about the characteristics of scientific articles; (2) VOSviewer, a software developed by Leiden University; and (3) Biblioshiny, an online R-based data analysis framework that captures the basic bibliometric algorithm [28]. The first two tools, which are extensively utilized globally for document analysis and review, facilitate the construction and visualization of relatively dependable survey networks. The remaining one presents the data overview, which includes a visualization of the most influential authors, the most cited documents, the most relevant sources, and trend topics.

Step 3, the study analyzes the content of 30 research publications drawn from Step 1 to identify the research gaps related to STT.

15.3 RESEARCH RESULTS

15.3.1 Regarding the Number of Relevant Published Articles

When gauging the present state of STT research publications, the quantity of articles published is a crucial metric. Figure 15.2 depicts a chronological view of the number of articles published on the research problem. According to analytical data, there were variations in the quantity of articles published annually throughout a ten-year period from 2013 to 2023. There were only 19 articles in 2017, 24 in 2018, and 29 in 2019. The number of articles soared from 2020 onwards, with 50 to 62 articles from 2020 to 2022. More remarkably, at the statistical time (December 21, 2023), 116 articles will have been published in 2023 alone.

These results are also illustrated in Figure 15.3. We found that STT has been significantly developed and widely implemented in multiple countries worldwide,

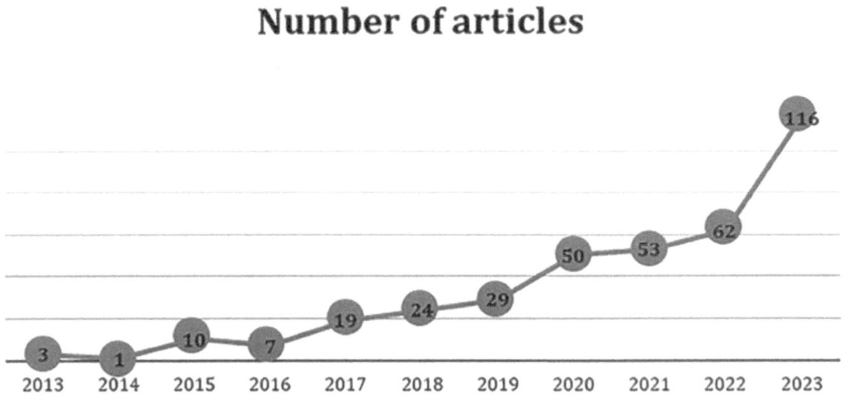

FIGURE 15.2 Number of Articles Published by Year (Source: Chart Made by the Author Team Based on ScienceScape.)

including China, which released the highest number of documents (84 articles), followed by Spain (43 articles), Korea (40 articles), Italy and the United Kingdom (23 articles each). Developing countries like India (21 articles), Taiwan (18 articles), Malaysia (16 articles), etc., are also strongly employing technology in their tourism industries. As a result, more studies are being conducted in this area.

15.3.2 Authors, Groups Of Authors, Organizations, and Countries That Have a Great Impact on STT

VOSviewer was used to analyze authors, organizations, countries and sources for the data of 374 articles to show which authors, organizations, countries and sources greatly influence STT.

15.3.2.1 Ten Groups of Authors with the Most Influential Research

When generating author data based on the co-authorship map, the threshold value was set at the highest level of an author having one article. The study findings identify influential researchers in the field of STT. Accordingly, out of a total of 374 authors, 365 authors have published one article.

In general, the citation index of the studies by the group of ten authors analyzed in Table 15.1 is extremely high, ranging from 124 to 962 citations. This indicates that the study conducted by the ten author groups has a significant influence on STT research and primarily focuses on countries like China, South Korea, Iran, and the United States.

Figure 15.4 shows circles representing the level of influence of the author groups on the STT. The citation index increases with the size of the circle. According to analytical data, the maximum and minimum threshold for the number of articles per group of authors is 2 and 1, respectively. Figure 15.4's visual picture findings depict

FIGURE 15.3 Map of Countries with a Large Number of Articles Published on STT. (Source: Figure Made by the Author's Team Based on Mapchart.com.)

TABLE 15.1
Top Ten Authors with the Highest Citation Index

RO	Author	Document	PY	AC	TLS	Country
Top 1	Gretzel et al.	1	2015	962	0	South Korea
Top 2	Chung et al.	1	2015	260	0	South Korea
Top 3	Buhalis	1	2020	227	0	Iran
Top 4	Wang et al.	1	2013	221	0	United States
Top 5	Vecchio et al.	1	2018	214	0	Italy
Top 6	Huang et al.	1	2018	188	0	United States
Top 7	Jeong & Shin	1	2017	138	0	China
Top 8	Gursoy et al.	1	2022	135	0	South Korea
Top 9	Dorcic et al.	1	2019	133	0	South Korea
Top 10	Pencarelli	1	2020	124	0	China

Source: Authors' Team.
Note: RO: Ranking order; AC: Average citations; PY: Average publication; TLS: Total link strength.

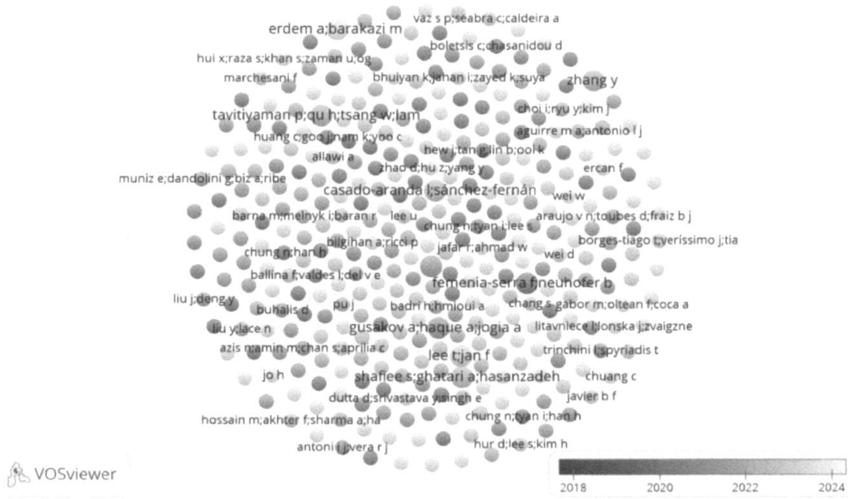

FIGURE 15.4 The Network Visualization of the Co-Authorship Network on STT. (Source: Figure Made by the Authors' Team Based on VOS.)

circular spots that occur separately without any correlation. The dots' color indicates when the research was published, changing over time from 2018 to 2023. This image is logically compared to the strong linkage indexes in Table 15.1, all of which are 0.

This analysis puts forward the conclusion that the author groups conducted research separately and lacked connectedness.

15.3.2.2 Ten Organizations with the Most Influence

The results are shown in Table 15.2 when the research team chose a threshold of one article per organization. Ten organizations with the highest citation index are filtered out of a total of 357 organizations after removing irrelevant information. The citation indices range from 164 to 962, which supports the conclusion that the top ten organizational names on this table have the greatest effect on STT research.

15.3.2.3 Ten Institutions with the Strongest Research Collaborations

As indicated in Figure 15.5, the circular spots representing research institutions on STT are mostly small to medium in size and have separate locations. It can be inferred that cross-institution collaborations in this field remain novel and newly emerged. The results demonstrated in Table 15.5 show that the highest TLS of organizations range from 8 to 31. These organizations are Italy, Malaysia, China, and Dubai, with the year of publications of collaborated articles between 2021 and 2023.

15.3.2.4 Ten Most Influential Countries

In the total of 374 articles, the search threshold was 1 article, from which 219 countries appeared, divided into 219 clusters. Citation indicators help to reflect the level of influence of ten countries on STT, as shown in Table 15.4.

TABLE 15.2
Top Ten Organizations with the Most Influence on STT Research

RO	Organization	Document	AC	TLS	Country
Top 1	Department of Management Information Systems	1	962	1	South Korea
Top 2	Guangdong University of Science and Technology	1	962	1	China
Top 3	Department of Gastronomy and Culinary Arts	1	589	1	Malaysia
Top 4	School of Performing Art	1	446	1	China
Top 5	Daulat Ram College	1	297	3	India
Top 6	Key Laboratory of Spatial Data mining and Information sharing of MOE	1	242	2	China
Top 7	The University of Manchester	1	217	1	United Kingdom
Top 8	School of Electrical and Computer Engineering	1	188	7	United Kingdom
Top 9	Information Sciences	1	169	0	Portugal
Top 10	Modul University Vienna	1	164	1	Austria

Source: Authors' Team.
Note: RO: Ranking order; AC: Average citations; TLS: Total link strength.

TABLE 15.3
Top Ten Organizations Collaborating in Research on STT

RO	Organization	Document	AC	TLS	Year	Country
Top 1	Department of Management	6	3	31	2023	Italy
Top 2	School of Management	7	0	27	2021	Malaysia
Top 3	School of Tourism Management	4	30	17	2023	China
Top 4	College of Hospitality and Tourism Management	3	0	17	2023	Malaysia
Top 5	School of Business and Economics	2	11	15	2023	Malaysia
Top 6	Geography Research Unit	2	3	15	2023	Malaysia
Top 7	Department of Computer Science	3	118	8	2021	China
Top 8	Department of Service Management and Service Studies	1	6	8	2023	Italy
Top 9	Faculty of Business Administration and Center for Cognitive and Brain Sciences	1	0	8	2023	Malaysia
Top 10	School of Business Administration	5	0	8	2021	Dubai

Source: Authors' Team.
Note: RO: Ranking order; AC: Average citations; APY: Average publication; TLS: Total link strength.

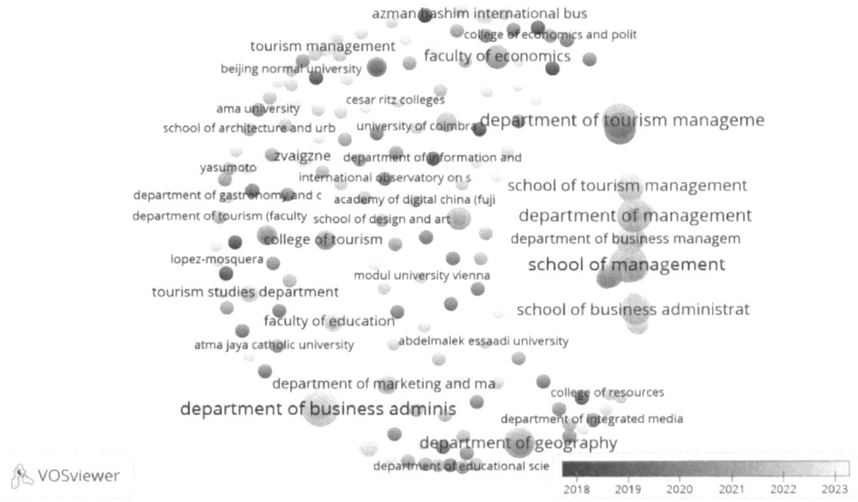

FIGURE 15.5 Network Visualization of Institutional Research Collaborations on STT. (Source: Figure Made by the Authors' Team Based on VOS.)

TABLE 15.4
Top Ten Countries with the Most Influence on STT Research

RO	Country	Document	AC	TLS
Top 1	China	84	1075	33
Top 2	India	21	262	30
Top 3	Italy	23	504	33
Top 4	Malaysia	16	228	27
Top 5	Portugal	18	238	24
Top 6	South Korea	40	2104	153
Top 7	Spain	43	762	90
Top 8	Taiwan	18	145	53
Top 9	United Kingdom	23	931	134

Source: Authors' Team.
Note: RO: Ranking order; AC: Average citations; TLS: Total link strength.

TABLE 15.5
Top Ten Countries with the Strongest Research Collaborations on STT

RO	Country	Documents	AC	TLS
Top 1	China	84	141	66
Top 2	South Korea	40	0	64
Top 3	Spain	43	68	49
Top 4	United States	30	476	49
Top 5	India	21	27	43
Top 6	Australia	10	50	39
Top 7	England	8	141	32
Top 8	Malaysia	16	1	30
Top 9	France	7	38	28
Top 10	Canada	6	7	27

Source: Authors' Team.
Note: RO: Ranking order; AC: Average citations; TLS: Total link strength.

15.3.2.5 Ten Countries with the Strongest Research Collaborations in STT

There are many outstanding strong countries, represented by large round labels on the map. In terms of connections between countries, China is strong, with 84 articles and a link strength of 66 links connecting with countries such as the UK, Taiwan, USA, and Korea. Next is Spain, which has 43 articles and a link strength of 49. It has nine links with Portugal, Korea, Italy, Greece, the United Kingdom, the USA, India, Australia, and Brazil. Korea is a new country that has recently emerged in this research topic, with 40 articles, a link strength of 18, and six links. This country also cooperates frequently with Spain, the US, Australia, and the UK, as shown in Table 15.5.

15.3.2.6 Top Ten Journal Sources for Publishing Research

The appropriate source of journals to publish research results is the basis for researchers to submit articles and publish their publications. A total of 374 articles have been published on the issue of STT in 200 journals. In Figure 15.6, the top ten journals publishing articles on this issue include *Sustainability* (19 articles), Sustainability (Switzerland) (19 articles), *Information Technology and Tourism* (15 articles), Asia *Pacific Journal of Tourism Research* (ten articles), *Journal of Hospitality and Tourism Technology* (ten articles), *Tourism Review* (ten articles), *Advances in Intelligent Systems and Computing* (nine articles), *Worldwide Hospitality and Tourism Themes* (eight articles), *Springer Proceedings in Business and Economics* (six articles), and the *International Journal of Contemporary Hospitality Management* (six articles).

As seen in Figure 15.6, these are top-tier, prestigious journals with a focus on tourism that researchers are interested in. Meanwhile, researchers are encouraged to submit publications under this category. Among these, STT-related publications have often been published in the top ten journals that publish on research concerns.

15.3.3 30 Publications That Have A Great Importance And An Impact on Research Issues

Among the 374 articles, 30 articles were selected. They will have a significant impact and will be a crucial document for researchers to address the subject at hand. Gretzel et al.'s (2015) "Smart Tourism Foundation Developments" has the most citations with 962 citations, and 50 links, as Table 15.6 shows. This author group is regarded as one of the leading ones when it comes to publishing research in this field. "Tourists' Intention to Visit a Destination: The Role of Augmented Reality (AR) Application for a Heritage Site" by Chung et al. (2015) has 260 citations and six links. Wang et al. (2013) have 221 citations and 17 links with the article name "China's "smart tourism destination" Initiative: A Taste of the service-dominant Logic".

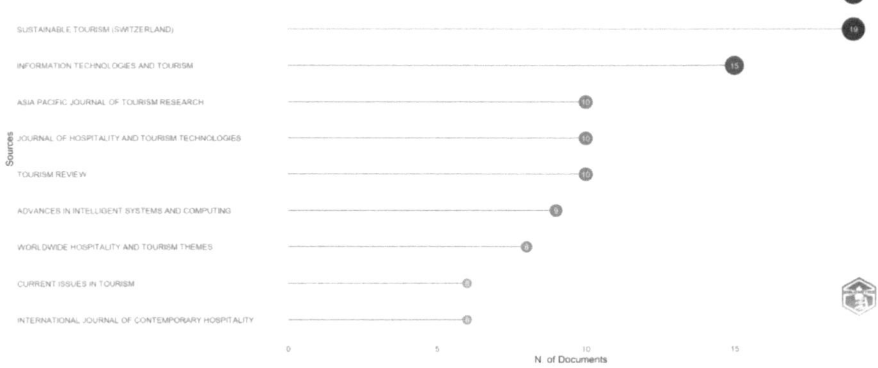

FIGURE 15.6 Top Ten Relevant Journals. (Source: Chart Made by the Authors' Team Based on Biblioshiny.)

TABLE 15.6
30 Publications that have Great Importance on STT

	Author	Article	AC	Source	SJR	H-index
1	Gretzel, Sigala, Xiang, & Koo (2015)	Smart tourism: foundation developments	962	Electronic Markets	Q1	49
2	Chung, Han, & Joun (2015)	Tourists' intention to visit a destination: The role of augmented reality (AR) application for a heritage site	260	Computers in Human Behavior	Q1	226
3	Buhalis (2020)	Technology in tourism-from information communication technologies to eTourism and smart tourism towards ambient intelligence tourism: a perspective article	227	Tourism Review	Q1	48
4	Liberato, Alen, & Liberato (2018)	Smart tourism destination triggers consumer experience: the case of Porto		European Journal of Management and Business Economics	Q1	25
5	Vecchio, Mele, Ndou, Secundo (2018)	Creating value from Social Big Data: Implications for Smart Tourism Destinations	214	Information Processing and Management	Q1	144
6	Huang, Goo, Nam, Yoo (2017)	Smart tourism technologies in travel planning: The role of exploration and exploitation	188	Information & Management	Q1	182 152
7	Jeong & Shin (2020)	Tourists' Experiences with Smart Tourism Technology at Smart Destinations and their Behavior Intentions	138	Journal of Travel Research	Q1	159
8	Dorcic, Komsic, & Markovic (2019)	Mobile technologies and applications towards smart tourism – state-of-the-art	133	Tourism Review	Q1	48
9	Pencarelli (2020)	The digital revolution in the travel and tourism industry	124	Information Technology & Tourism	Q1	30

(continued)

TABLE 15.6 (Continued)
30 Publications that have Great Importance on STT

	Author	Article	AC	Source	SJR	H-index
10	Yi, Yuan, & Yoo (2020)	The effect of the perceived risk on the adoption of the sharing economy in the tourism industry: The case of Airbnb	122	Information Processing & Management	Q1	144
11	Gursoy, Malodia, & Dhir (2022)	The metaverse in the hospitality and tourism industry: An overview of current trends and future research directions	0	Journal of Hospitality Marketing & Management	Q1	67
12	Go & Kang (2023)	Metaverse tourism for sustainable tourism development: Tourism Agenda 2030	0	Tourism Review	Q1	48
13	Wei (2022)	Gemiverse: The blockchain-based professional certification and tourism platform with its own ecosystem in the metaverse	0	International Journal of Geo Heritage and Parks	Q1	13
14	Lee (2022)	Tourism Using Virtual Reality: Media Richness and Information System Successes	0	Sustainability (Switzerland)	Q1	136
15	Suanpang, Niamsorn, Pothipassa, Chunhapataragul, Netwong, & Jermsittiparsert (2022)	Extensible Metaverse Implication for a Smart Tourism City	0	Sustainability (Switzerland)	Q1	136
16	Filimonau, Ashton, & Stankov (2022)	Virtual spaces as the future of consumption in tourism, hospitality, and events	0	Journal of Tourism Futures	Q1	27
17	Monaco & Sacchi (2023)	Travelling the Metaverse: Potential Benefits and Main Challenges for Tourism Sectors and Research Applications	0	Sustainability (Switzerland)	Q1	136

TABLE 15.6 (Continued)
30 Publications that have Great Importance on STT

	Author	Article	AC	Source	SJR	H-index
18	Chu (2023)	Deep resource allocation for a massively multiplayer online finance of tourism gamification in metaverse		Information Technology & Tourism	Q1	30
19	Sang, Wu, Liu, & Fang (2022)	Digital Twin Platform Design for Zhejiang Rural Cultural Tourism Based on Unreal Engine	0	International Conference on Culture-oriented Science and Technology (CoST 2022)		
20	Litavniece, Kodors, Adamoniene, & Kijasko (2023)	Digital twin: an approach to enhancing tourism competitiveness	0	Worldwide Hospitality and Tourism Themes	Q2	28
21	Ivanov, Nikolskaya, Radchenko, Sokolinsky, & Zymbler (2020)	Digital Twin of City: Concept Overview	0	Global Smart Industry Conference (GloSIC)		
22	Baker, Nam, & Dutt (2023)	A user experience perspective on heritage tourism in the metaverse: Empirical evidence and design dilemmas for VR	0	Information Technology and Tourism	Q1	30
23	Yuce (2023)	Digital twins and sustainable developments in the tourism and hospitality industry	0	Handbook of Research on Applications of AI, Digital Twin, and Internet of Things for Sustainable Development		
24	Rahmadian, Feitosa, & Zwitter (2023)	Governing Digital Twin technology for smart and sustainable tourism: A case study in applying a documentation framework for architecture decisions	0	Handbook on the Politics and Governance of Big Data and Artificial Intelligence		

(continued)

TABLE 15.6 (Continued)
30 Publications that have Great Importance on STT

	Author	Article	AC	Source	SJR	H-index
25	Zaman, Koo, Abbasi, Raza, & Qureshi (2022)	Meet Your Digital Twin in Space? Profiling International Expats' Readiness for Metaverse Space Travel, Tech-Savviness, COVID-19 Travel Anxiety, and Travel Fear of Missing Out	0	Sustainability	Q1	136
26	Buhalis, Leung, & Lin (a) (2023)	Metaverse as a disruptive technology revolutionizing tourism management and marketing	0	Tourism Management	Q1	236
27	Buhalis, Lin, & Leung (b) (2023)	Metaverse as a driver for customer experience and value co-creation: implications for hospitality and tourism management and marketing	0	International Journal of Contemporary Hospitality Management	Q1	113
28	Ioannidis & Kontis (2023)	Metaverse for tourists and tourism destinations	0	Information Technology & Tourism	Q1	30
29	Dwivedi et al. (2023)	Metaverse marketing: How the metaverse will shape the future of consumer research and practice	0	Psychology & Marketing	Q1	130
30	Koo, Kwon, Chung, & Kim (2023)	Metaverse tourism: conceptual framework and research propositions	0	Current Issues in Tourism	Q1	94

Source: Authors' Team.
Note: C: Average citations; GS Rank: Google Scholar Rank; SJR: SC Imago Journal Rank..

Notably, articles on digital twins and metaverse, which have few citations, were selected. A separate analysis of 69 documents on the subject reveals that the majority of these documents—60 in total—were published in the 2022–2023 edition.

In addition, 30 publications are all in Q1 journals, and the H-Index is comparatively high, ranging from 48 to 226, as shown in Table 15.6. Identify the journals or published articles that are most closely related to the research problems.

15.3.4 Topic Analysis

Researchers gain a more thorough comprehension of research trends by analyzing the co-occurrences. According to some authors, the more keywords a topic has, the

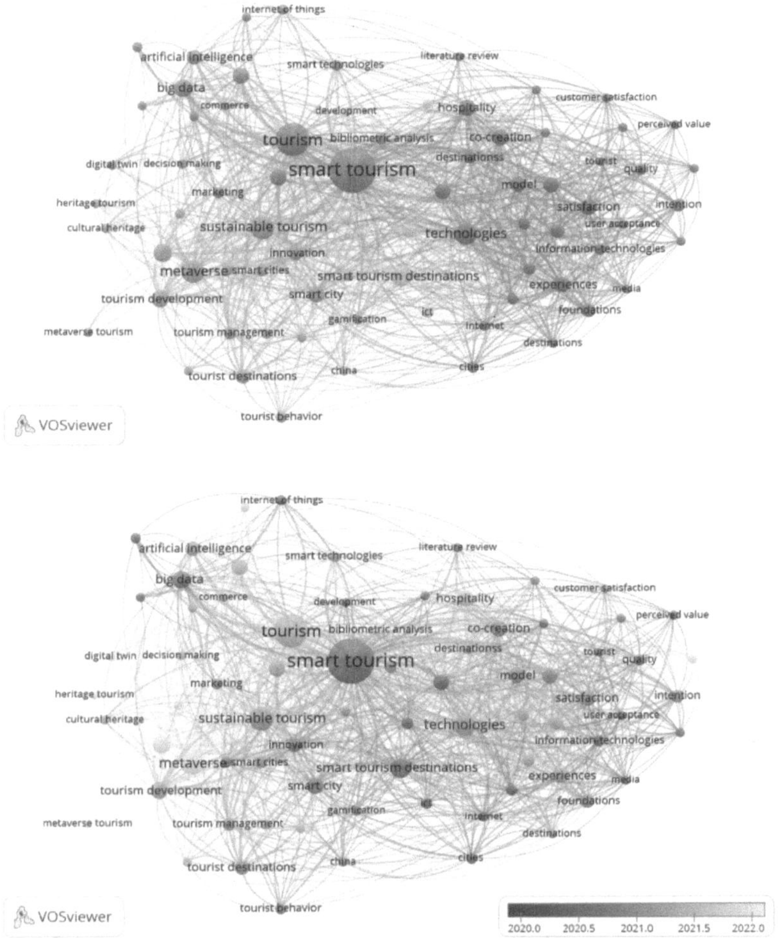

FIGURE 15.7 Network Visualization with 76 Most Common Co-Keyword on STT. (Source: Figure Made by the Authors' Team Based on VOS.)

Note: (a) Co-Keyword Network Visualization Was Based on Occurrences; (b) Co-Keyword Overlay Visualization Was Based on the Occurrences and Average Publication Per Year Score.

closer and more significant the association between them is [29]. There are a total of 2,035 keywords related to STT. The threshold criterion was set for seven occurrences of each keyword, resulting in 76 keywords showing up and five clusters, as shown in Figure 15.7.

From the results of the STT visual network in Figure 15.7, the research team selected eight clusters in Table 15.7 (smart tourism technologies; destinations image;

TABLE 15.7
Thematic Analysis of Keywords on the STT

Cluster		Sub-theme
1	Smart tourism Link:73 TLS: 503; O:176 Avg.Pub.Y:2020.20	Technology adoption: smart technologies; AI; big data; IoT; social networking (online); digital twin; gamification; metaverse tourism; virtual reality; AR; Topic: bibliometric analysis; COVID-19; cultural heritage; decision-making; heritage tourism; innovation; sustainable tourism; destination image; user experience;
2	Smart tourism technologies Link:38 TLS: 71; O:17 Avg.Pub.Y:2021.29	Smart tourism; attributes; tourism; information management; marketing; tourism industry; development trends; marketing; tourism industry; smart cities; competitiveness; experience; acceptance; intention; revisit intention; loyalty; quality; travel; customer satisfaction; determinants; travel
3	Destinations image Link:20; TLS: 30; O:8 Avg.Pub.Y:2021.88	Smart tourism; satisfaction; technologies; perceived value; loyalty; service; memorable tourism expericences; travel; intention
4	Tourism experiences Link:46 TLS: 50; O:9 Avg.Pub.Y:2020.89	Smart tourism
5	Risk perception Link:53 TLS: 63; O:6 Avg.Pub.Y:2021	Smart tourism; trust; risk perception; consumer; behavior;
6	Gamification Link:20; TLS: 28 O:7; Avg.Pub.Y:2020.29	Smart tourism; engagement; metaverse; tourism, model
7	Metaverse Link:37; TLS: 90; O:43; Avg.Pub.Y:2022.66	Media richness theory; COVID-19; sustainable tourism; technologies; tourism market; tourism; gamification; AR; virtual reality; hospitality; model; authenticity; blockchain
8	Digital twin Link:42; TLS:46; O:7 Avg.Pub.Y:2023	Smart tourism; tourism; decision-making

Source: Author group compiled.

smart tourism; tourism experience; risk perception; gamification; metaverse; digital twin) and their sub-themes to be a basis for content analysis.

The bibliometric analysis shown in Table 15.7 provides that clusters (1,2,3,4,5,6,8) are both commonly connected to the keywords: "smart tourism". This indicates a high level of interest in this topic from researchers in recent years.

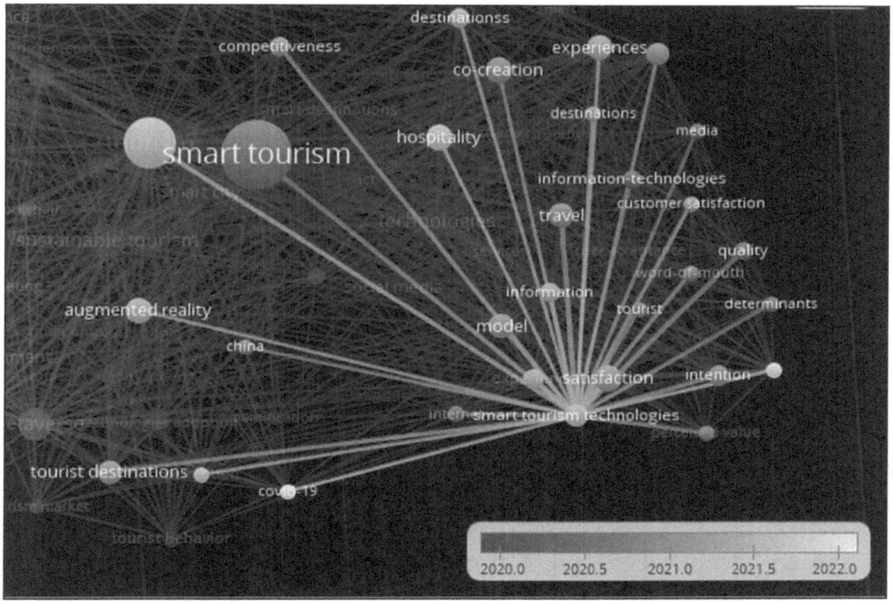

FIGURE 15.8 The Correlation Network Visualization of "Smart Tourism Technology". (Source: Figure Made by the Authors' Team Based on VOS.)

One similarity among the topic group of specific STT applications (gamification, metaverse, and digital twin) is that they all emerged in the 2020–2023 timeframe. In particular, digital twin is a new keyword that has appeared most recently (Avg. Pub.Y:2023), and metaverse has the highest link strength of 90. When triangulated to "Sub-themes" in Table 15.7, the number of keywords associated with these topics is still limited.

Along with the strong ST research trend, some clusters have started to emerge evaluating the influence of STT on this industry. Examples of these topics are "destination image", "tourist experiences", and "risk perception", shown in Figure 15.8 as tiny circles.

Thus, researchers studying this subject should closely follow this tendency from now on.

15.4 CONTENT ANALYSIS

This study inherits 30 publications of great importance and impact on the research problem to overview research on STT (Table 15.6) and eight keyword phrases analyzed by topic (Table 15.7). Close analysis reveals that these studies focus on two issues: (1) Experiencing smart travel technology enhances the tourist experience; (2) Challenges and risks involved when using STT during tourists' travel.

15.4.1 The Use of Smart Tourism Technology Enhances the Tourist Experience

Some authors proposed that technological readiness, visual appeal, and favorable conditions affect the usefulness and ease of augmented reality (AR) use, thereby influencing tourists' intention to use AR and visit the destination through AR [8]. Meanwhile, many authors concluded that the use of STT has a significant impact on tourists' overall travel experience and intention to return to the destination in the future [30]–[31]. Conversely, positive experiences with STT lead to higher intent to return to the destination, proving the importance of providing a seamless and engaging technological experience for tourists. At the same time, using STT helps to strengthen tourists' engagement with the destination, providing a richer and more personalized travel experience. AR can also bring visitors valuable experiences without needing a tour guide [8]. In fact, the media richness of VR content enhances the perceived usefulness and perceived enjoyment, leading to higher levels of satisfaction. For example, VR technology enables tourists to virtually explore destinations and attractions, providing a sense of presence and immersion that can enhance satisfaction [14]. It is also like the Geniverse platform, which focuses on building rich experiences for travelers and provides unique and engaging ways to explore destinations [32].

Regarding STT's use to improve the tourist experience, emphasis must be placed on the Digital twin and metaverse. The expanded metaverse system (divided into three layers of creating labels, characters, and VR) has enhanced the travel experience by creating new and engaging virtual reality experiences for tourists in smart destinations. It enables the development of personalized travel programs tailored to traveler preferences with travel program features, delivering a relevant and engaging experience [33]. Metaverse will change consumer behavior by providing opportunities for trip planning, interaction, and engagement [21]. Digital twins can enable the creation of personalized experiences for travelers by analyzing their preferences and behavioral patterns, thereby providing tailored recommendations and services [23].

More specifically, the metaverse provides rich experiences by converging the physical and digital universes, enabling users to move seamlessly between them [21]. Thus, users can explore a vibrant environment to work, study, trade, explore interests, and interact with others in the metaverse. Through the metaverse, users can travel in time, experience ancient encounters, and explore space or dangerous natural phenomena, such as volcanic eruptions, which creates a sense of immersion and authenticity. The gaming ecosystem has demonstrated effective interactions in the metaverse, where users can participate in immersive experiences together. At the same time, metaverse enhances destination awareness by providing virtual experiences that enable users to explore destinations virtually and gain a deeper insight into them. More importantly, users can interact with digital twin destinations in the metaverse, providing real-time information and insights about their location, history, culture, and other relevant details, enhancing already rich experiences. This is a key difference between this technology and previous ones. These rich experiences in the metaverse motivate users to explore and learn more about different places, potentially inspiring real-life travel experiences and delivering a personalized experience for tourists [14], [23], [10].

In parallel, digital replicas of destinations in the metaverse provide real-time information and insights about physical locations, further bridging the gap between the physical and digital worlds. Users can move seamlessly between the physical and digital universes in the metaverse, which blurs the boundaries and enables them to engage in virtual encounters, space exploration, and natural phenomena. Metaverse also has the potential to transform tourism marketing and management, creating opportunities for trip planning, interaction, and engagement, effectively transforming consumer behavior [21]. Similar studies point to the significant advantages of Digital twin and metaverse in enhancing tourist experience, such as [34]–[41].

15.4.2 Challenges and Risks Involved When Using Smart Tourism Technology in Tourists' Travel Process

Some research said that there are concerns about user security and privacy. This may negatively affect how smart travel technology is used [30]. Similarly, some research found that privacy and financial risks are involved when a tourist desires to use the sharing economy, particularly in the setting of Airbnb [42]. In contrast, physical and performance risks were found to have a positive impact on the intent to use the sharing economy. This risk paradox can be explained by the disruptive innovation of the sharing economy and the characteristics of risk-taking in tourism.

As analyzed above, despite its outstanding advantages, the application of technology (such as digital twins and metaverse) also has many downsides that pose long-term challenges. Metaverse offers promising solutions to long-standing challenges in the industry, such as concerns about trust and security, decision-making confidence, and the risks that travelers face throughout their vacation and travel planning process [22]. More particularly, video games are linked to mental health disorders, and excessive participation in virtual spaces can contribute to obsessive behavior [11]. The design and features of virtual spaces, such as rewards, achievements, and social interactions, can trigger compulsive behaviors and reinforce addictive patterns. Virtual currencies and in-game economies in virtual spaces can create a sense of urgency, thus driving compulsive shopping and spending behaviors.

From there, some authors provide recommendations to minimize the downsides of technology for users [11]. For example, "mindful" experiences can be designed in virtual spaces to ensure they are not overused and do not cause obsession. Additionally, it is important to consider the needs and expectations of different socio-demographic groups of Level 2 consumers in cyberspace, such as Generations Y and Z, to identify relevant content and features that can be acquired to attract them without leading to obsession. Implementing processes and protocols to manage deviant behavior, such as cyberbullying and harassment, in cyberspace in order to create a safe and healthy environment for consumers is also advisable. At the same time, it is important to develop policies that protect Level 2 consumers from potential cyber addiction, including guidelines on usage limits, privacy protection, and financial transactions.

15.5 CURRENT RESEARCH GAPS, NEW AND DEVELOPING RESEARCH AREAS

Through content analysis, it can be inferred that the technology applications and related relationships—represented by keywords—clearly illustrate how the various aspects relate to one another. However, the results shown in Figure 15.8 point out that "STT" has no correlation with "Gamification", "Metaverse", and "Digital Twin".

Moreover, as Figure 15.9 highlights, the relationship between STT applications and tourism destination image is a huge gap that needs to be filled. This includes analyzing the attributes of STT and how they relate to perceptions of experience, satisfaction, and value. Meanwhile, the keyword "Destination image" has very little association with "ST" or "STT", and the writers have not discovered a relationship between them.

Although the keyword "Gamification" (7 appearances) first surfaced in 2020, it has become a new research trend in Figure 15.10, particularly when it comes to the metaverse and Virtual Reality. Moreover, it also links with other keywords like Experience, Travel, Co-creation, Tourist behavior, etc. In the future, researchers should pay more attention to these other STT-related research concerns.

Moreover, the correlation between "Destination image" and keyword analysis for "Smart Travel", "Smart Travel Technology", "Digital Twin", "Metaverse", or "Gamification" is minimal. Besides, the visual images in Figure 15.9 and Figure 15.11 show that the keywords, "destination image", "Digital Twin", and "Metaverse", all

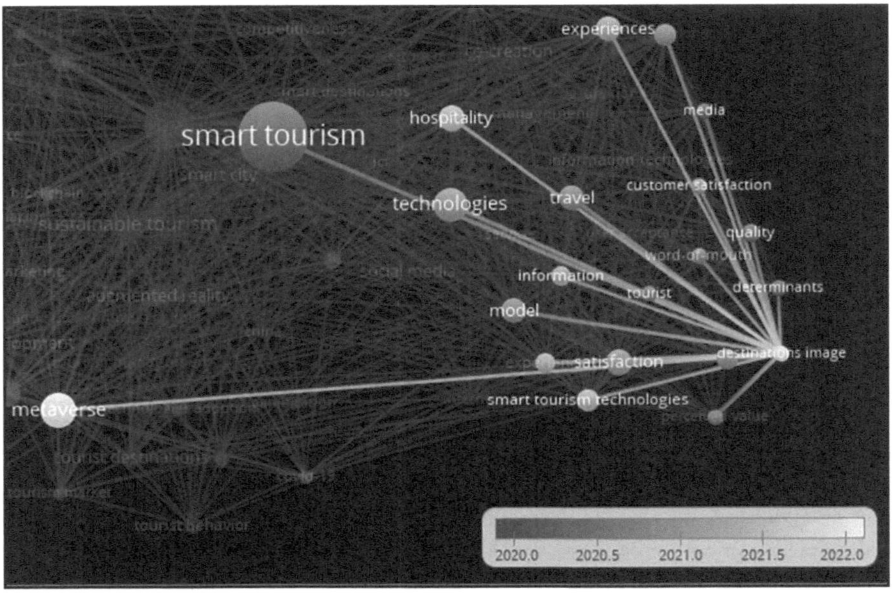

FIGURE 15.9 The Correlation Network Visualization of "Destination Image". (Source: Figure Made by the Author's Team Based on VOS.)

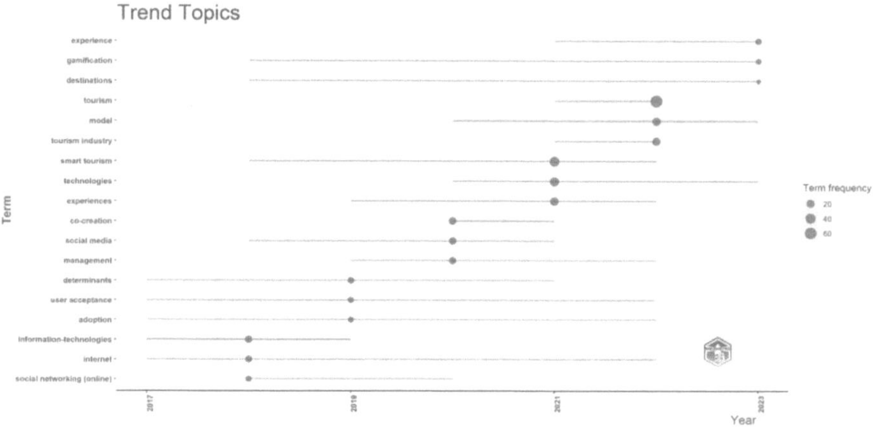

FIGURE 15.10 Trends Topics Related to STT. (Source: Chart Made by the Authors' Team Based on Biblioshiny.)

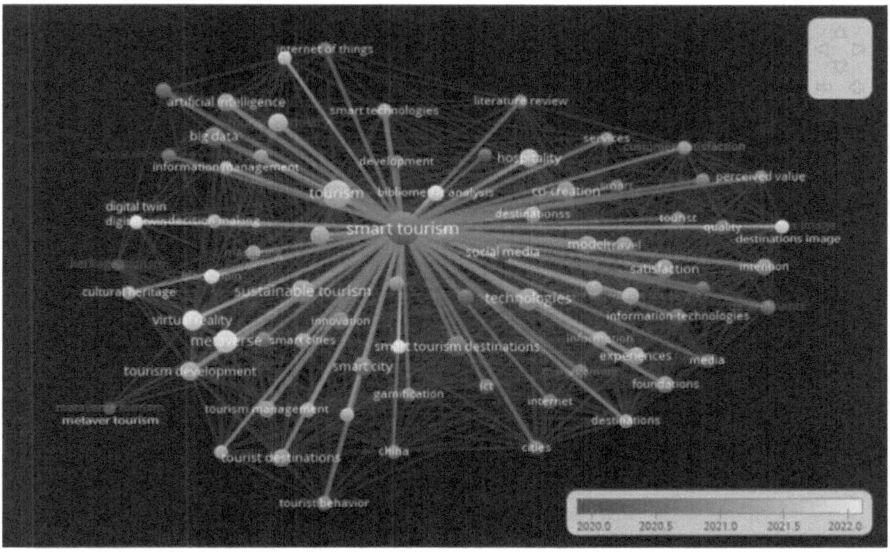

FIGURE 15.11 The Keywords "Digital Twin" and "Metaverse" Have Only Recently Appeared in Negligible Numbers, Represented by Small Round Labels. (Source: Figure Made by the Author Team Based on VOS.)

appear with small circles. This represents a significant research gap as well as a trend that would be of interest to researchers. Specifically, several studies have demonstrated that employing ST technologies may improve tourist destination image and increase the possibility that they will plan to return. Thus, researchers must bridge this gap as quickly as possible by identifying and analyzing aspects of this issue.

15.6 CONCLUSION

Research results show that the number of articles published has steadily increased annually and has been stable in the last 5 years. In recent years, researchers have become widely interested in the issue of STT, especially related to digital twins and metaverse. This is also consistent with the development and application of the trend of smart technology in tourism. At the same time, the cooperation of countries around the world is increasingly taking place on a broader scale, mainly with developed countries, especially countries with substantial strength, such as China, Spain, Korea, etc. Thus, to conduct a more thorough study on STT, it is necessary to increase collaboration with strong countries like China, Spain, Korea, the United States, etc. Additionally, some territories and countries, like Taiwan, Malaysia, India, etc., are coming up in this issue's investigation.

The study also shows publications with a significant impact on STT, including author groups, articles, citations, and journals that are most closely associated with the research topic in terms of ranking and H-Index. These publications have only appeared in recent years, from 2020 to 2022. Furthermore, the presence of digital twins and metaverse in publications only began in 2022 and 2023, supporting the assertion that research on the topic of STT is in its infancy.

At the same time, the study shows the strengths of researchers around the world who are paying close attention to the following aspects: smart tours, tourism, digital twin, metaverse tourism, satisfaction, AR, co-creation, tourist destination, information, intention, customer satisfaction, experience, quality, loyalty, competitiveness, perceived value, etc. Notably, current research has not focused much on connecting STT with destination image. Although the scarcity of the virtual appearance of the keyword "Tourism destination image" creates a gap in current research, it facilitates investigation from pertinent angles in future research.

Furthermore, research has indicated which journals have the most articles on these issues. This finding grants scholars options of credible and appropriate journal sources to submit for publication and announce their research results. On the other hand, the results of the article also help to provide more information, especially an understanding of the overall research situation on the issue of STT in the period 2013 – 2023. This article can be a reference for scholars, learners, or interested librarians as an updated document on STT.

REFERENCES

[1] F. Ercan, "Smart tourism technologies: Applications in hotel business," *Theory and Practice in Social Sciences,* ST. Kliment Ohridski University Press Sofia, Chapter 43, pp. 528–546, 2019.

[2] M. Naramski, "The application of ICT and smart technologies in polish museums—towards smart tourism," *Sustainability,* vol. 12, no. 21, p. 9287, 2020.

[3] U. Gretzel, L. Zhong, and C. Koo, "Application of smart tourism to cities," *International Journal of Tourism Cities,* vol. 2, no. 2, 2016, doi: 10.1108/IJTC-04-2016-0007.

[4] A. Erdem and F. Şeker, "Tourist Experience and Digital Transformation," in *Handbook of Research on Digital Communications, Internet of Things, and the Future of Cultural Tourism*: IGI Global, 2022, pp. 103–120.

[5] C.-K. Pai, Y. Liu, S. Kang, and A. Dai, "The role of perceived smart tourism technology experience for tourist satisfaction, happiness and revisit intention," *Sustainability,* vol. 12, no. 16, p. 6592, 2020.

[6] A. Mitala, A. Konstantoglou, D. Folinas, and T. Fotiadis, "Smart Tourism Technologies: How Effective Are They for Millennial Tourists?," in *Transcending Borders in Tourism Through Innovation and Cultural Heritage: 8th International Conference, IACuDiT, Hydra, Greece, 2021*: Springer, 2022, pp. 339–351.

[7] T. Pencarelli, "The digital revolution in the travel and tourism industry," *Information Technology & Tourism,* vol. 22, no. 3, pp. 455–476, 2020.

[8] N. Chung, H. Han, and Y. Joun, "Tourists' intention to visit a destination: The role of augmented reality (AR) application for a heritage site," *Computers in Human Behavior,* vol. 50, pp. 588–599, 2015.

[9] D. Buhalis, "Technology in tourism-from information communication technologies to eTourism and smart tourism towards ambient intelligence tourism: A perspective article," *Tourism Review,* vol. 75, no. 1, pp. 267–272, 2020.

[10] E. Rahmadian, D. Feitosa, and A. Zwitter, "Governing Digital Twin technology for smart and sustainable tourism: A case study in applying a documentation framework for architecture decisions," in *Handbook on the Politics and Governance of Big Data and Artificial Intelligence,* vol. 0, p. 105, 2023.

[11] V. Filimonau, M. Ashton, and U. Stankov, "Virtual spaces as the future of consumption in tourism, hospitality and events," *Journal of Tourism Futures,* vol. 10, no. 1, pp. 110–115, 2022.

[12] S. Ivanov, K. Nikolskaya, G. Radchenko, L. Sokolinsky, and M. Zymbler, "Digital Twin of City: Concept Overview," in *2020 Global Smart Industry Conference (GloSIC)*: IEEE, 2020, pp. 178–186.

[13] Z. Wei, J. Zhang, X. Huang, and H. Qiu, "Can gamification improve the virtual reality tourism experience? Analyzing the mediating role of tourism fatigue," *Tourism Management,* vol. 96, p. 104715, 2023.

[14] U.-K. Lee, "Tourism using virtual reality: Media richness and information system successes," *Sustainability,* vol. 14, no. 7, p. 3975, 2022.

[15] F. Xu, J. Weber, and D. Buhalis, "Gamification in Tourism," in *Information and Communication Technologies in Tourism 2014: Proceedings of the International Conference in Dublin, Ireland, January 21–24, 2014*, 2013: Springer, pp. 525–537.

[16] E. Özkul, E. Uygun, and S. Levent, "Digital Gamification in the Tourism Industry," in *Handbook of Research on Smart Technology Applications in the Tourism Industry*: IGI Global, 2020, pp. 169–203.

[17] E. Bartoli, B. Elmi, D. Pascuzzi, and A. Smorti, "Gamification in tourism," *Psychology & Behavioral Science,* vol. 8, no. 3, pp. 93–95, 2018.

[18] D. Liberato, P. Liberato, and M. Nunes, "Gamification and Storytelling Enhancing Successful Wine and Food Tourism Products," in *International Conference of the International Association of Cultural and Digital Tourism*: Springer, 2022, pp. 433–452.

[19] Y. Prihastomo *et al.*, "Gamification for E-Tourism Based on Virtual Reality Study on Indonesia Tourism," in *2022 International Conference on ICT for Smart Society (ICISS)*: IEEE, 2022, pp. 01–05.

[20] L. Xin, G. Qiao, H. Song, Z. Huang, and M. Ye, "Gamification innovation: Understanding industrial tourism products and tourist response," *Asia Pacific Journal of Tourism Research,* vol. 28, no. 3, pp. 274–291, 2023.

[21] D. Buhalis, D. Leung, and M. Lin, "Metaverse as a disruptive technology revolutionizing tourism management and marketing," *Tourism Management,* vol. 97, p. 104724, 2023.
[22] S. Ioannidis and A.-P. Kontis, "Metaverse for tourists and tourism destinations," *Information Technology & Tourism,* vol. 25, no. 4, pp. 483–506, 2023.
[23] L. Litavniece, S. Kodors, R. Adamoniene, and J. Kijasko, "Digital twin: An approach to enhancing tourism competitiveness," *Worldwide Hospitality and Tourism Themes,* vol. 15, no. 5, pp. 538–548, 2023.
[24] A. L. Porter, A. Kongthon, and J.-C. Lu, "Research profiling: Improving the literature review," *Scientometrics,* vol. 53, pp. 351–370, 2002.
[25] R. Vogel and W. H. Güttel, "The dynamic capability view in strategic management: A bibliometric review," *International Journal of Management Reviews,* vol. 15, no. 4, pp. 426–446, 2013.
[26] D. Moher, A. Liberati, J. Tetzlaff, D. G. Altman, and P. Group, "Preferred reporting items for systematic reviews and meta-analyses: The PRISMA statement," *Annals of Iinternal Medicine,* vol. 151, no. 4, pp. 264–269, 2009.
[27] P. Mongeon and A. Paul-Hus, "The journal coverage of Web of Science and Scopus: a comparative analysis," *Scientometrics,* vol. 106, pp. 213–228, 2016.
[28] M. Aria and C. Cuccurullo, "bibliometrix: An R-tool for comprehensive science mapping analysis," *Journal of Informetrics,* vol. 11, no. 4, pp. 959–975, 2017.
[29] J. E. Raffaghelli, S. Cucchiara, F. Manganello, and D. Persico, "Different views on Digital Scholarship: Separate worlds or cohesive research field?," *Research in Learning Technology,* vol. 24, pp. 1–17, 2016.
[30] C. D. Huang, J. Goo, K. Nam, and C. W. Yoo, "Smart tourism technologies in travel planning: The role of exploration and exploitation," *Information & Management,* vol. 54, no. 6, pp. 757–770, 2017, doi: 10.1016/j.im.2016.11.010.
[31] M. Jeong and H. H. Shin, "Tourists' Experiences with smart tourism technology at smart destinations and their behavior intentions," *Journal of Travel Research,* vol. 59, no. 8, pp. 1464–1477, 2019, doi: 10.1177/0047287519883034.
[32] D. Wei, "Gemiverse: The blockchain-based professional certification and tourism platform with its own ecosystem in the metaverse," *International Journal of Geoheritage and Parks,* vol. 10, no. 2, pp. 322–336, 2022.
[33] P. Suanpang, C. Niamsorn, P. Pothipassa, T. Chunhapataragul, T. Netwong, and K. Jermsittiparsert, "Extensible metaverse implication for a smart tourism city," *Sustainability,* vol. 14, no. 21, p. 14027, 2022.
[34] H. Go and M. Kang, "Metaverse tourism for sustainable tourism development: Tourism agenda 2030," *Tourism Review,* vol. 78, no. 2, pp. 381–394, 2023.
[35] D. Buhalis, M. S. Lin, and D. Leung, "Metaverse as a driver for customer experience and value co-creation: Implications for hospitality and tourism management and marketing," *International Journal of Contemporary Hospitality Management,* vol. 35, no. 2, pp. 701–716, 2022.
[36] D. Gursoy, S. Malodia, and A. Dhir, "The metaverse in the hospitality and tourism industry: An overview of current trends and future research directions," *Journal of Hospitality Marketing & Management,* vol. 31, no. 5, pp. 527–534, 2022.
[37] U. Zaman, I. Koo, S. Abbasi, S. H. Raza, and M. G. Qureshi, "Meet your digital twin in space? Profiling international expat's readiness for metaverse space travel, tech-savviness, COVID-19 travel anxiety, and travel fear of missing out," *Sustainability,* vol. 14, no. 11, p. 6441, 2022.

[38] C. Koo, J. Kwon, N. Chung, and J. Kim, "Metaverse tourism: Conceptual framework and research propositions," *Current Issues Tourist,* vol. 26, no. 20, pp. 3268–3274, 2023.

[39] Y. K. Dwivedi *et al.*, "Metaverse marketing: How the metaverse will shape the future of consumer research and practice," *Psychology & Marketing,* vol. 40, no. 4, pp. 750–776, 2023.

[40] A. Yuce, "Digital Twins and Sustainable Developments in the Tourism and Hospitality Industry," in *Handbook of Research on Applications of AI, Digital Twin, and Internet of Things for Sustainable Development*: IGI Global, 2023, pp. 461–472.

[41] J. Baker, K. Nam, and C. S. Dutt, "A user experience perspective on heritage tourism in the metaverse: Empirical evidence and design dilemmas for VR," *Information Technology & Tourism,* vol. 25, no. 3, pp. 265–306, 2023.

[42] J. Yi, G. Yuan, and C. Yoo, "The effect of the perceived risk on the adoption of the sharing economy in the tourism industry: The case of Airbnb," *Information Processing & Management,* vol. 57, no. 1, p. 102108, 2020.

16 Customized Cancer Care Using Digital Twins

The Case of Breast Cancer in Women

Ananya Chenat and Divya S.

16.1 INTRODUCTION

While numerous initiatives are underway to incorporate digital twin strategies across many sectors, including healthcare, there remains a lack of knowledge on effectively deploying digital twins in intricate systems, particularly in the realm of sensitive and complex medical data. It is also crucial for this type of system to be user-friendly, catering to the lack of expertise in computer science or biomedical engineering, such as doctors and nurses, as these professionals play an important role in developing the system.

Examining data from populations of digital twins can contribute to enhancing strategies for cancer screening, prediction, prevention, detection, treatment, and research investments. This paper investigates and analyzes the various real-time uses of digital twin-based treatment, specifically in the context of breast cancer, utilizing both conceptual and statistical analyses. Furthermore, the paper presents an innovative framework illustrating various ailments where digital twin technology can be implemented for future research. Focusing on user-friendly, accessible, and intuitive ways for end users with no technical expertise on to how to interact with digital twin systems in their industry is another crucial priority, along with the timely and correct update and replication of data from a physical twin onto the digital twin. To demonstrate the potential of digital twins, this paper also addresses the challenges in implementing digital twin technology in cancer diseases, followed by a discussion of several promising research directions in this field.

16.2 BREAST CANCER – SYMPTOMS AND TRADITIONAL DIAGNOSIS

Breast cancer is a type of cancer that, in most cases, spreads to other organs further away, such as the bones, lungs, brain, and liver, which makes the process of curing it difficult. Understanding the condition is crucial before delving into the specifics

DOI: 10.1201/9781003465904-20

of diagnosis and treatment procedures. Breast cancer starts in the cells of the breast. Most people are unaware that men and women can both fall prey to breast cancer, although it is more likely to be found in women. There are three basic components of a breast: connective tissue, ducts, and lobules. Typically, breast cancer develops in the lobules, which are responsible for the production and transportation of milk. Breast cancer cells begin to multiply and expand, eventually forming a tumor or lump. It can even spread to other areas of the body outside of the breasts via the lymphatic or blood systems. The spread of cancer cells to different parts of the body is known as metastatic cancer.

Taking care of your health requires understanding how your body feels and looks. Any unusual changes in your body should be noted. The secret to successful treatment is early detection. The following are indicators and symptoms that may serve as warning signs for breast cancer:

- A painless lump
- Nipple inversion
- Breast swelling
- Nipple discharge
- Dimpling of the skin
- Alterations in the skin on the breast or nipple
- A swelling or lump under the arm or on the collarbone

16.3 WHO IS AT RISK?

Around 5% to 10% of breast cancer cases are believed to have a hereditary component, stemming from gene abnormalities (referred to as mutations) passed down from a parent. Women with immediate blood relatives who have had breast cancer face a greater risk of developing it themselves. If a woman has a direct relative, either a sister, daughter or a mother, who has had breast cancer, then her risk is almost doubled. Studies show that Caucasian women have a slightly higher chance of developing breast cancer compared to African American women. However, African American women are at a greater risk of encountering more aggressive forms of breast cancer, and they are often diagnosed at a younger age and in more advanced stages. A higher risk of developing breast cancer is seen in women who exhibit denser breast tissue on a mammogram, which means they have a higher proportion of glandular tissue and reduced fatty tissue.

- Old Age: Age increases the risk of breast cancer. According to reports, many breast cancer cases are reported by women over the age of 40.
- Genetic Changes: Breast and ovarian cancer risk is increased in females who have hereditary changes or mutations to specific genes, such as BRCA1 and BRCA2.
- Reproductive backgrounds: Women are exposed to hormones for a longer amount of time when their menstruation starts before the age of 12 or hit

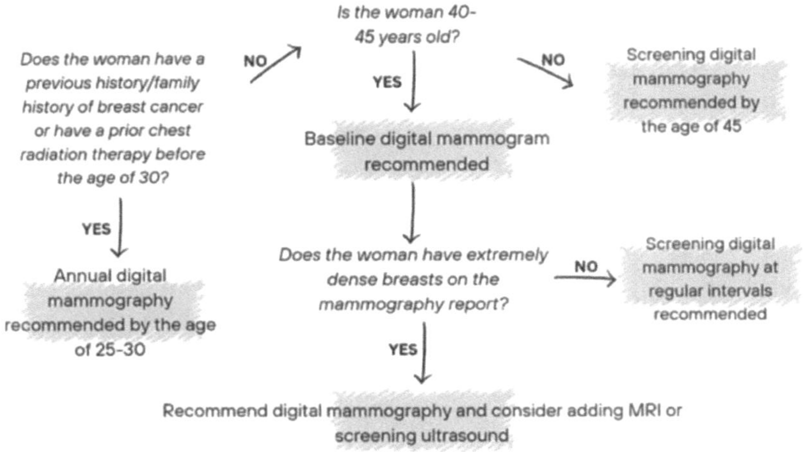

FIGURE 16.1 Medical Imaging Decision Tree.

menopause after 55 years of age, which increases their likelihood of developing breast cancer.

- Denser Breasts: As connective tissue predominates over fatty tissue in dense breasts, tumors may sometimes be difficult to detect on a mammogram. Thus, breast cancer is more likely to affect women who have dense breasts.
- A previous history of breast cancer or a particular non-cancerous breast condition: Breast cancer is more likely to recur in women who have already had it. Women suffering from breast conditions, such as atypical hyperplasia and lobular carcinoma, are also at risk.
- Previous radiation therapy treatment: Breast cancer is more likely to develop in women at some point in their lives if they have undertaken radiation therapy to the chest or breasts.

16.3.1 Diagnosis

If breast cancer is detected early, then treatment can be fairly effective, with an 84% survival rate. Typically, a triple assessment approach is used by doctors to determine a breast cancer diagnosis. Having a physical examination is the first step. A clinical breast examination (CBE) is performed on the patient to look for lumps or any changes in the size or shape of the breasts. Imaging is the next step if any physical anomalies are discovered. Breast magnetic resonance imaging scans (MRIs), breast ultrasounds, and mammograms are a few of the popular and efficient imaging methods. Low-dose X-rays called mammograms are used to identify breast cancer in its earliest stages, when it is still small and even before a lump can be felt. This is when it can be treated most simply. However, it is less successful for individuals under the age of 40 and those with dense breast tissue. It also exhibits lower sensitivity to tiny tumors (less than 1 mm, or approximately 100,000 cells) and does not

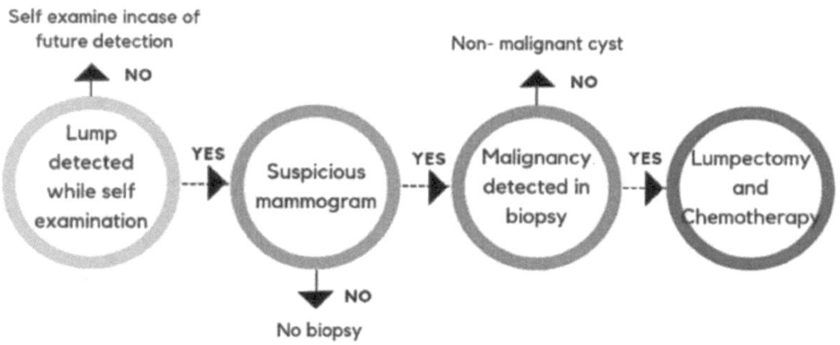

FIGURE 16.2 Risk Model.

predict how the illness will progress over time. Breast ultrasounds, which use sound waves, are a non-invasive imaging method that produces images of the breast tissue. It can spot alterations such as fluid-filled cysts, which mammograms struggle to detect. An MRI scan is advised for females who have a high risk of developing breast cancer. To produce finely detailed images of the breast tissue, the MRI uses a magnetic field and radio waves. The third step is a biopsy, which entails removing a tiny sample of breast tissue for microscopic inspection. A biopsy is required to determine whether a suspicious region is malignant. Breast biopsies are typically carried out in addition to screening methods to differentiate between malignant and benign tissues, but this procedure is costly and thus calls for experienced and trained individuals. After completing these stages, the doctor can assess the breast cancer's stage (extent), which aids in recommending the most effective treatment options.

16.3.2 Treatment

The treatment approach for breast cancer is greatly influenced by its biological characteristics and behavior. Some tumors may be small but exhibit rapid growth, while others, though larger, progress more slowly. Personalized treatment recommendations hinge on various factors, including:

- The stage of the tumor
- Patient-specific factors, such as age, health, menopausal status
- The presence of known mutation genes, such as BRCA1 or BRCA2 in inherited breast cancer, as indicated by genetic testing results

The cancer's stage, the size and location of the tumor, the features of the cancer cells, and the patient's general health all influence the sort of treatment that is used.

- Surgery: The first prescribed course of treatment is surgery. The three primary surgical procedures for breast cancer are a mastectomy (complete removal of the breast), lumpectomy (excision of the tumor, along with some adjacent tissue), and lymph node removal.
- Radiation therapy: Radiation treatment employs high-energy X-rays or other radiation forms to kill cancer cells. Typically administered post-surgery, it eradicates any remaining cancer cells and reduces the risk of recurrence.
- Chemotherapy: Drugs are used in chemotherapy to kill the cancer cells. It is typically administered either before surgery to reduce the size of the tumor and facilitate removal, or after surgery to eradicate any cancer cells that may have survived.
- Targeted therapy: Targeted therapy is a more recent form of cancer treatment that specifically targets certain proteins or genes associated with the development and spread of cancer cells. HER2-positive breast cancers are the ones most frequently treated with it.

16.3.3 The Role of AI and ML in Breast Cancer Diagnosis

In the past five years, ML and the AI revolution have been used to detect and diagnose several serious diseases. This increases the patient's probability of survival by paving the way to an early diagnosis and course of therapy. Health outcomes and patient experiences can be improved by using ML algorithms to process medical data quickly and assist medical practitioners in identifying trends and traits that could be signs of breast cancer.

Radiologists can employ computer-aided detection (CAD) systems as a sort of technology to help them find anomalies. These technologies analyze medical images using algorithms based on ML and AI to identify potential areas for more research.

Radiomics is a branch of medical imaging that analyzes and extracts quantitative data from medical pictures, including mammograms, ultrasound images, and MRI scans, to find patterns and traits that are suggestive of malignancy.

Using ML and AI algorithms, a technology called AI-powered ultrasound can analyze ultrasound images and spot patterns and features that could be signs of breast cancer. Although this technology is still in its infancy, it has the potential to completely revolutionize how breast cancer is detected and treated.

16.4 DIGITAL TWINS

A digital twin is a virtual model made with real-time data and simulation technology to represent a physical thing accurately. Typically, they are developed by combining sensor data, computational models, and ML techniques. The ML algorithms and computational models are used to simulate and anticipate the behavior of the physical system or process, while the sensor data is utilized to record real-time information about it. Although the concept of a digital twin has been around for some time, new technological developments have made it possible to build more sophisticated and effective digital twins. The integration challenge between IoT and data analytics can be addressed by the digital twin by establishing a linked physical and virtual twin.

The idea of a digital twin typically consists of three essential elements: a physical object existing in the real world, a virtual representation existing in a digital environment, and the exchange of information between these two components. In the context of our discussion, the physical entity specifically pertains to an individual, commonly a patient.

16.4.1 Digital Twins in Healthcare

Digital twins have revolutionized the industrial sector by changing how products are developed, and equipment is utilized. By leveraging the data on an object's evolution, digital twins enable us to foresee and enhance its performance. The healthcare field is currently undergoing a shift toward customized medicine through digital transformation. The digital twin stands out as a crucial instrument in driving this transformation. Healthcare analysis uses digital twins, often known as medical twins, in the hospital setting. They have the potential to deliver individualized medical treatments and enhance patient recovery.

- Disease Modeling: Digital twins can be used to simulate disease behaviors and forecast how quickly they will spread throughout a patient's body.
- Surgical Planning: By using digital twins to construct 3D models of a patient's anatomy, surgeons can plan and simulate procedures before carrying them out on actual patients.
- Drug Development: By simulating how drugs behave in the human body using digital twins, researchers may more efficiently create and test new medications.
- Monitoring: Real-time monitoring of a patient's health via digital twins enables medical professionals to quickly identify alterations in the patient's condition and take prompt action.

Several research studies have proposed and built digital twin models for various health issues, such as cardiovascular diseases, etc. This is provided in detail in Table 16.1.

In the realm of healthcare, the digital twin is seen as a catalyst for incremental and evolutionary advancements, signifying its role in accelerating existing trends rather than instigating a revolutionary shift. Presently, challenges in technology, regulations, and ethics have impeded a unanimous agreement on the possibility of a huge impact over the next decade. This paper aims to outline the current uses of digital twins in healthcare, focusing on breast cancer, to identify both the prospects and obstacles that either encourage or impede their broader and swifter adoption.

TABLE 16.1
Digital Twins in Healthcare

S No	Fields in Healthcare	Impact	Author References
1	Cardiovascular disease	The algorithms acquire data from many sources and integrate them to find patterns predict alternative tests, and overall enhance the survival of the patient.	Coorey et al., 2022
2	Key areas of ICU medicine	Using supportive technologies, the digital twin is used to meet the in-demand goals of ICU care.	Geoffrey Chase et al., 2021
3	Radiopharmaceutical therapies	Using the results of radiobiological reports, radiopharmaceutical therapies are customized using digital twins.	Abdollahi & Rahmim, 2022
4	Personalize medicine	A digital twin is made for each patient on which different medications are tested so as to obtain the most suited option for the patient.	Björnsson et al., 2020
5	The potential in surgery	For the patient to test different treatment options, phenotypic copies are created using digital twinning.	Ahmed & Devoto, 2020
6	Pneumonia patient treatment	Using digital twins, a three-step approach is used. First, the urgent cases are identified, after which ventilation and critical care are assigned to those in need. Later, ventilation can be discontinued after optimal time. This three-step approach allows smooth flow of patients in and out of ITUs.	Chakshu & Nithiarasu, 2022
7	Mechanical stability of vertebrae in cancer patients	Deep convolutional generative adversarial network (DCGAN) optimization algorithms are used to virtually reconstruct the trabecular bone microstructure.	Ahmadian et al., 2022
8	Detecting the severity of carotid stenoses from head vibration	A semi-active digital twin model is used to find out the severity of carotid stenosis.	Chakshu et al., 2019
9	The emergency room service (ERS)	The detection and diagnosis of various diseases or disorders using digital twins with good accuracy, even better than experienced physicians.	Aluvalu et al., 2023

TABLE 16.1 (Continued)
Digital Twins in Healthcare

S No	Fields in Healthcare	Impact	Author References
10	Cancer care	Digital twins help to understand the case of each patient individually and thus help to prescribe personalized treatment and therapy with precision.	Kaul et al., 2023
11	Intelligent radiotherapy	Intelligent network platform consisting of radiotherapy preparation and quality control (QC) for post-treatment follow-up using digital twins.	Guangpeng Chen et al., 2022
12	Smart healthcare	Use digital twins to keep a check on a person's health and positively impact their therapy.	Volkov et al., 2021
13	Vertebra and predicting its fracture response	AI-based framework to predict the risk of vertebrate fracture in a cancer patient having spinal metastasis.	Radchenko et al, 2021
14	Predicting neurological complications of pediatric cancers and their treatment	Identify causes of neurological difficulties.	Thiong'o GM, Rutka JT, 2022
15	Based on the patient's age, to prescribe a customized therapy based on the patient's age.	Duration of applying the commercialized fentanyl patches.	Flora Bahrami et al., 2022

16.4.2 Digital Twins in Breast Cancer

Narrowing our research to breast cancer, digital twinning technology enables healthcare professionals to have a virtual clone of the patient's breast. This virtual model can be used to test various treatment alternatives and simulate various scenarios, enabling personalized and effective treatment programs for specific patients. Doctors can prescribe the best imaging approach to identify the existence of cancer and personalize treatment options for patients by modeling the activity of cancer cells in a virtual environment.

Through this review paper, the objective is to explore the potential of digital twinning in the realm of breast cancer diagnosis.

16.5 DIGITAL TWIN ARCHITECTURE

The architecture of digital twins in the healthcare industry involves a comprehensive procedure, including data collecting, storage, analysis, and communication with patients and healthcare providers. An outline of the main components and steps in the architecture is provided below:

1. Gathering of Data:
   - National Cancer Databases and Clinical Records: To obtain a report of a patient's medical record, treatment options, and results, preliminary data is gathered from national cancer databases and clinical records.
   - Wearable Technology and Sensors: Continuous physiological data is provided via real-time data from wearable technology and sensors, such as temperature sensors and health monitoring gadgets. Vital signs, exercise levels, and other pertinent health measures are included in this.
2. Integration and Storage of Data:
   - Data integration is the process of combining information from several sources, such as real-time sensor data, clinical records, and national databases, to produce a comprehensive and cohesive picture of a patient's health profile.
   - Cloud Computing Infrastructure: Large data sets can be managed effectively with cloud storage, which also guarantees analytical accessibility.
3. ML and AI Algorithms for Data Analysis:
   - AI and ML: The integrated data is analyzed using advanced algorithms. These algorithms can identify trends, patterns, and possible correlations that conventional analysis techniques could miss.
   - Formulation of Diagnoses: Based on the examined data, ML and AI algorithms are used to create diagnoses and forecasts.
4. Creation of Digital Twins:
   - Digital twins or virtual representations are made possible in part by the outcomes of data analysis. By combining real-time and historical data, these digital twins mimic the patient's health condition and present it in a dynamic, changing manner.
   - Behavioral Modeling: New developments in digital twins aim to make these virtual representations more lifelike to more closely mimic the behavior of actual people. This could involve simulating reactions to different stimuli, treatment plans, and adjustments to medical conditions.
5. Interaction and Output:
   - Patient and Physician Interaction: Patients and medical professionals receive the output from the digital twins, which includes diagnosis, treatment suggestions, and health insights.
   - Secure Communication Channels: Patient portals can be used to give people immediate access to their digital twin data, and communication channels are secured to safeguard critical health information.

6. Database Updating and Learning:
   - Access to Past Data: To gradually enhance their capacity for learning and diagnosis, the digital twin's algorithms may have access to past data kept in databases.
   - Ongoing Learning: The digital twins' ML and AI algorithms are always picking up fresh information from which to improve their models and adjust for variations in a patient's condition.
7. Security and Regulatory Compliance:
   - Compliance Measures: A key component of the architecture is strict adherence to healthcare standards, particularly data protection legislation such as the Health Insurance Portability and Accountability Act of 1996 (HIPAA). Compliance procedures are put in place to guarantee the moral use of patient data.
   - Data Security: Throughout the digital twin's lifetime, security measures are used to safeguard the confidentiality and integrity of patient information.

16.5.1 Data Collection

A fundamental puzzle piece in crafting digital models lies in the acquisition of real-world data. As digital health technology is developing, the deployment of medical data from laboratories and clinics has become less challenging. The better the quality and quantity of data are, the more the digital model's performance and functionality will be enhanced.

Data is collected from Surveillance, Epidemiology, and national databases such as The Cancer Genome Atlas Program (TCGA), the National Cancer Database (NCDB) and the End Results Program (SEER). The most often used data source, however, is data from patients in clinics and medical records. Data from generative adversarial network (GAN) machines, CAD, temperature sensors, and wearable health devices, such as bracelets, watches, or headbands, are also common sources of information. The data types employed are computed tomography (CT) scans, MRIs, positron emission tomography (PET) scans, images from histopathological microscopes, computing image processing (CIP), mammograms, and ultrasounds.

There are different types of data collected. Diagnostic data helps to identify the extent or stage of cancer. The survival rate and the recurrence of cancer in the patient are determined using prognostic data. Predictive data can tell how a patient will respond to the particular type of treatment undertaken, while treatment monitoring data is used to build AI models or train the existing ones.

16.5.2 ML Techniques Used

The aim of AI, specifically the subfield of ML, revolves around making algorithms and models that enable computers to use data without explicit programming and predict useful insights. Healthcare is just one of the many industries that use ML, which has become one of the fastest-growing topics in computer science.

ML is used in digital twinning to train models to learn from real-world data and use that information to forecast or decide on how the system being modeled will

TABLE 16.2
Database and Data Type to Build Digital Twin Models

Sr No	Data type	Database
1	CT scan, MRI, PET scan, and microscopic images obtained through histopathology	Cancer care databases such as the Surveillance, Epidemiology, and End Results Program (SEER), The TCGA, National Cancer Database (NCDB)
2	Data from patients using GAN machines	Blood samples and medical report information of patients diagnosed with breast cancer
3	Thermography for breast cancer detection	Data was taken from all temperature sensors to identify any anomalies
4	CAD models	Bracelets, watches or headbands, and other wearable health devices
5	MRI scans	Data from patients with breast cancer
6	CIP	Digital health records and medical history of patients
7	Mammography	Data from patients with breast cancer
8	Ultrasound imaging	Data from patients with breast cancer
9	Blood samples and demographic, anthropometric, and clinical data	Digital Health records and medical history of patients

TABLE 16.3
ML Algorithms Used in Digital Twins

Sr No	ML Techniques	Objective	Research Studies
1	MLLR	Use ML techniques to create real-time and reliable digital twins of breast cancer for diagnostic and therapeutic purposes.	• Moztarzadeh et al., 2023 • Cancer Digital Twins in Metaverse December 2022
2	DTR	Using advanced ML methodologies, the aim is to develop dynamic and dependable digital replicas of breast cancer in real time. These digital twins are designed to serve both diagnostic and therapeutic objectives with a high degree of accuracy and reliability.	• Moztarzadeh et al., 2023 • Cancer Digital Twins in Metaverse December 2022

TABLE 16.3 (Continued)
ML Algorithms Used in Digital Twins

Sr No	ML Techniques	Objective	Research Studies
3	RFR	Using advanced ML methodologies, the aim is to develop dynamic and dependable digital replicas of breast cancer in real time. These digital twins are designed to serve both diagnostic and therapeutic objectives with a high degree of accuracy and reliability.	• Moztarzadeh et al., 2023 • Cancer Digital Twins in Metaverse December 2022
4	GBA	Using advanced machine learning (ML) methodologies, the aim is to develop dynamic and dependable digital replicas of breast cancer in real time. These digital twins are designed to serve both diagnostic and therapeutic objectives with a high degree of accuracy and reliability.	• Moztarzadeh et al., 2023 • Cancer Digital Twins in Metaverse December 2022
5	CIP in face recognition	The proposed model, incorporating digital twins in the ERS, accelerates patient care through a streamlined, fast-track service, thereby minimizing the duration of stay in the ER.	• Aluvalu et al., 2023
6	ML algorithms to extract data from sensors	Detect cancer through skin temperature analysis.	• L. Béthencourt et al., 2021 • Τηε στατε οφ τηε αρτ οφ διγιταλ τωινσ ιν ηεαλτηχαρε, Toward τηε φιρστ υσε χασεσ, Ρομι Δεχραενε, submitted to Ghent University, 2021–22

TABLE 16.3 (Continued)
ML Algorithms Used in Digital Twins

Sr No	ML Techniques	Objective	Research Studies
7	Ant colony algorithm (ACO)	Examine the clinical utility of employing intelligent algorithms to analyze MRI features in the diagnosis and treatment of breast cancer. This aims to offer an effective reference for the assessment of breast cancer diagnoses.	• Shuang Liu et al., 2021
8	Simulated heat transfer model that uses thermal properties of the tissues present in the breast	A decision tree approach is implemented for the early detection of breast cancer using data gathered from smart devices.	• Safa Meraghni, et al., 2021
9	CNN algorithm	Examine the diagnostic efficacy of multimodal images using AI target detection algorithms in the early detection of breast cancer.	• Meiping Jiang et al., 2022

behave. The accuracy of the digital twin can be increased using real-time data from sensors.

Four traditional ML techniques—decision tree regression (DTR), gradient boosting algorithm (GBA), ML linear regression (MLLR), and random forest regression (RFR)—are the focus of many digital twin breast cancer screening methods. LR is a fundamental ML technique used to model the relationship between a dependent variable and one or more independent variables. On the other hand, a decision tree is a graphical representation that analyzes the response to queries as either true or false with the purpose of categorizing or regressing data. When depicted visually, it takes on the appearance of a tree structure. The blend of simplicity and precision of the RFR is its primary advantage. Operating as an iterative algorithm, the GBA is renowned for its efficacy in solving intricate problems, making it the backbone of medical analysis.

Another method employed is CIP for patient face recognition. These days, ML algorithms to extract data from sensors connected to smart devices, such as headbands, watches, or bracelets, are also popular.

Convolutional neural networks (CNNs) belong to a category of deep learning algorithms. They are frequently used for segmenting, classifying, and recognizing images and videos. The capacity of CNNs to automatically learn complex data representations without the need for manual feature engineering is one of their main features. Additionally, they can work with high-dimensional inputs, such as mammograms and MRI pictures, and they are resistant to changes in input, such as changes in lighting and perspective.

Some models make use of the ant colony optimization (ACO) algorithm, which can quickly find excellent solutions even in large solution spaces. There have also been algorithms developed to extract data that is simulated in the human body by a heat transfer model.

Other methods employed by digital twin models include the Internet of Medical Things (IoMT), blockchains, extended reality, parallel medical imaging, artificial imaging systems and augmented reality.

16.6 AI IN DEVELOPING HEALTHCARE DIGITAL TWINS

The recent advancements in fields such as AI and IoT have largely helped in the development of healthcare digital twins and also formed the primary instigators for research in the future. We can already see the implementation of AI around us to capture data, such as IoT devices and wearable sensors. Each individual patient is distinct and unique and will need a different combination of treatments. There is a need to assess the medical history data of a patient in order to prescribe the best-suited course of care. The combination of AI with digital twins helps enable a system that can make accurate, personalized, and precise decisions.

16.6.1 Developing the Digital Twin

The data collected from various sources needs to be processed. Outliers need to be identified, missing values need to be removed, and filtration needs to be performed.

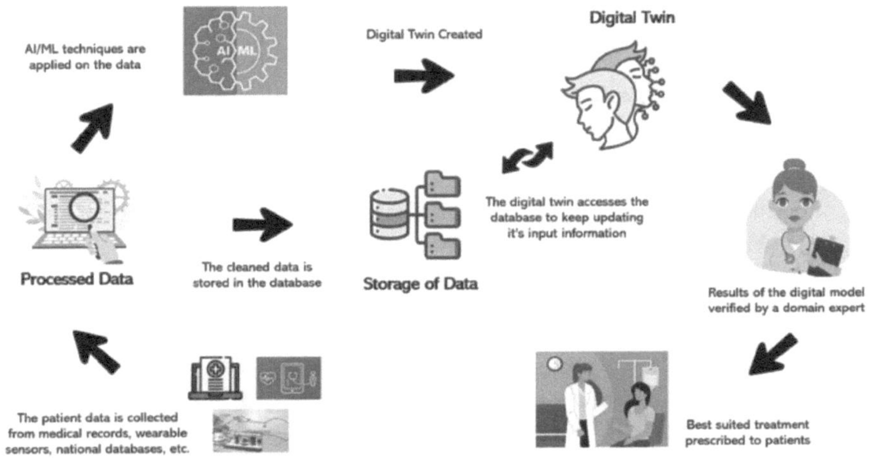

FIGURE 16.3 Digital Model Development.

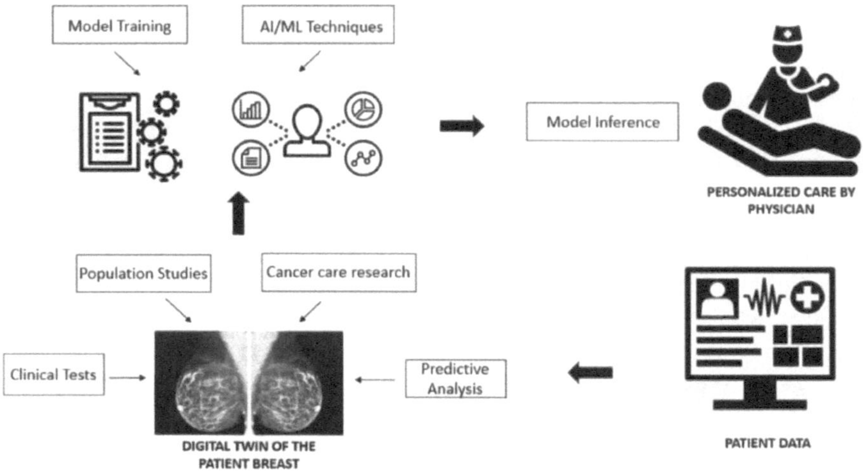

FIGURE 16.4 Implementation of Digital Twins in Breast Cancer Care.

This is the first step in creating the digital model. The next step is to use AI algorithms to define the gathered data and apply ML techniques. The last and final step is to assure the quality of the model, and this is done by experts with domain knowledge. When new data is received from the real world, it is integrated into the existing digital model. As we can see, there is a continuous interplay between the digital and physical realms. It is interesting to note that recent studies have shown that optimization of cancer treatment using digital twins has quite significantly enhanced the rate of survival of the patient. Digital twins thus help health care professionals to prescribe the best treatment for each patient.

16.7 DIGITAL TWIN IMPLEMENTATION

The aim of the model is that an individual's best course of treatment should no longer be determined by their present or past conditions, but rather by a prediction of their future condition. However, creating a real-time digital representation that incorporates the physical, mental, and various attributes of an actual person is a challenging task. Integrating this information into a digital twin and generating accurate summaries pose significant hurdles. Overcoming these challenges is essential for successful implementation. These difficulties include the need for computational capacity, cyber security concerns, data sharing problems, a wide range of ethical restrictions, and obstacles to doctors using AI-based illness management decision tools.

A key priority is focusing research on user-friendly, accessible, and intuitive ways for users who are not familiar with high-level technology to make use of digital twins in their respective industries is another crucial priority. Equally important is ensuring the accurate and timely recreation of data taken from a physical body, such as a human organ, onto its digital counterpart.

Accelerating the resolution of issues inherent in a physical system, enhancing the precision of 3D visualizations, and expediting the provision of data sets essential for constructing digital twins are crucial advancements in physics-based modeling.

To conclude, digital twin research includes research studies that demonstrate the application of strategies that are woven from research data. These exemplify the objectives of precise and customized medicine.

16.8 CHALLENGES

Because of variations in formats, standards, and quality, integrating data from several sources (including genetic data, wearable technology, imaging studies, and electronic health records) can be difficult. Inadequate standardization and interoperability between various healthcare systems, gadgets, and data sources can impede the smooth integration of data required for generating precise and all-encompassing digital twins. There have been many speculations about data ownership. A patient's medical records and data are extremely sensitive, and it is important to monitor access to this information. This calls for difficult data protection rules.

The challenge of funding the digital twin is another issue. Not all hospitals can finance this technology, and purchasing a new system is very expensive. Patients and healthcare experts need to be proficient in using this technology, too. Hence, the user-friendliness of this interface also emerges as another concern.

It is difficult to create precise ML and AI algorithms for digital twins that detect and diagnose breast cancer. For these algorithms to be accepted in clinical practice, they must be dependable, comprehensible, and clinically validated. There are ethical questions about obtaining informed consent for the development and usage of digital twins. To create and update digital representations of their health, patients must be informed about this process and provide their consent.

Despite its foundation in AI, the progress in this field presents practical challenges that extend beyond the context of breast cancer. Specifically, it relies on various human factors and a breadth of multidisciplinary scientific expertise. Collaboration between healthcare professionals, technologists, and patients is necessary to address these issues. We can thus say that personalized medicine using digital twinning technology still remains in its infancy. There is a void to fill, particularly in the context of breast cancer.

16.9 CONCLUSION

In light of the escalating pressure on healthcare experts to derive relevant information from the extensive medical data, within a limited consultation window, it becomes essential to equip experts with effective decision-making solutions. This paper thus argues that employing digital twins for the same reason strikes a harmonious balance between precision and personalization.

Despite the challenges discussed above, it is clear that digital twins have immense potential in the area of cancer. As technology advances and our comprehension of cancer biology grows, we expect that the combined use of digital twins alongside conventional diagnostic methods will result in more precise, individualized, and efficient strategies to combat breast cancer, and cancer in general. This study marks a substantial stride toward enhancing patient results and, ultimately, preserving lives. Additionally, the integration of digital twin technology into routine clinical workflows requires careful consideration of regulatory and ethical frameworks. A detailed analysis of existing strategies will enable researchers to identify their shortcomings and create more powerful substitutes, making humanity less susceptible to the dangers linked with this expanding issue.

BIBLIOGRAPHY

[1] Siegel, R.L., Miller, K., D., & Jemal, A. Cancer Statistics, 2017. *CA Cancer J Clin.* 2017;67:7–30.

[2] Washbrook, E. Risk factors and epidemiology of breast cancer. *Women's Health Med.* 2006;3:8–14.

[3] Brewer, H.R., Jones, M.E., Schoemaker, M.J. et al. Family history and risk of breast cancer: An analysis accounting for family structure. *Breast Cancer Res Treat.* 2017;165:193–200.

[4] Hellquist, B.N., Czene, K., Hjälm, A., Nyström, L., & Jonsson, H. Effectiveness of population-based service screening with mammography for women ages 40 to 49 years with a high or low risk of breast cancer: Socioeconomic status, parity, and age at birth of first child. *Cancer.* 2012;118:1170–1171.

[5] Onega, T., Goldman, L.E., Walker, R.L., Miglioretti, D.L., Buist, D.S., Taplin, S., Geller, B.M., Hill, D.A., & Smith-Bindman, R. Facility mammography volume about breast cancer screening outcomes. *J Med Screen.* 2016;23:31.

[6] Schneble, E.J., Graham, L.J., Shupe, M.P., Flynt, F.L., Banks, K.P., Kirkpatrick, A.D., Nissan, A., Henry, L., Stojadinovic, A., Shumway, N.M., et al. Future directions for the early detection of recurrent breast cancer. *J Cancer.* 2014;5:291–300.

[7] Yen, T.W.F., Li, J., Sparapani, R.A., Laud, P.W., & Nattinger, A.B. The interplay between hospital and surgeon factors and the use of sentinel lymph node biopsy for breast cancer. *Medicine.* 2016;95:e4392.

[8] Abel, E.J., Carrasco, A., Culp, S.H., Matin, S.F., Tamboli, P., Tannir, N.M., & Wood, C.G. Limitations of preoperative biopsy in patients with metastatic renal cell carcinoma: Comparison to surgical pathology in 405 cases. *BJU Int.* 2013;189:1692.

[9] Sporn, M.B. Approaches to prevention of epithelial cancer during the preneoplastic period. *Cancer Res.* 1976;36:2699–2702.

[10] Elizalde, P.V., Cordo Russo, R..I, Chervo, M.F. et al. ErbB-2 nuclear function in breast cancer growth, metastasis and resistance to therapy. *Endocr Relat Cancer.* 2016;23:243–257.

[11] Meraghni, S., Benaggoune, K., Al Masry, Z., Terrissa, L.S., Devalland, C., & Zerhouni, N. (2021). Towards Digital Twins Driven Breast Cancer Detection. In: Arai, K. (eds) *Intelligent Computing. Lecture Notes in Networks and Systems*, vol. 285. Springer, Cham.

[12] Popa, E.O., van Hilten, M., Oosterkamp, E. et al. The use of digital twins in healthcare: socio-ethical benefits and socio-ethical risks. *Life Sci Soc Policy.* 2021;17:6.

[13] Coorey, G., Figtree, G.A., Fletcher, D.F. et al. The health digital twin to tackle cardiovascular disease—A review of an emerging interdisciplinary field. *npj Digit Med.* 2022;5:126.

[14] Geoffrey Chase, J., Zhou, Cong, Knopp, Jennifer L., Shaw, Geoffrey M., Näswall, Katharina, Wong, Jennifer H.K., Malinen, Sanna, Moeller, Knut, Benyo, Balazs, Yeong Shiong Chiew, & Thomas Desaive. Digital twins in critical care: What, when, how, where, why?. *IFAC-PapersOnLine*, 2021;54: 310–315.

[15] Abdollahi, H., & Rahmim, A. Digital Twins for personalized healthcare: Application to radiopharmaceutical therapies. *Front Biomed Technol.* 2022;10(1):1–5.

[16] Björnsson, B., Borrebaeck, C., Elander, N. et al. Digital twins to personalize medicine. *Genome Med.* 2020;12;4.

[17] Ahmed H, Devoto L. The potential of a Digital Twin in surgery. *Surg Innovation.* 2021;28(4):509–510.

[18] Chakshu, N.K., & Nithiarasu, P. An AI based digital-twin for prioritizing pneumonia patient treatment. *Proc Institut Mech Eng Part H J Eng Med.* 2022;236(11):1662–1674.

[19] Chakshu, N.K., Carson, J., Sazonov, I., & Nithiarasu, P. A semi-active human digital twin model for detecting severity of carotid stenoses from head vibration—A coupled computational mechanics and computer vision method. *Int J Numer Meth Biomed Engng.* 2019;35:e3180.

[20] Aluvalu, Rajanikanth, Mudrakola, Swapna, Uma Maheswari, V., Kaladevi, A.C., Sandhya, M.V.S., & Rohith Bhat, C. The novel emergency hospital services for patients using digital twins. *Microprocess Microsyst.* 2023;98: 104794.

[21] Kaul, R., Ossai, C., Forkan, A. R. M., Jayaraman, P. P., Zelcer, J., Vaughan, S., & Wickramasinghe, N. The role of AI for developing digital twins in healthcare: The case of cancer care. *WIREs Data Mining Knowledge Discovery.* 2023;13(1):e1480.

[22] Flora Bahrami, René Michel Rossi & Thijs Defraeye. Predicting transdermal fentanyl delivery using physics-based simulations for tailored therapy based on the age. *Drug Delivery.* 2022;29:1, 950–969.

[23] Guangpeng Chen, Jianxiong Cui, Jindong Qian, Jianbo Zhu, Lirong Zhao, Bangyu Luo, Tianxiang Cui, Liangzhi Zhong, Fan Yang, Guangrong Yang, Xianlan Zhao, Yibing Zhou, & Mingying. Rapid progress in intelligent radiotherapy and future implementation. *Cancer Investigation,* 2022;40:425–436.

[24] Volkov, I., Radchenko, G. & Tchernykh, A. Digital Twins, internet of things and mobile medicine: A review of current platforms to support smart healthcare. *Program Comput Soft.* 2021;47:578–590.

[25] Moztarzadeh, O., Jamshidi, M., Sargolzaei, S., Jamshidi, A., Baghalipour, N., Malekzadeh Moghani, M., & Hauer, L. Metaverse and healthcare: Machine learning-enabled Digital Twins of cancer. *Bioengineering.* 2023;10:455.

[26] Aluvalu, Rajanikanth, Mudrakola, Swapna, Uma Maheswari, V., Kaladevi, A.C., Sandhya, M.V.S., & Rohith Bhat, C. The novel emergency hospital services for patients using digital twins. *Microprocess Microsyst.* 2023;98:104794, ISSN 0141–9331.

[27] Croatti, Angelo, et al. On the integration of agents and digital twins in healthcare. *Journal of Medical Systems,* 2020;44:161.

[28] Meraghni, S., Benaggoune, K., Al Masry, Z., Terrissa, L.S., Devalland, C., & Zerhouni, N. (2021). Towards Digital Twins Driven Breast Cancer Detection. In: Arai, K. (eds) *Intelligent Computing. Lecture Notes in Networks and Systems,* vol. 285. Springer, Cham.

[29] Foy, K.C., Fisher, J.L., Lustberg, M.B., Gray, D.M., DeGraffinreid, C.R., & Paskett, E.D. Disparities in breast cancer tumor characteristics, treatment, time to treatment, and survival probability among African American and white women. *NPJ Breast Cancer.* 2018;4(1):7.

[30] Ivanov, D. Conceptualization of a 7-element digital twin framework in supply chain and operations management. *Int J Prod Res,* 2024;62(6):1–13.

[31] Tao, F., Xiao, B., Qi, Q., Cheng, J., & Ji, P. Digital twin modeling. *J Manufact Syst.* 2022;64:372–389.

[32] World Health Organization. Cancer. 2022. www.who.int/news-room/fact-sheets/detail/cancer

[33] Kaul, Rohit, et al. The role of AI for developing digital twins in healthcare: The case of cancer care. *Wiley Interdiscipl Rev Data Mining Knowledge Discov.* 2023;13(1):e1480.

[34] Moztarzadeh, Omid, et al. Metaverse and healthcare: Machine learning-enabled digital twins of cancer. *Bioengineering.* 2023;10(4):455.

[35] Konopik, Jens, Wolf Larissa, & Schöffski Oliver. Digital twins for breast cancer treatment–An empirical study on stakeholders' perspectives on potentials and challenges. *Health Technol.* 2023;13: 1–8.

[36] Thiong'o, Grace M., & James T. Rutka. Digital twin technology: The future of predicting neurological complications of pediatric cancers and their treatment. *Frontiers in Oncology.* 2022;11: 781499.

17 Digital Twin in Water Utilization

A Case Study to Estimate the Prediction of Water Usage

E.B. Priyanka, S. Thangavel, S.K. Devatharani, A. Jose Anand and Saravanan Krishnan

17.1 INTRODUCTION

The world economy relies significantly on the availability of fresh water, a natural blessing. Having enough water is crucial for activities such as farming, drinking, industry, and recreation. Ironically, the presence of natural or human-induced contaminants can take away this blessing, making our world more challenging. It is widely recognized that having clean water is essential for our health. Unfortunately, as technology and industry grow, the global supply of clean water is at risk. Around one-sixth of the world's population faces challenges due to the shortage of freshwater [1]. Urban areas in developing countries are increasingly encountering water scarcity issues, and the situation could worsen due to rapid changes in the hydro-environment, including climate and land-cover alterations. Traditionally, scholars have commonly approached the urban water supply issue as having two separate elements: a resource challenge and a delivery challenge.

Various solutions, such as desalination, water markets, interbasin transfers and wastewater recycling and reuse, have been proposed to tackle the resource-related aspect of the problem. However, the problem is not due to the scarcity of resources. On the contrary, the issue arises from mismanagement in the delivery process. Water utilities often fail to accurately meter the water usage. A growingly popular alternative, with numerous advantages across different domains, is grounded in the Internet of Things (IoT) concept [2]. IoT has played a pivotal role in diverse sectors, particularly in its integration into smart water management systems to design and analyze digital twins, as shown in Figure 17.1. In residential settings, the implementation of IoT involves installing sensors for real-time data collection, enabling continuous monitoring of data. Traditionally, water demand forecasting relied on historical data and statistical models, often proving limited in capturing dynamic fluctuations and local nuances. Our approach, however, leverages the real-time flow rate data streamed through low power wide area networking (LoRaWAN) sensors. These devices directly

DOI: 10.1201/9781003465904-21

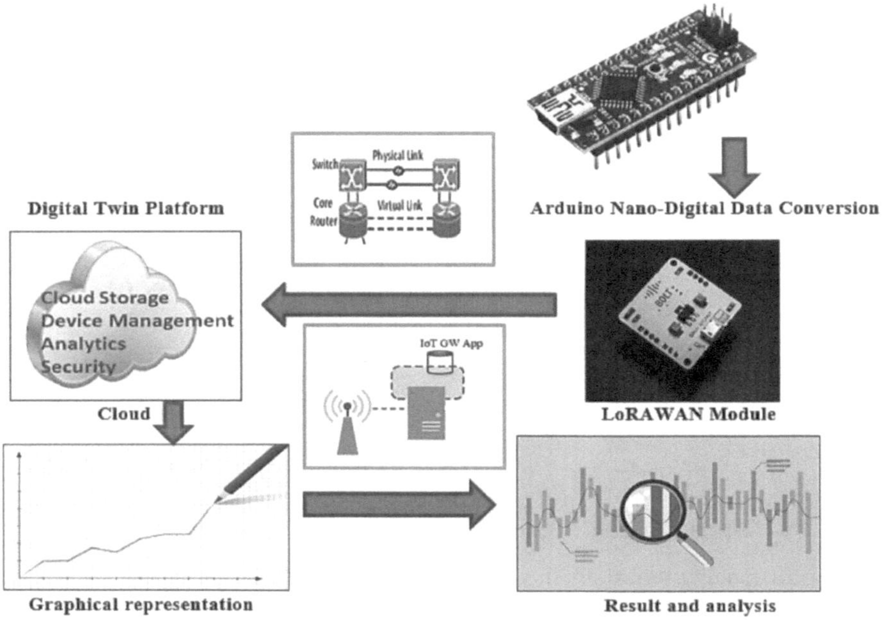

FIGURE 17.1 Digital Twin Architecture with IoT and Cloud Management.

capture water usage at strategic points, providing a granular and dynamic picture of consumption patterns [3].

Machine learning (ML) plays a pivotal role in revolutionizing water demand prediction by leveraging advanced algorithms to analyze complex datasets. ML enhances water demand prediction by unraveling complex relationships, adapting to changing conditions, and recognizing patterns within diverse datasets. Its ability to integrate data from various sources, achieve precision, and continuously improve makes it a valuable tool for optimizing resource allocation and decision-making in water management. This technology enables the identification of patterns and correlations within factors influencing water consumption. By employing techniques such as regression analysis, neural networks, and decision trees, ML models can provide accurate forecasts, aiding water utilities and policymakers in proactive resource planning [4]. The integration of ML enhances the precision of predictions, enabling more effective management of water resources and the development of sustainable strategies to meet evolving demands. Water utilization rates, a key metric in assessing the demand and distribution of water resources, are influenced by a myriad of factors. Among these, flow rate, representing the volume of water moving through a system over time, holds particular significance. Understanding the intricate relationship between flow rate and overall water utilization is crucial for effective decision-making in water resource planning. The development of a digital twin platform is known for its ability to handle complex datasets and uncover nonlinear relationships, becomes a valuable tool in this context [5]. By employing historical data that includes flow rate as a key input parameter, the digital twin with ML algorithms can construct a real-time decision model that unveils patterns and dependencies contributing to variations in water utilization rates.

17.1.1 Existing Methodologies on the Analysis of Water Utilization Rate

The pioneering research of the water demand analysis suggests that the most crucial characteristics for identifying leaks in pipelines are the pressure and velocity fields. Water demand analysis can be done by several methods, including software methods and hardware methods. Zubaidi et al. [6] introduce a new method for predicting monthly municipal water demand, accounting for different weather scenarios. It combines noise-removing "wavelet" technology with feature selection and model optimization, proving more accurate than traditional methods. Tested in Melbourne, Australia, it predicted demand within 5.5% on average, offering water utilities a powerful tool to reduce waste and optimize resource allocation [6]. Chen et al. [7] tackle this challenge with a clever "transfer learning" approach. Imagine learning from one city's water usage patterns and applying that knowledge to predict demand in another city, by using "multi-head attention" technology to focus on the most relevant factors, such as weather and city features. Tested on real data, this method reduced prediction errors by 50%, offering water utilities a powerful tool to anticipate demand, reduce leaks, and ensure sufficient water for everyone [7]. Emami et al. [8] investigated effectively to predict county-level irrigation water demand in California, with X-Boost achieving the highest accuracy. Climate data, crop types, and even socioeconomic factors influence demand, highlighting the need for comprehensive models. This approach can optimize water resource management in agriculture, promoting sustainability and efficiency [8]. Papacharalampous and Langousis [9] introduced a new method for forecasting water demand with probabilities, not just averages. Using clever "quantile regression" algorithms can tell you not just how much water a city might need but also the range of possibilities, such as a weather forecast. This probabilistic approach helps water managers to prepare for varying demand scenarios, from mild days to heatwaves, ensuring a reliable supply for everyone [9]. Hao et al. [10] investigated to predict water demand accurately for both the short and long term. This method compared different models, including neural networks and statistical methods. By incorporating weather data and historical water usage patterns, the models could predict daily water demand within 1% and weekly demand within 5%. This information helps Milan to optimize water distribution and prevent shortages, ensuring a reliable supply for its residents [10].

Bhushan[11] explored the power of long short-term memories (LSTMs) in water demand forecasting and analysis. By analyzing water consumption data alongside factors such as weather, time of day, and holidays, LSTMs can learn hidden relationships and make accurate predictions [11]. Xu et al.[12] analyzed the new ground by using a team of ML models (similar to Sherlock Holmes and his crew) to predict complex usage patterns. They combine weather data, holidays, and even population trends, each model focusing on different clues [12]. Nemčić-Jurec et al. [13] examined the five chemical parameters over ten years; they found nitrate levels increasing while pH and organic matter levels decreased. Rainfall impacts water quality, but its effects are complex and depend on local factors such as topography [13]. The pre-statistics emphasizes the need for multivariate statistical techniques and ML to understand long-term trends and predict future contamination risks. The impact of the water utilization rate of the current water pipeline system was mainly examined through the use of commercial simulation software. Hence, the proposed chapter outlines the application of digital twin technology in future to predict the water demands using modern CART algorithm on real-time datasets to enhance the distribution rate of municipal sectors.

17.2 SMART LORAWAN COMMUNICATION IN REAL-TIME ACQUISITION OF WATER PIPELINE DATA

Pipelines stand out as highly practical means for transporting water over extended distances. Over the past few decades, the global pipeline system has experienced rapid growth due to its safety, cost-effectiveness, and efficient transport capabilities. In addition to their use in water supply, pipelines find widespread application in industries and power production centers. The primary risk associated with this method of water transportation is the potential for pipeline bursts, often attributed to pressure surges and corrosion. Pipeline leaks are a common occurrence in any transporting system and can result from factors such as abrupt pressure changes, cracks, corrosive processes, pipe defects, poor workmanship, external damage, and insufficient maintenance [14]. These leakages not only cause wastage of water but also contamination. This contamination was traced back to the use of damaged subterranean pipelines following a cyclone.

The previous research investigated the fluid flow characteristics to understand how the geometry of leaks in a pipe wall affects velocity and pressure distributions within the pipe. The findings suggest that in the proximity of leaks, both velocity and pressure magnitudes exhibit rapid variations, with the shape of the cracks exerting a maximum impact. Pressurized water supply lines commonly encounter partial blockages over their operational lifespan. These blockages typically initiate with a minor increase in wall roughness, which progressively grows due to physical or chemical processes, potentially obstructing a significant portion of the pipe's cross-sectional area. Such blockages lead to energy wastage, a decrease in pipe-carrying capacity, and an elevated risk of contamination [15]. Furthermore, heavily restricted flows from blockages can cause flow imbalances in the pipe network, potentially resulting in pipe overpressure and the emergence of leaks. Detecting blockages promptly is crucial to address these issues effectively.

An IoT solution can be designed for real-time monitoring of water quality. The implemented system integrates Arduino Mega and LoRAWAN module target boards with various sensors, achieving successful interfacing. A real-time algorithm has been developed to monitor water quality effectively. To track parameters such as pH value, water turbidity, tank water level, and temperature and humidity of the surrounding environment, a web-based application, specifically an online operator interface through Thing Speak open-source software data is utilized. This enables for continuous monitoring through a web server, as shown in Figure 17.2.

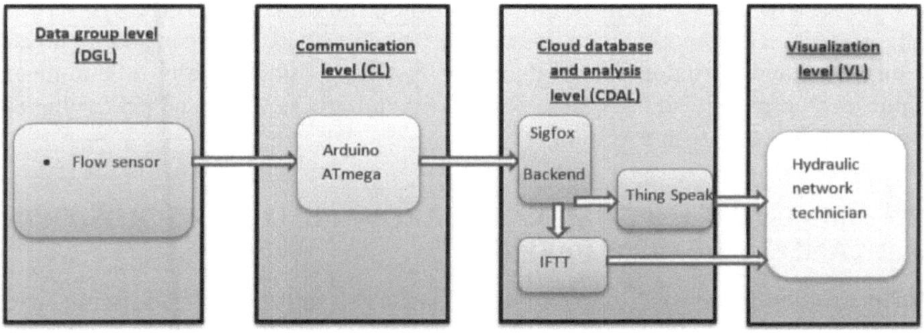

FIGURE 17.2 Digital Twin Based Data Collection and Mining.

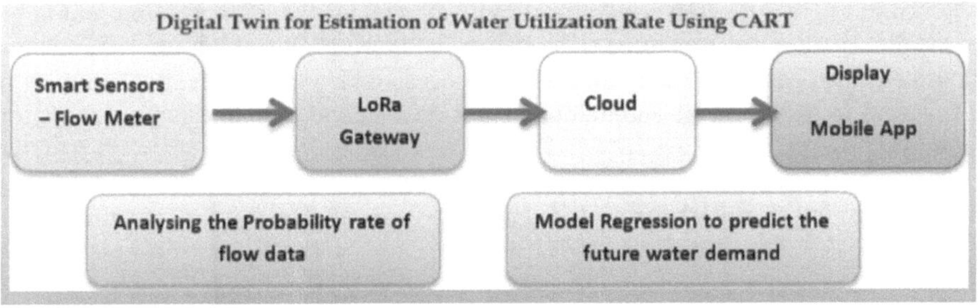

FIGURE 17.3 Proposed Framework of Smart Automation of Water Pipeline System.

An experimental exploration and prototype development of a water pipeline leakage monitoring system is shown in Figure 17.3. The system is built around a cost-effective microcontroller, Arduino, and incorporates modern smart sensors supporting LoRa-based spectrum bands. The primary function of the system is to monitor the flow rate of water before it is pumped into the household water tank. In the event of any distribution mismatch rate, as already predicted by the digital twin model, the user receives notifications through either SMS or Telegram notifications. The variation of flow rate and demand rate is facilitated by water flow with parallel computing by the cloud server.

This chapter enumerates the incorporation of intelligent automation such as digital twin into public water distribution systems, utilizing IoT and ML to improve efficiency and tackle crucial issues. In the initial phase, IoT devices are strategically placed throughout the water distribution network to detect and monitor key parameters such as pressure, flow rate, and temperature. These IoT sensors continuously collect real-time data on these parameters, transmitting it to a centralized system. This abundant data forms the basis for implementing ML algorithms, which are integral to the decision-making process in the water distribution system [16]. These algorithms can analyze water consumption patterns over time, providing insights into water utilization rates. Such information is crucial for optimizing the distribution network, ensuring efficient water resource utilization while minimizing wastage. By scrutinizing deviations from normal patterns in flow rate data, the system can autonomously identify demand variations. The algorithm can then initiate alerts or automated responses to promptly address these issues, preventing water loss and preserving the integrity of the distribution system.

17.3 REAL-TIME EXPERIMENTAL HARDWARE DATA ACQUISITION AND DATA MONITORING

The main station consists of two pipes that are connected in succession, each including a gate valve (GV 1), an I/P converter, a pressure gauge (PG), centrifugal pumps, and a pressure transmitter (PT). The pump, flow transmitter (FT) with orifice flow meter unit, I/P converter, and gate valves (GV 2) are located in a separate area. Pressure and flow control valves, which are primarily connected and operated by the I/P (current to pressure converter) signals activated by the local control unit remotely, are the primary control components, as shown in Figure 17.4. A three-way drain valve integrated into a manual hand valve at the process tank's output terminal enables it to function in the event of anomalous conditions such as pipeline leaks, bursts, or fractures.

The primary station of the fluid transportation system is thought to be a scaled-down experimental lab configuration that includes boosting and final delivery substations.

FIGURE 17.4 Lab-Scale Setup Utilized for Fetching Real-Time Sensor Data for Modeling.

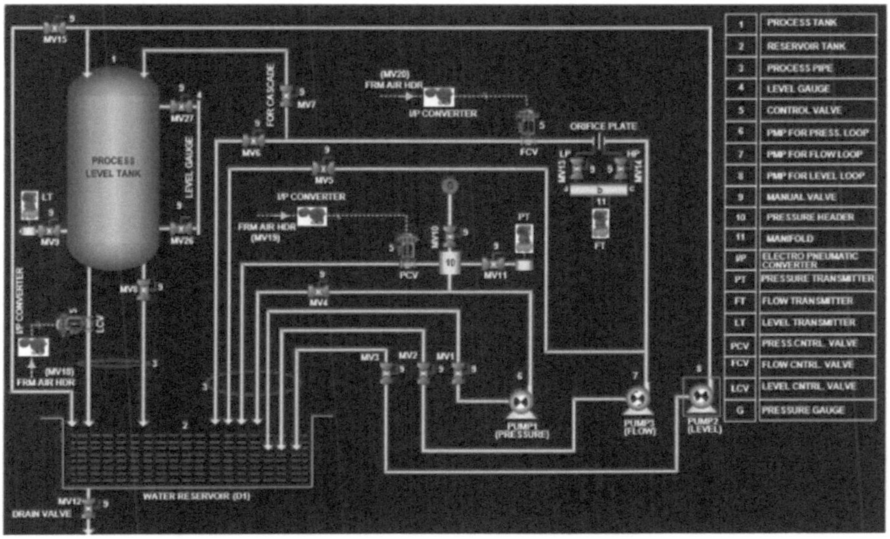

FIGURE 17.5 SCADA Interface for Virtual Visualization of Regulating Flow Rate in the Water Pipeline.

A supervisory control and data acquisition (SCADA) shows the run and stops operation status in addition to the on and off-pump status. The reservoir tank fills to 20% of its capacity when the experimental setup is ready to operate, at which point the electric pump is turned on to suction the fluid out of the tank and distribute it to each area. Through the Centum/IP net protocol, the relevant pressure and FT sends data about the current pressure and flow rate to the I/O hub module station at the field control station as the fluid begins to pass through the pipes, as displayed in Figure 17.5. The analog current output signal of 04–20 mA is concurrently sent to the main control unit by the flow and pressure transmitters. The proportional-integral-derivative (PID) controller on the Centum VP platform remotely controls the corresponding I/P converter of the pressure and flow control valve based on these two transmitter measurements during transportation. The pressure and flow I/P converter receives an analog current control signal with a range of 04–20 mA, which is fed in and converted to a calibrated pressure signal with a range of 0–3 psi. The control valve adjusts its opening and shutting of control valves till the water reaches the process tank by receiving a matching pressure signal.

High-level engineers can access control blocks, such as (PIC) and (FIC), which are positioned in the main control panel and monitor the fluid driven from the reservoir tank to the process tank. The engineers are unable to operate the (PCC) and (FCC) that make up the controlling section. These components are shown on the local control panel. The Centum VP platform's controller remotely adjusts these FCC and PCC to process control signals sent to the appropriate I/P convertor of pressure and flow control valves; therefore, regulating the flow and pressure of the fluid being delivered is modeled by representing in Figure 17.6.

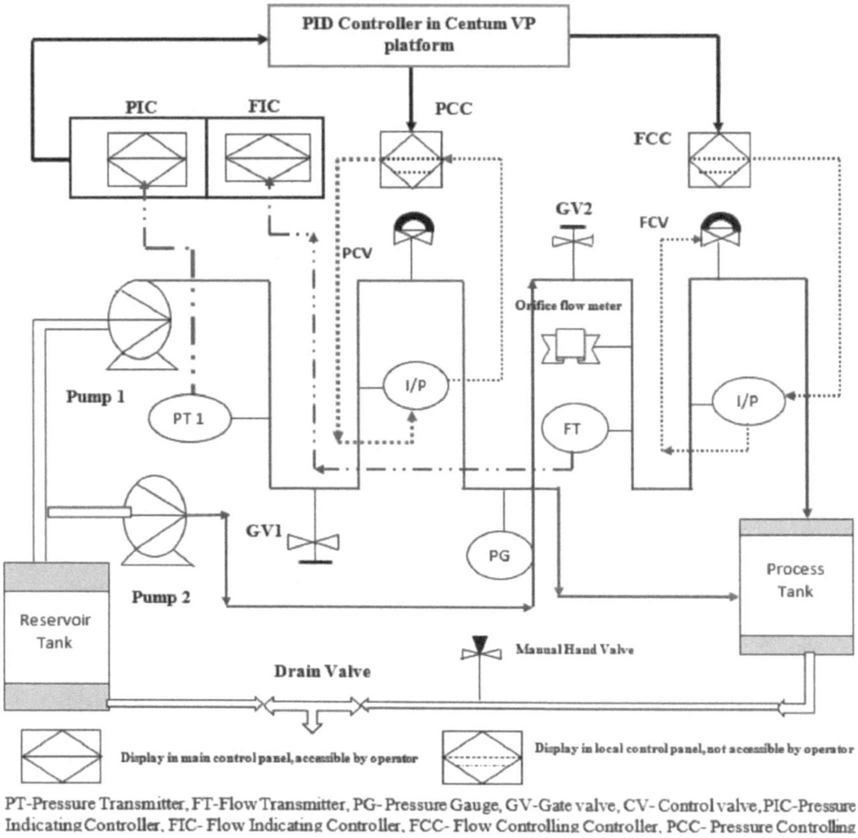

FIGURE 17.6 Digital Twin Virtual Modeling of Sensor Position to Monitor Flow Rate to Examine the Water Demand.

When making manual conclusions, the PG places the manual hand valve and drain valve in the proper locations for safe transportation and shows the pressure as bar scale readings. In the event that any risky events, such as leaks or cracks/bursts in the water pipelines during transit, occur, the emergency push button on the control panel is utilized to turn off the whole system. Using virtual case analysis, if the water demand rate has to be increased or decreased, then it can be attained by modeling in this digital twin platform by adjusting the parameters such as pressure or velocity of the water to identify the exact actuation in real-time by skipping real-time experimentation trial and error rate.

17.4 IMPLEMENTATION OF CART ALGORITHM ON WATER UTILIZATION RATE PREDICTION

In order to compute classification rules, the majority of statistical learning approaches, including CART, assume independent samples. When estimating algorithmic quantities and evaluating the asymptotic characteristics of estimators, this assumption proves to be quite useful. The data being studied in many ecological or environmental applications are samples of certain regionalized variables that may be modeled as spatially dependent random fields. Direct application of supervised classification algorithms results in biased discriminant rules when the sampling technique is highly irregular. This can be because of the potential oversampling of some areas. With an emphasis on water utilization rate analysis applications, the CART method is modified for the situation of spatially dependent samples [17].

Data is treated as independent samples of random variables $(S^1,\ldots,S^P,K)$, where S^m represents predictor variables, and K is the qualitative variable under consideration. CART utilizes a method based on rules, where it builds a binary tree by employing recursive partitioning. This process involves splitting a subset (referred to as a leaf) of the dataset into two subsets, forming sub-leaves within the tree structure. The criterion for this split is determined by minimizing a heterogeneity measure calculated based on the sub-leaves that emerge as a result. Every division is determined by a single variable, and certain variables might be employed multiple times, whereas others may not be utilized at all. Subsequently, each sub-leaf undergoes additional divisions guided by independent rules [18]. Let's contemplate a decision tree, denoted as T, with one of its leaves identified as l. In mathematical terms, T is a mapping that associates a leaf l with each sample $(S_i^1,\ldots.S_i^P)$, where i will be the sample index. Let p(a | l) represent the proportion of a specific class j within a given leaf l.

The entropy index is

$$E_l = \sum_a \mathrm{p}(\mathrm{a} \mid \mathrm{l}) \log\{P(a \mid l)\} \tag{17.1}$$

The expression x log x = 0 holds true when x = 0. The Gini index is

$$D_l = \sum_{i \neq a} p(i|l)\, p(a \mid l) = 1 - \sum_i p(i \mid l)^2 \tag{17.2}$$

The indices specified are minimized when a leaf l contains only one class, and they reach their peak when all classes are represented in equal proportions within the leaf. In the CART method, the concept involves dividing a leaf l into two sub-leaves (l-) and (l+) using a variable criterion. This split is done in a way that maximizes the favorable contrast in the heterogeneity of the original leaf and the overall heterogeneity that arises within the two resulting sub-leaves. The process continues until further splitting is not feasible. Each leaf is then assigned to the most common class within it. However, the final tree might overly fit the available data, resulting in a typically large prediction error is represented as

$$R(T) = P\{T(S^1,\ldots,S^P) \neq K\} \tag{17.3}$$

To address this, when creating a classification tree, the primary aim is to create a tree (T) from the data with the smallest possible probability of prediction error (R(T)). In a later phase, the tree T undergoes "pruning" to produce a subtree T' with an anticipated risk lower than R(T). Since the distributions of K and $\left(S^1,\ldots,S^p\right)$ are not known, pruning relies on the empirical risk ($\hat{R}$(T)) calculated through cross-verification. The primary objective of the CART pruning algorithm is to achieve a balance between optimistic estimates of empirical risk. This equilibrium is attained by incorporating a complexity term, which serves to penalize larger subtrees [19]. When handling samples that are independent of each other, we estimate class proportions, denoted as p(a I l), using the formula

$$\hat{p}(a \mid l) = \frac{m_{al}}{m_l} \tag{17.4}$$

Here, m_{al} represents the number of samples in leaf l belonging to class a and m_l is the total number of samples in leaf l. The objective to reduce is expressed as

$$\frac{D_l - \left(m_{l-}D_{l-} + m_{l+}D_{l+}\right)}{m_l} \tag{17.5}$$

for the Gini index, and likewise for the entropy index. The empirical risk, denoted as $\hat{R}$(T), is calculated as 1/n times the sum over all samples (indexed by ∝) of the indicator function $I\{T\left(S_\propto^1,\ldots,S_\propto^p\right) \neq K_\propto$, where n represents the overall number of samples.

$$\hat{R}(T) = \frac{1}{n}\sum_{\propto=1}^{n} I\{T\left(S_\propto^1,\ldots,S_\propto^p\right) \neq K_\propto \tag{17.6}$$

In environmental or ecological applications, where the assumption of independence is untenable, the data is often a sample of some regionalized variable. This prompts a shift in perspective, considering that the samples $\left\{S^1(s_\propto),\ldots,S^p\left(s_\propto\right), K\left(s_\propto\right)\right\}$where ∝ ranges from one to n, originate from random fields $\left\{S^1(.),\ldots,S^p(.),K(.)\right\}$on a domain D in R^2. It becomes essential to explicitly account for the dependence structure among these fields. The initial strategy involves adjusting the weights assigned to individual samples. The aim is to assign lower weights to closely clustered or adjacent data points compared to those that are more spread out. This approach is based on the recognition that in scenarios where there exists a dependence structure within random fields, closely located data points are likely to exhibit similar values. Consequently, these nearby data points may contain somewhat repetitive or superfluous details regarding the correlation between the response variable Y and the set of explanatory variables $(S^1,\ldots,S^P)$. By assigning reduced weights to closely situated data, the objective is to mitigate the potential overemphasis on information due to proximity, thereby promoting a more balanced representation of the inherent dependence structure in the spatial data.

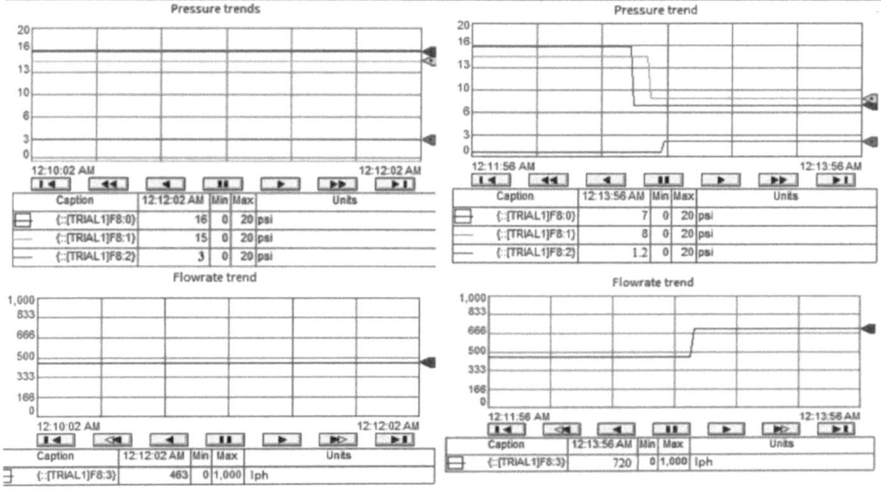

FIGURE 17.7 Regulation of Flowrate at Desired Range by Modeling of Sensor Data Using Digital Twin Technology.

From Figure 17.7, it is inferred that the data model generated by the digital twin platform making an efficient effort to optimize the flow rate data in order to observe the flow rate of the undertaken system with minimum error tolerance with fine-tuning of the cost function of the developed data model by the CART algorithm. Hence, smart decision-making can be obtained once the model developed by the digital twin platform can be implied in real time will give better efficiency as compared with other modeling and virtualization analysis concepts in the current industrial water transportation necessities [20]. Even multiple data-based modeling can be opted for based on the need for the prediction of the water utilization rate of the desired locations.

17.4.1 Real-Time Prediction of Water Utilization Rate Using CART Algorithm

The data warehouse of flow rate has been created using LoRa technology to continuously fetch real-time data from the designed experimental setup. The regulation of flow rate by the data modeling is validated through the SCADA screen, and the desired control mechanism is actuated based on the virtual analysis. Once the experimental analysis is over, the model is further developed to automatically predict the water utilization rate based on the current trend utilization. The prediction will help to make the desired water resource allocation in the upcoming time based on the real-time model prediction.

The need for water is produced by dynamic processes that are always changing and are based on interactions at several scales between natural systems and human actors. As a result of this realization, the creation and application of dynamic models have increased, as seen in Figure 17.8. The majority of demand functions are

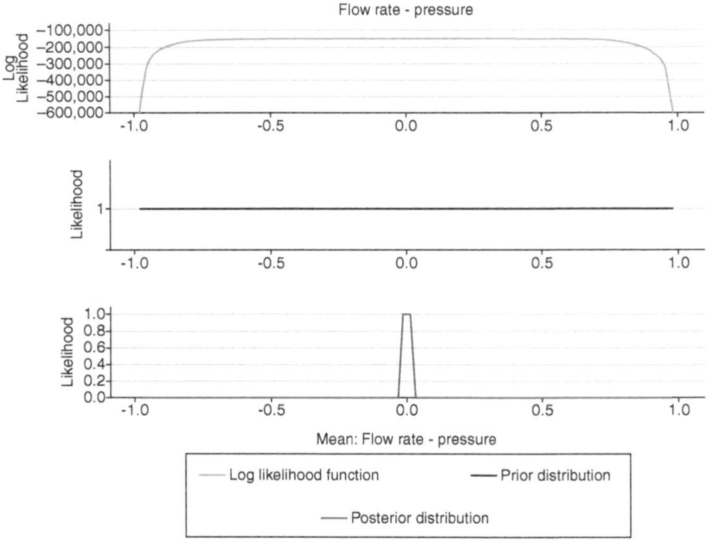

FIGURE 17.8 Automatic Estimation of Water Utilization Rate Using Likelihood Function.

built to be static, yet studies have indicated that historical water consumption has a significant impact on present water use. Therefore, compared to the conventional approaches, the accuracy and reliability of traditional estimations of parameters may be improved by developing a dynamic model that takes into account this link between past and present use. Initially, time series models and econometric models based on linear multivariate regression were employed to analyze urban water demand. These methods could be carried out with a small amount of processing power and only required a small number of datasets. These early approaches had a limited scope, concentrating on the creation of precise forecasting techniques to maximize water supply infrastructure and lower expenses and risks assumed by water providers. Furthermore, the initial water demand models ignored differences in water usage within the geographic focal region since they were essentially spatial in nature. The inability of existing software to handle and retain vast volumes of geographical data resulted in a constraint on modeling and analysis. Progress has been made in identifying and quantifying relationships among various social, climate, and water consumption variables as a result of increased data richness. However, methods that can integrate these diverse types of data and utilize this rich information to clarify relationships at multiple scales still need to be developed [21]. The CART algorithm is one technique that is moving in this direction, but there is still an opportunity to enhance the variable interactions and the underlying system structure's transparency. In order to make water utility managers' jobs easier, it is necessary to have some water utilization rate forecasting techniques that are not overly complex but can include spatially explicit data that is already readily accessible in water demand modeling.

17.5 DIGITAL TWIN BASED PREDICTION OF MAINTENANCE OF EXECUTION ON WATER PIPELINE SYSTEM

One of the crucial technological tools for ensuring and enhancing the normal water pipeline distribution system is predictive maintenance, which is built to facilitate the exchange of information and resources. The digital twin is incorporated into the system as a crucial supporting technology to enhance the precision of fault diagnosis and prediction and facilitate the creation of a maintenance plan with greater accuracy and dependability [22]. Within this framework, an integrated mathematical programming model is established, taking parameter uncertainty into account, and a boosted linear regression algorithm is employed to solve it. Additionally, an adjustment strategy for a maintenance plan is discussed in response to the dynamic features of the actual maintenance environment. Historically, modeling techniques such as manual, deterministic, stochastic, and artificial intelligence have been used to assess the performance of operational infrastructure. The manual assessment of infrastructure performance and conditions takes a lot of time and requires subject matter specialists [23]. Concerns have been raised over the accuracy of the projections for future scenarios by the deterministic, stochastic, and artificial intelligence-based methodologies, which mostly depend on previous data to anticipate infrastructure performance. In addition, the performance and conditions of the infrastructure are adversely affected by the ever-increasing consumption brought on by urbanization and changing weather patterns. The uncertainty in the estimations is further increased by the fact that the effect of these factors is not taken into account when utilizing the historical data [24]. Hence, in the undertaken experimental setup, the following real-time pressure data are measured at two faulty conditions, such as blockage and leakage and compared with the normal performing pipeline for modeling of the water distribution system given in Table 17.1.

TABLE 17.1
Real-Time Fault Data for Mathematical Predictive Model Analysis

Sl. No	Percentage of control valve opening (%)	Water Distribution System-Pressure 1, 2, and 3 readings								
		Normal operating condition			Crack scenario			Blockage scenario		
		P1	P2	P3	P1	P2	P3	P1	P2	P3
1	10	17.9	15.9	2.9	0.41	5.12	0.39	26.01	17.12	9.01
2	20	16.8	14.8	2.6	1.32	3.56	0.62	18.87	18.87	6.38
3	30	14.6	13.9	1.95	9.21	3.27	0.71	32.65	16.88	4.81
4	40	14.2	11.7	0.92	5.67	0.91	0.19	20.34	17.49	3.99
5	50	11.9	11.2	0.91	6.79	4.85	0.08	24.15	19.21	8.02
6	60	10.9	9.65	0.9	5.02	3.41	0.07	31.27	27.29	6.54
7	70	9.89	7.9	0.91	8.99	1.24	0.81	27.36	22.52	5.99
8	80	9.90	7.95	0.91	3.29	1.02	0.21	31.65	18.74	6.89
9	90	7.89	5.92	0.89	5.96	1.32	0.63	24.21	22.44	9.93
10	100	6.99	5.84	0.89	1.18	3.57	0.81	28.64	23.08	9.53

In order to evaluate the best attribute for the maintenance execution, the risk rates of the pressure parameters have to be verified by modeling the real-time data as given in Equation 17.6.

$$P_m = \sum_{i=1}^{N}\sum_{q=1}^{m} p_{iq}^{mk} \left\| a_i - c_q \right\|^2 + Q_{iq} \tag{17.6}$$

The partition array P indicates which clusters the values in the data range belong to, and k indicates how faulty the system is (usually N = 2) [25]. The following conditions must be fulfilled by each element of the P array:

$$P = \left\{ \sum_{q=1}^{m} p_{iq}^{m} \in [0,2] \mid \sum_{q=1}^{m} p_{iq} = 1, \forall\, y \text{ and } 0 < \sum_{i=1}^{N} p_{iq} < N, \forall c' \right\} \tag{17.7}$$

The periodic behaviors of the infrastructure, such as state transition and performance, are represented by the behavioral model. Behavior simulation, which includes a range of models, such as state models and dynamics models, is a difficult procedure. Finite-state machines can be used to generate these models. State modeling includes both state diagrams and activity diagrams. The former defines the dynamic behaviors of infrastructure throughout the course of its existence, whereas the latter outlines the steps required to carry out an activity [26]. Dynamics modeling covers both fluid and rigid body motion. In behavioral modeling, algorithms and data optimization technologies are used extensively. As a consequence, the distance for multidimensional feature space may be determined using the boosted linear regression (BLR) equation below:

$$Q_{iq}^2 = 1 - N(c_m, q_n) = 1 - e^{\left(\frac{\left\| c_m - q_n \right\|^2}{\mu}\right)} \tag{17.8}$$

The BLR follows the center moving average mechanism with a smoothing technique to pre-process the data for prediction with other cost functions μ with the feature of forecast its most possible nearer slope on the best-fit line. Due to its boosted level, the best-fit error values will be reduced by probability comparison with the current error to get a better prediction of the risk data [27]. The obtained risk scatter plot for the pressure data is presented in Figure 17.9.

Based on the inference from Figure 17.9., the model developed starts to analyze the risk and normal functioning pressure data to make decisions on the maintenance of the water pipeline infrastructure. In a digital twin, the service layer serves as the kernel. In accordance with the intended data structure, this layer aims to combine all accessible data sources. The real-time data processing and analysis features in the service architecture would update the physical infrastructure assets' current status. The service architecture provides application assistance to its clients for forecasting, monitoring, and simulation, all of which help with infrastructure management

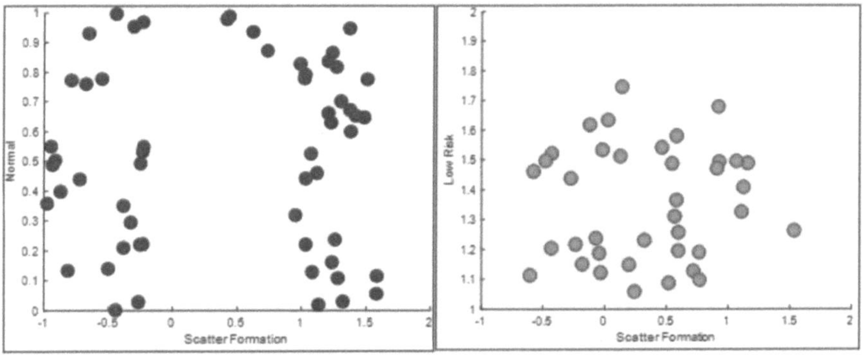

FIGURE 17.9 Analysis of Rik Data to Predict the Maintenance Action on the Pipeline Using Pressure Data.

decision-making. A range of external services, including data services, knowledge services, and algorithms, are needed to build a functional service layer. Additionally, the digital twin's service component has to be consistently supported by a range of platform services that might make it possible to develop bespoke software and provide services. Predicting the necessary inspection or control action at a particular time based on real-time historical data is crucial, in addition to monitoring and controlling anomalous pressure generated by cracks and leaks in the water pipeline system. Since pressure data from the past is kept in the cloud, data analysis may be done using the IoT front-end operator interface.

The rate of fault occurrences in the future is anticipated by the use of BLR time series forecasting, as given in Figure 17.10. According to the results of the regression statistical analysis, multiple R indicates the strongest connection at 0.939, and R square values of 0.901 further indicate that the future estimated model matches the forecast the best. Higher F values with a significance range less than alpha (3.26281E-08 < 0.05, respectively) in the analysis of variance (ANOVA) result guarantee the statistical significance of the projected coefficient. As a result, an inspection operator can solve issues with oil pipeline deformation by making proactive decisions. As a result, the pipeline's lifespan may be extended; maintenance time and expenses can also be reduced by performing pigging at the suitable station at the desired time rather than on a regular basis.

As a result, an operator engineer may determine and respond as to whether the pipeline system needs cleaning to improve production performance in the future or requires emergency pipeline replacement. From this anticipated time series, one may also calculate the risk occurrence rate resulting from cracks and blockages in the water pipeline system. It is possible to calculate the detrimental effect of pipeline tolerance status on future performance using this forecast. As a result, by anticipating potential problems before they arise, this forecasting enables ideal proactive decision-making. It also lengthens the pipeline's lifespan by timely oversight and analysis of risk occurrences.

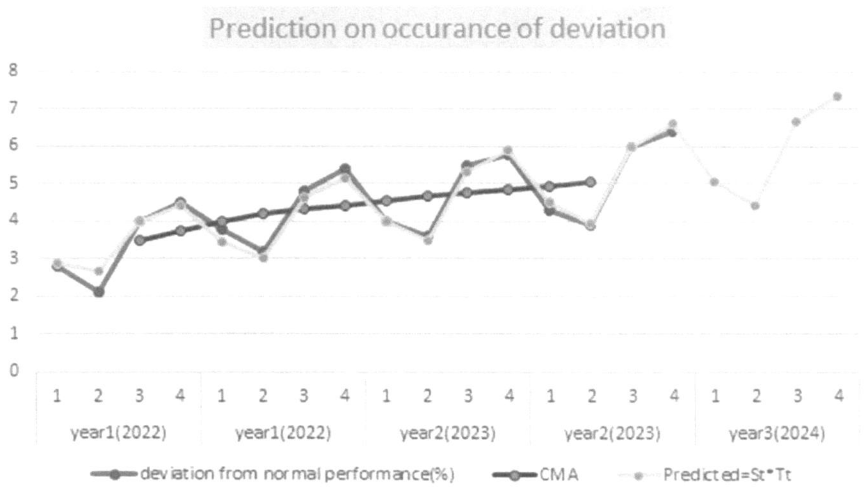

FIGURE 17.10 Prediction of Deviation in Pressure Data to Decide Maintenance in Water Pipeline Infrastructure.

17.6 CONCLUSION

The many benefits of the digital twin primarily serve as a counterbalance to the criticisms leveled against the traditional narrative literature review. Most significantly, it is extensive in nature; that is, the meta-analyst attempts to incorporate all empirical investigations that offer adequate data for statistical analysis when gathering research. Conversely, a standard literature review's sample of papers is often chosen at random by the author, which carries a significant bias of leaving out certain essential studies and increases the possibility of subjectivity. The increase in statistical efficiency that comes from combining primary studies is a second significant benefit. In reality, by merging inconclusive data, the meta-analysis technique may produce significant conclusions. The use of disparate specifications, data, and methodologies makes it difficult to compare and draw clear conclusions. To address these challenges, meta-analysis is based on a statistical process known as meta-regression, which makes it easier to synthesize empirical studies and explains why estimates of outcomes vary. The digital twin's simulation capability aids in obtaining more precise maintenance needs and even supports the process simulation of a maintenance plan. It is possible to forecast additional process data to help with more precise maintenance decision-making. Based on this framework, associated application services for predictive maintenance decisions may be designed and deployed in a cloud platform, relieving the technical and financial pressures on municipal water distribution systems at the engineering application level.

REFERENCES

1. Pelorosso, Raffaele. "Modeling and urban planning: A systematic review of performance-based approaches." *Sustainable Cities and Society* 52 (2020): 101867.
2. Zubaidi, Salah L., Sandra Ortega-Martorell, Hussein Al-Bugharbee, Ivan Olier, Khalid S. Hashim, Sadik Kamel Gharghan, Patryk Kot, and Rafid Al-Khaddar. "Urban water demand prediction for a city that suffers from climate change and population growth: Gauteng province case study." *Water* 12, no. 7 (2020): 1885.
3. Nieuwenhuis, Eva, Eefje Cuppen, Jeroen Langeveld, and Hans de Bruijn. "Towards the integrated management of urban water systems: Conceptualizing integration and its uncertainties." *Journal of Cleaner Production* 280 (2021): 124977.
4. Kakwani, Nikita S., and Pradip P. Kalbar. "Review of circular economy in urban water sector: Challenges and opportunities in India." *Journal of Environmental Management* 271 (2020): 111010.
5. Ma, Yongchi, Yong Jiang, and Stephen Swallow. "China's sponge city development for urban water resilience and sustainability: A policy discussion." *Science of the Total Environment* 729 (2020): 139078.
6. Zubaidi, S. L., K. Hashim, S. Ethaib, N. S. S. Al-Bdairi, H. Al-Bugharbee, and S. K. Gharghan. "A novel methodology to predict monthly municipal water demand based on weather variables scenario." *Journal of King Saud University-Engineering Sciences* 34, no. 3 (2022): 163–169.
7. Chen, Z., C. Deng, F. Che, and Y. Li. (2022, December). "Predicting the Urban Water Demand Based on Transfer Learning Method with Multi-Head Attention." In *GLOBECOM 2022–2022 IEEE Global Communications Conference* (pp. 3760–3765). Rio De Janeiro, Brazil: IEEE.
8. Emami, M., A. Ahmadi, A. Daccache, S. Nazif, S. F. Mousavi, and H. Karami. "County-level irrigation water demand estimation using machine learning: Case study of California." *Water* 14, no. 12 (2022): 1937.
9. Papacharalampous, G., and A. Langousis. "Probabilistic water demand forecasting using quantile regression algorithms." *Water Resources Research* 58, no. 6 (2022): e2021WR030216.
10. Hao, W., A. Cominola, and A. Castelletti. "Comparing predictive machine learning models for short-and long-term urban water demand forecasting in Milan, Italy." *IFAC-PapersOnLine* 55, no. 33 (2022): 92–98.
11. Bhushan, S. (2022, September). "The Use of LSTM Models for Water Demand Forecasting and Analysis." In *Proceedings of 3rd International Conference on Machine Learning, Advances in Computing, Renewable Energy and Communication: MARC 2021* (pp. 247–256). Singapore: Springer Nature Singapore.
12. Xu, Z., Z. Lv, J. Li, and A. Shi. "A novel approach for predicting water demand with complex patterns based on ensemble learning." *Water Resources Management* 36, no. 11 (2022): 4293–4312.
13. Nemčić-Jurec, J., D. Ruk, V. Oreščanin, I. Kovač, M. Ujević Bošnjak, and A. S. Kinsela. "Groundwater contamination in public water supply wells: Risk assessment, evaluation of trends and impact of rainfall on groundwater quality." *Applied Water Science* 12, no. 7 (2022): 172.
14. Zhao, Lihong. "Prediction model of ecological environmental water demand based on big data analysis." *Environmental Technology & Innovation* 21 (2021): 101196.

15. Vásquez-Lavín, Felipe, Leonardo Vargas, José I. Hernández, and Roberto D. Ponce Oliva. "Water demand in the Chilean manufacturing industry: Analysis of the economic value of water and demand elasticities." *Water Resources and Economics* 32 (2020): 100159.
16. Wang, Qiang, and Xiaowei Wang. "Moving to economic growth without water demand growth–A decomposition analysis of decoupling from economic growth and water use in 31 provinces of China." *Science of the Total Environment* 726 (2020): 138362.
17. Guo, Lishuo, Wenbin Zhu, Jiaxing Wei, and Lifang Wang. "Water demand forecasting and countermeasures across the Yellow River basin: Analysis from the perspective of water resources carrying capacity." *Journal of Hydrology: Regional Studies* 42 (2022): 101148.
18. Dzimińska, Paulina, Stanisław Drzewiecki, Marek Ruman, Klaudia Kosek, Karol Mikołajewski, and Paweł Licznar. "The use of cluster analysis to evaluate the impact of COVID-19 pandemic on daily water demand patterns." *Sustainability* 13, no. 11 (2021): 5772.
19. Darby, Zachary, Neelam Chandra Poudyal, Adam Frakes, and Omkar Joshi. "Economic analysis of recreation access at a lake facing water crisis due to municipal water demand." *Water Resources Management* 35 (2021): 2909–2920.
20. Carpenter, Abhay, and Mahendra Kumar Choudhary. "Water demand and supply analysis using WEAP model for Veda River basin Madhya Pradesh (Nimar region), India." *Trends in Sciences* 19, no. 6 (2022): 3050–3050.
21. Priyanka, E. B., C. Maheswari, S. Thangavel, and M. Ponni Bala. "Integrating IoT with LQR-PID controller for online surveillance and control of flow and pressure in fluid transportation system." *Journal of Industrial Information Integration* 17 (2020): 100127.
22. Ramos, Helena M., Alban Kuriqi, Mohsen Besharat, Enrico Creaco, Elias Tasca, Oscar E. Coronado-Hernández, Rodolfo Pienika, and Pedro Iglesias-Rey. "Smart water grids and digital twin for the management of system efficiency in water distribution networks." *Water* 15, no. 6 (2023): 1129.
23. Mücke, N. T., P., Jain, S., Bohté, S. M. and Oosterlee, C. W. A probabilistic Digital Twin for leak localization in water distribution networks using generative deep learning. *Sensors* 23, no. 13 (2023): 6179.
24. Priyanka, E. B., S. Thangavel, X. Z. Gao, and N. S. Sivakumar. Digital twin for oil pipeline risk estimation using prognostic and machine learning techniques. *Journal of Industrial Information Integration* 26 (2022): 100272.
25. Brahmbhatt, Parth, Abhilasha Maheshwari, and Ravindra D. Gudi. "Digital twin assisted decision support system for quality regulation and leak localization task in large-scale water distribution networks." *Digital Chemical Engineering* 9 (2023): 100127.
26. Wu, Zheng Yi, Alvin Chew, Xue Meng, Jianping Cai, Jocelyn Pok, Rony Kalfarisi, Kah Cheong Lai, Sock Fang Hew, and Jia Jie Wong. "High fidelity Digital Twin-based anomaly detection and localization for smart water grid operation management." *Sustainable Cities and Society* 91 (2023): 104446.
27. Li, Xiaoming, Jian Luo, Yulei Li, Weixi Wang, Wuyang Hong, Mingwei Liu, Xingyi Li, and Zhihan Lv. "Application of effective water-energy management based on digital twins technology in sustainable cities construction." *Sustainable Cities and Society* 87 (2022): 104241.

Index

A

B

C

D

O

P

R

S

T

U

V

W